Nonequilibrium Statistical Mechanics I

Theoretical and Mathematical Physics

This series, founded in 1975 and formerly entitled (until 2005) *Texts and Monographs in Physics* (TMP), publishes high-level monographs in theoretical and mathematical physics. The change of title to *Theoretical and Mathematical Physics* (TMP) signals that the series is a suitable publication platform for both the mathematical and the theoretical physicist. The wider scope of the series is reflected by the composition of the editorial board, comprising both physicists and mathematicians.

The books, written in a didactic style and containing a certain amount of elementary background material, bridge the gap between advanced textbooks and research monographs. They can thus serve as a basis for advanced studies, not only for lectures and seminars at graduate level, but also for scientists entering a field of research.

Carlos Mejía-Monasterio ·
Lamberto Rondoni

Nonequilibrium Statistical Mechanics I

Foundations and Modern Applications

Springer

Carlos Mejía-Monasterio
School of Agricultural, Food, and Biosystems Engineering
Technical University of Madrid
Madrid, Spain

Grupo Interdisciplinar de Sistemas Complejos (GISC)
Madrid, Spain

Lamberto Rondoni
Dipartimento di Scienze Matematiche "Giuseppe Luigi Lagrange"
Politecnico di Torino
Turin, Italy

Istituto Nazionale di Fisica Nucleare
Turin, Italy

ISSN 1864-5879 ISSN 1864-5887 (electronic)
Theoretical and Mathematical Physics
ISBN 978-3-031-98248-4 ISBN 978-3-031-98249-1 (eBook)
https://doi.org/10.1007/978-3-031-98249-1

C M-M work is supported by the Spanish Government Grant No. PID2021-127795NB-I00 from MCIN/AEI/10.13039/501100011033 and FEDER, UE.

This Springer imprint is published by the registered company Springer Nature Switzerland AG
The registered company address is: Gewerbestrasse 11, 6330 Cham, Switzerland

To Laura and Carlotta

To Gabriella, Elena Hilary, and Letizia Kylie

Preface

This book attempts to provide a modern view of the foundations of nonequilibrium physics, together with a set of modern applications in the fields of physics, chemistry, and biology. Despite the many good books written on nonequilibrium, our goal has been to present in a classic and concise fashion the general foundational methods of the physics of the nonequilibrium, and their application to modern problems of current interest such as anomalous transport, nonequilibrium fluctuation theorems, stochastic lattice gases, cell kinetics, to mention a few.

Nonequilibrium statistical mechanics has become a broad field that it is difficult if not impossible to cover it fully in a book. This makes the decision on the contents of the book difficult. Unavoidably, the list of subjects in this book has been impacted by keeping the length of the book reasonable, as well as by our personal choice. In the end our approach has been to include the subjects that we consider necessary to explore current advances in the physics of nonequilibrium through the series of recent applications that we present. To account for the material and methods that we have intentionally not included, each chapter contains a final brief section covering other current research and open questions.

As an introduction of the book we start discussing the crucial importance of the choice of space and time scales, of the observables of interest, of measurement tools and of protocols, and how this choice makes our statistical theories not subjective. This naturally leads to the hydrodynamic approach presented in Chap. 2. Chapter 3 is devoted to the theory of Brownian motion that led to the modern atomistic theory of matter. We continue in Chaps. 4 and 5 the fundamental concepts of energy equipartition and the virial theorem, and of Legendre transforms and their modern counterpart of large deviation theory. Up to here, the book chapters are amenable to be read independently. We next focus on nonequilibrium fluctuations, discussing the theory of linear response in Chap. 7, the theory of Onsager-Machlup in Chap. 8, and the nonequilibrium fluctuation relations in Chap. 9. All these chapters rely on Chap. 6 on probabilities in phase space, where the Boltzmann equation is also discussed.

Students and researchers with a background in dynamical systems will find Chap. 6 particularly useful. Various appendices complete the text, giving technical details of subjects more informally treated in the main text, as well as illustrating side issues that are, however, strictly connected with the development of a theory of nonequilibrium phenomena.

The importance of scales to choose a model or interpret the resulting theories is a conceptual message recalled throughout this book. For this, fluctuations play the central role, and they serve as a guiding line of our argumentation.

We assume that the reader has familiarity with equilibrium statistical physics. Also, since most interesting problems are covered in the applications, the book does not contain an extra set of problems or exercises. Nevertheless, we hope this book will be an interesting teaching reference for courses at graduate and postgraduate levels.

A non-exhaustive list of recent applications that will be contained in Volume 2 of this book is:

- Transport Phenomena in Low Dimensions
- Anomalous Transport in Disordered and Complex Environments
- Emergent Collective Phenomena
- Stochastic Thermodynamics
- The living Matter
- Nonequilibrium Quantum Systems
- Probabilistic learning
- Thermodynamics of long-range systems.

Madrid, Spain
Turin, Italy
May 2025

Carlos Mejía-Monasterio
Lamberto Rondoni

Acknowledgements In the past decades, we have enjoyed the privilege of discussing the matter with established outstanding scientists, as well as with young active collaborators and students. We greatly profited from all the exchanges we had with them. We express our gratitude to those whose remarks have more deeply shaped our views.

In alphabetic order, we thank Erik Aurell, Santo Banerjee, Eli Barkai, Giancarlo Benettin, Olivier Bénichou, Michael Berry, Ricardo Brito, Giulio Casati, Salvatore Calabrese, Sergio Chibbaro, Giovanni Ciccotti, Sergio Ciliberto, E. G. D. Cohen (much missed since he passed away in 2017), Matteo Colangeli, Stephan De Bievre, Paolo De Gregorio, David Dean, Abishek Dhar, Bob Dorfman, Jean-Pierre Eckmann, Denis Evans, Maurizio Ferraris, Roman Frigg, Giovanni Gallavotti, Pierre Gaspard, Claudio Giberti, Sheldon Goldstein, Jim Hudspeth, Alberto Imparato, Owen G. Jepps, Paul Krapivsky, Antti Kupiainen, Hernan Larralde, Joel Lebowitz, Stefano Lepri, Roberto Livi, Valerio Lucarini, Satya Majumdar, Ralf Metzler, Gary Morriss, Paolo Muratore-Ginanneschi, Sergei Nechaev, Jordan Orchard, Gleb Oshanin, Juan M. Parrondo, Luca Peliti, Giovanni Pistone, Antonio Politi, Amilcare Porporato, Tomaž Prosen, Sid Redner, David Ruelle, Stefano Ruffo, Keiji Saito, Debra Searles, Udo Seifert, Herbert Spohn, Ray Streater, Tamas Tel, Jürgen Vollmer, Angelo Vulpiani, Hong Zhao, Paul Zweifel.

Competing Interests The authors have no competing interests to declare that are relevant to the content of this manuscript.

Contents

Reflections on Method

1

A book should open old wounds, even inflict new ones. A book should be a danger.

Drawn and Quartered, Emil M. Cioran

How immoderate we mortals are! Delight in watching the fluctuations of the contest is our true lot. Uber die Entwicklung der Methoden der theoretischen Physik,

Ludwig Boltzmann

Since philosophical thought, including religious beliefs, is documented, humans have tried to explain the natural world in terms of the smallest possible number of fundamental elements. Anaximenes proposed air as such a fundamental element; Tales proposed water. Concerned about the great complexity of the world, Empedocles thought that the elements must be four: air, water, earth, and fire. An immediate consequence of dealing with a limited set of elements implies the possibility to develop a unitary description, or a theoretical framework, for all phenomena.

Aristotle envisaged a hierarchy of approaches to represent the world, stating that different "sciences" reveal different kinds of causes, but metaphysics brings them together. In modern times, this idea is echoed by the French Encyclopédists, by Oken's thought about the systematic unity of natural philosophy, by the methodological unity of the Vienna School, and still more recently, by the cybernetic views and the science of complexity.

The major step in this direction was taken in the fifth century B. C. by Leucippus and Democritus, who suggested that the most variegated and dynamic natural phenomena can be understood in terms of countless, small, indivisible, eternal, and invisible objects called *"atoms"*. Atoms were so fundamental to Democritus that he ventured to state that phenomena commonly considered objective, such as sweet

C. Mejía-Monasterio and L. Rondoni, *Nonequilibrium Statistical Mechanics I*,
Theoretical and Mathematical Physics,
https://doi.org/10.1007/978-3-031-98249-1_1

or bitter taste, hot or cold sensations, were just conventions. Varying in size and shape, atoms had to be the only objective reality, capable of producing all the rest while incessantly moving and interacting in the empty space. It is striking that such a theory could be developed so long ago, when no technological tool could provide insights on its validity, but considering enough evidence for the existence of atoms the gradual erosion of bodies over time. In fact, the idea remained scarcely appreciated in ancient times, apart from important acolytes such as Epicurus and Lucretius. Be that as it may, the distinction between subjective although persistent perceptions and the objective reality of atoms has been eventually confirmed, so that taste and temperature can at last, more than 2000 years later, be even quantitatively explained in terms of atoms, of their interactions and of their motion.

Statistical mechanics and, in particular, its branch devoted to the study of nonequilibrium phenomena, developed precisely to provide an understanding, i.e., a unitary picture, of macroscopic phenomena, given that matter is made of atoms, and that atoms move in space obeying a few fundamental laws. Speaking of fluids in standard conditions, these rules appear consistent with classical mechanics; treating the transport properties of solids or phenomena that are not common at the scale of our daily life experience, quantum mechanics ought to be invoked. In both cases elementary principles apply, and yet the subject remains particularly subtle, to the point that it appears elusive to many and that various contrasting points of view can be found in the literature.

Despite this, the tools developed in Statistical Mechanics to investigate the connection between the microscopic and macroscopic levels of reality, or to investigate states of aggregation of matter at other scales, have been very successful. In particular, Nonequilibrium Statistical Mechanics finds application in the whole range spanning from the realm of macromolecular structures to that of cosmological scales, in addition to nuclear and particle colliders physics.

Before entering the wide set of concepts and techniques of this field let us reflect on the meaning of physical theories, from a point of view relevant for the subject of this book. First of all, we have to remark an implicit assumption in the above: the concrete existence of both observable objects and their elementary constituents. While some thinkers question such an assumption, developing fascinating and intriguing arguments, we find it hard to follow that line of thought and to reconcile it with the success of Physics, in particular, in describing nature. Therefore, we will not discuss further this issue and we will consider Physics as referring to objectively existing realities, although they may be perceived differently by different individuals.

Now, suppose that we have prepared a cup of good Italian coffee, which is a small amount of hot water and coffee, and that we have placed it on a wooden table for you. The table is in a dining hall in your city, which is on Earth, which is a planet of the Solar system, that belongs to the Milky Way, and so on. If you do not drink coffee and leave the cup on the table for a while, your senses will reveal that it gradually becomes less and less hot. In half an hour, you will realize that the coffee is not hot at all. After a few days, the coffee is no longer there: it evaporated, leaving dark stains on the bottom of the cup. Within several thousand years or less, the table has deteriorated, and maybe the cup like your city has been destroyed; whether war, bugs, earthquakes, or any other event produced this effect does not matter; the previous situation has drastically changed. But this is nothing compared to what will happen in the next billion years: the Earth will most likely be engulfed by the expanding Sun; still later, the Milky Way will undergo a destructive collision with the Andromeda galaxy.

Question: Has our coffee ever been in a stationary state? a state in which its physical properties do not change in time? Has any of the small or large objects that we mentioned ever been stationary?

We ask such questions because thermodynamics, hence statistical mechanics, is by and large concerned about such states. In particular, they concern *equilibrium* states, which are a special kind of stationary states. Thermodynamics is a most successful branch of physics that describes first of all objects and phenomena occurring at the scale of our daily life. Physics, in turn, is the realm of quantitative—hence mathematically expressible—theories meant to describe and predict the being and becoming of natural phenomena. There is an irresistible drive that motivates humans to understand the world and to give life to new realities, which passes through the development of theories, i.e., intellectual frameworks capable of explaining what we observe, in terms of a reduced number of fundamental elements. In the case of physics, this means mathematical frameworks. The formulation of a correct theory requires and inspires experiments (not just experiences) that involve measurements of relevant quantities, in order to test its validity. Measurements should provide objective data that become intelligible within the mathematical framework that constitutes the theory. Objective means that data do not depend on the observer, so that their significance stands, for example, for a passer-by, whether that person is aware or not of the measurement. If water in a pot is $90^o C$ hot, anyone sticking a hand in it will be burned, whether they knew or did not know the value of the temperature. Temperature is an example of thermodynamic property.

The goal of thermodynamics is to investigate the consequences of fundamental principles such as the conservation of energy, the impossibility of perpetual motion, etc. Its second law has been described with the following words:

> The second law of thermodynamics is, without a doubt, one of the most perfect laws in physics. Any reproducible violation of it, however small, would bring the discoverer great riches, as well as a trip to Stockholm. The world's energy problems would be solved in one stroke. It is not possible to find any other law (except, perhaps, for super selection rules such as charge conservation) for which a proposed violation would bring more skepticism than this one. Not even Maxwell's laws of electricity or Newton's law of gravitation are so sacrosanct, as each has measurable corrections coming from quantum effects or general relativity. The law has caught the attention of poets and philosophers and has been called the greatest scientific achievement of the nineteenth century (Lieb and Yngvason 1999).

Describing the behavior of matter at the scale of our daily life, or what we call the macroscopic realm, does not suffice to characterize thermodynamics. One should further specify the kinds of measurement, the tools, and the protocols being considered. In fact, measurement tools, which we may view as an extension of our senses, and the protocols followed in using them necessarily affect the picture of reality one gets. In the first place, different measurement tools probe different properties of reality, hence yielding different descriptions of a given object. It is like wearing glasses that transmit only a given kind of radiation; around 510–530 THz makes the whole world look yellow or black. Other glasses give a different picture. Moreover, we may decide to cover both eyes or only one with the glasses, and why should not we like to smash the glasses on the object of interest to test its rigidity? In fact, Hofstadter ponders the possibility that the encounter between a meteorite and a digital disk containing Bach's music might be an intelligent event (Hofstadter 1979).[1]

Although at times these facts are taken as the basis of subjectivity in science, denying the Galilean paradigm, objectivity persists, making possible our ever growing control over natural phenomena.

If measurement tools and protocols are chosen by someone who wants to investigate a given phenomenon rather than another, one may conclude that the objectivity of data is problematic. Indeed, the first and most familiar measurement tools are our senses: amazingly complex structures, more accurate and dependable than many artificial sophisticated tools. However, they are also imperfect and may badly lead us into error through hallucinations or misjudgments. Even when such disturbing situations do not happen, our senses may still induce different people to draw contradictory conclusions about the very same phenomenon: *Nattō* rotten beans are considered a delicacy in Japan and totally disgusting by many westerners.

Now, measurement tools extend our senses in various ways, but the increase in accuracy and resolution that they afford cannot wipe out that kind of "subjectivity." Presently, the duration of a second can be split in 10^{16} parts, the length of a forearm in 10^{10} parts, and the mass of a pollen grain in 10^{18} parts. Although this allows different operators to agree on a large number of digits concerning certain measurements, full agreement will never be reached. But more crucial is that the observer shall subjectively decide which property of the object of interest is to be measured, which tools to be used, which protocols to be adopted. Furthermore, not only different quantities, such as mass and temperature, require different tools to be measured, but even a single property needs different tools, depending on the scale of observation. One cannot weigh a star with a scale built on Earth, while one such scale can weigh potatoes and, if suited for potatoes, it is not suitable to weigh diamonds.

[1] Hofstadter considers possible definitions of intelligence, but his reasoning touches on questions of interest in Physics as well:

> To take an extreme example, consider a meteorite which, instead of deciphering the outer-space Bach record, punctures it with colossal indifference, and continues in its merry orbit. It has interacted with the record in a way that we feel disregards the record's meaning. Therefore, we might well feel tempted to call the meteorite 'stupid'. But perhaps we would do the meteorite a disservice. Perhaps it has a 'higher intelligence' that we in our Earth chauvinism cannot perceive, and its interaction with the record was a manifestation of that higher intelligence. Perhaps, then, the record has a 'higher meaning'—totally different from that which we attribute to it; perhaps its meaning depends on the type of intelligence perceiving it. Perhaps. Hofstadter (1979), p. 172.

Advancing technology may produce tools with wider ranges of applicability, but there are intrinsic limitations also to that.

The choice of observable quantities, tools and protocols is essential; at the same time: **(a)** it eliminates from observations aspects that could have been otherwise interesting, and **(b)** for the chosen properties, it selects a given range of space and time scales, that is limited by the resolution power of the tools. An optical microscope can measure neither the size of atoms, nor that of the Earth.

Even agreeing on the quantities to be evaluated and on the tools to be used, people can still get to totally different conclusions. Consider a cloud and use your eyes. A common experience of travelers is that the cloud looks like a white object with a definite shape and extension in the blue sky, but flying closer and closer to it, the cloud boundaries gradually vanish, and after a while there is only gray: no white, no blue, although no boundary has been crossed. Which picture shall we consider as correct, accurate, or revealing the *real* nature of the cloud? The white shape in the blue sky, or the gray that seems to have neither a beginning nor an end? Our descriptions do not always agree, much the same as our observations of the cloud from the ground and from inside it do not agree. The cloud, however, is what it is: it does not care what we think of it. The table on which our laptop rests is an objective reality that does not depend on our opinion of it. If we intend to measure its height, performing higher and higher resolution measurements, looking at finer and finer details, at some stage the top surface of the table vanishes, like the border of the approaching cloud, and measuring on that scale its distance from the floor stops making sense. On the scale of atomic nuclei we do not express the surface of the table very accurately; on that scale there is mostly empty space and no definite boundary between table and empty space. The ideas of "table length, width, and height" fade: they are lost for computing the occupied volume or the load it can carry, i.e., for relations that we commonly find useful and interesting. Of course, it is totally legitimate to investigate the nature of the table on microscopic scales, but the useful or interesting quantities that can be profitably related in some mathematical expression concern other aspects of the table. Alternatively, one may drastically change measurement tools, like in the detection of gravitational waves, but then "length" means quite a more elaborate phenomenon than the one obtained with a ruler. The table is what it is, while "useful" or "interesting" are notions that depend on us, and may change in time and from person to person. Thus, the point of a theory is not the height of the table, but what we call height, and why we care about it.

One could conclude that there are too many points of view on a very same object, hence plenty of room for subjectivity and no room for objective Galilean science, in spite of the fact it has led progress for centuries. But this is not the case. As Chinese culture teaches, apparently contrasting views can be harmonized to produce a superior account of reality. There is room for subjectivity in deciding what is interesting to measure: we may choose to classify the colors of objects, rather than their masses. There is objectivity in using the radiation wavelengths for colors and pounds for masses. The selection of interesting questions is largely a subjective issue, but digits read off a ruler are objective.

Think of the Rosetta space probe and its adventure in space[2] : it took eleven years to *objectively* reach the 67P-Churyumov-Gerasimenko comet, and to objectively land the probe Philae on it. Such an incredible enterprise was made possible by our understanding of objective and persistent properties of gravitating bodies, as well as by a variety of instruments developed thanks to our understanding of other objective physical realities. There is a sense in which measurements are objective! Alexandre Koyré, strongly expresses this view, highlighting Newton's contribution to science:

> Thus it seems to me that I have the right to assume that when we are speaking about Newton and Newtonianism we know more or less what we are speaking of. More or less! Somehow this very expression used in connection with Newton strikes me as improper, because it is possible that the deepest meaning and aim of Newtonianism, or rather, of the whole scientific revolution of the seventeenth century, of which Newton is the heir and the highest expression, is just to abolish the world of the "more or less," the world of qualities and sense perception, the world of appreciation of our daily life, and to replace it by the (Archimedean) universe of precision, of exact measures, of strict determination. [...] this characterization is very nearly equivalent to the mathematization (geometrization) of nature and therefore the mathematization (geometrization) of science. [...] This, in turn, implies the disappearance—or the violent expulsion - from scientific thought of all considerations based on value, perfection, harmony, meaning, and aim, because these concepts, from now on merely subjective, cannot have a place in the new ontology. Koyré (1968).

Our measurement tools show at once only a given facet of the phenomenon at hand. Imagine that we are looking at the size of the orbitals of one atom of an aluminum bar. We don't see the bar, and we are not able to measure its length. We must zoom out and lose resolution for that. Great resolution and accuracy make impossible the measurement of the length of the bar, but there is no paradox: on that scale the length of the bar is a notion of scarce physical relevance, or hard to use. Indeed, couldn't we like to increase further the measurement accuracy, reaching the internal structure of the nuclei of the atoms? And then the internal structure of nucleons, and so on? We can, but such a project is physically misguided: higher accuracy requires more energy, and when the energy is too large, the object of interest may even be destroyed. What is then its length, when the bar is vaporized? In addition, certain levels of resolution are impossible, even in principle. For instance, chaotic dynamics increase uncertainties at an exponential rate, therefore, accurate predictions on even relatively simple phenomena would rapidly need an accuracy on initial conditions requiring more energy than the object of interest can take, and even more than afforded by a finite Universe... with all due respect to certain cosmologies (Tipler 1995). Besides the choice of observable quantities, tools and protocols, also resolution, the scale of observation, is also essential and must be properly chosen by physicists. Once the property of interest has been selected, different measurement methods are expected to yield the same result *within the relevant scales*. One may use different kinds of thermometers to measure the thermodynamic temperature of a gravitational bar, including electronic devices that look at microscopic vibrations, and contact mercury

---

[2] See, e.g., https://www.nasa.gov/rosetta.

thermometers. The measurement is considered accurate if the results agree up to a certain number of digits, keeping in mind that many more digits would not correspond to thermodynamic temperature but to some other, perhaps equally interesting but different physical property.

After the observables of interest have been identified and the range of measurements has been delimited, a theory will connect the sets of measurements performed within those scales, hopefully constituting a satisfactory explanation of the phenomenon under consideration. If the scales are changed, hence the tools and the protocols, and perhaps the observables are changed, a new theory will be needed. This new theory may match with the previous one at the border between the two levels of description, or could be contained or contain the previous one. In principle, one can find a vocabulary to translate into each other the terms and statements of the different theories, but often, however, the different theories do not mathematically reduce to each other Chibbaro et al. (2014), extra assumptions, alien to them, are required to make a new theory mathematically agree with the old. The mathematical inconsistencies are only such: There is no claim that atoms disappear from matter observed at the macroscopic scale just because thermodynamics neglects them. The same happens with our senses: a cathedral is meaningfully described in a language that differs from that we might use to describe its stones, although the stones afford quite a detailed picture. The cathedral is made of stones, but so much detail overshadows what we intend as beauty. Our (present) knowledge of stones does not immediately reveal what makes a cathedral beautiful. Analogously, the (present) mathematical structures with which we describe atomic dynamics do not immediately lead to the second law of thermodynamics. Atomic dynamics and entropy growth pertain to (so far) irreducible mathematical representations of the same object, although the object itself does not care, like the apparently irreconcilable north and south sides of the same mountain are not self-contradictory. For a theory to be successful it suffices that the range of scales to which it successfully applies is not too narrow, so that different people bound to perform measurements in different conditions get compatible results. As noted above, it makes no sense to agree on all possible scales; on sufficiently separate scales, theories even turn "orthogonal" to each other. Take, as an illustration, the following function:

$$f(x) = 10^{-7} \sin x \,, \quad x \in [0 : 2\pi]$$

Figure 1.1 shows that, within a wide range of observations, the phenomenon f is correctly represented by a straight horizontal line; it looks wavy within intermediate scales, and eventually it appears as vertical lines. All these representations of f are legitimate and objective; the one that in addition is useful or meaningful depends on the scale one wants to explore. We could say that the horizontal straight line represents a macroscopic phenomenon, the wavy picture represents a mesoscopic phenomenon, and the vertical line a microscopic phenomenon. Yes, the macro and the microscopic phenomena may very well be orthogonal, i.e., irreducible to one another. There is no paradox in all that: the function is objectively what it is, and looking at it under strong magnification objectively shows vertical lines, while over

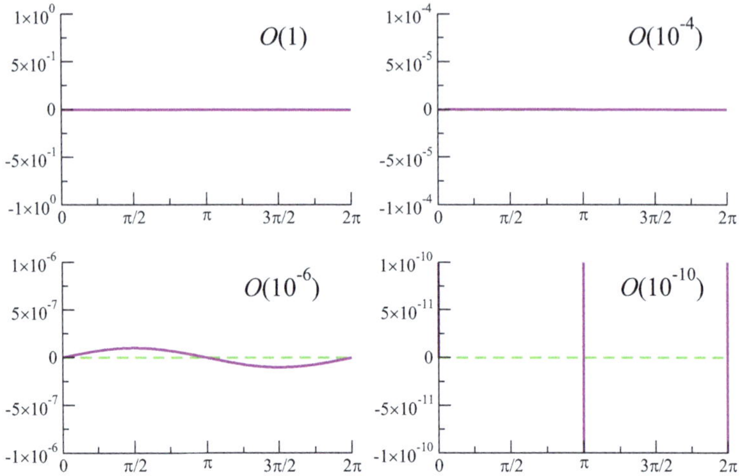

Fig. 1.1 Plots of $f(x) = 10^{-7}\sin x$ at various resolutions $O(10^{-k}), k = 0, -4, -6, -10$. For many levels of description, f can be treated as a straight horizontal line; for other resolutions it can be treated as several vertical straight lines. Only at intermediate resolutions does f look like a sinusoidal function. All representations are correct, but usefulness depends on our purposes

infinitely many scales it is a horizontal line for all observers. How interesting a rule would be that meaningfully translates "vertical" into "horizontal" depends on the problem at hand; it does not necessarily need to be a concern.

The choice of space and time scales, of the observables of interest, of measurement tools, and of protocols is crucial for the development of a physical theory and does not imply any subjectivity. Understanding more deeply a given phenomenon does not always mean to increase the resolution of measurements: ever more detailed information might not help. In general, a large fraction of information has to be discarded, in order to highlight the *relevant* information. Galileo expressed this view with his celebrated *"difalcare gli impedimenti,"* i.e., his invitation to dispose of inessential details, in order to highlight the pure phenomenon: the one which is of interest. This in itself constitutes the general program of Statistical Mechanics. The message of a movie is not transmitted by an accurate analysis of the pixels and sound bits of all frames of the film: it emerges when the frames pass before our eyes at a speed that allows us to grasp only a tiny fraction of the information in them. However, while this is obvious in the case of a movie, in general it may be far from trivial to distinguish essential from inessential details. Furthermore, once the essential questions have been identified, it may not be easy to select the most appropriate tools of investigation, and, indeed, different tools have variable degrees of success in open fields, such as nonequilibrium physics.

We will try to make this clear, together with the teaching that the mathematical formalism is as powerful and fundamental (the language of Nature, according to Galileo) as it can be misguiding, if not properly used. The Zermelo paradox is one celebrated example of that, particularly relevant for Physics. We will treat it in detail to show that something mathematically correct can be physically completely wrong.

Even a certain use of ensembles, the most typical tool of statistical mechanics, can lead to error because a number of conditions must be satisfied to obtain correct results. The fact that such conditions are quite generally satisfied in the study of thermodynamic phenomena often makes us forget that they are needed. We will see that for thermodynamic quantities to make sense, the elementary constituents of matter must properly interact (e.g., short-range repulsion) and must be very many. Merely using the mathematical formalism offered by the classical ensembles, one may derive the law of perfect gases for a single particle, when even the notion of mass density makes no sense, not to speak of pressure and temperature. The formalism is so robust that one may devise a thermodynamic formalism for potatoes falling from a truck. The formalism does not stop working; it does work and yields its result. Therefore, physical insight is required to separate what makes sense and what does not. In a profound paper of 1993, Vega, Uzer, and Ford intriguingly and convincingly argue that the physical relevance of polygonal billiards dynamics:

> has for decades been obscured by the perfections of pure mathematics. Vega et al. (1993).

Far from denying the usefulness, besides the beauty of the mathematics of polygonal billiards, we also think that keeping these facts in mind helps the investigation of phenomena exceeding our current knowledge. Present day science and technology reach scales alien to thermodynamics, yet thermodynamics is what we know best; hence, we like to continue to use its terminology. But, when the numbers of elementary constituents are not large, or when their interactions are complicated, etc., what shall we say with respect to the space and time scales of our daily life? We need proper tools to judge whether a new result, that appears paradoxical or revolutionary, isn't, in fact, merely the fruit of incorrect use of the available mathematics. At the same time, what looks like an ordinary result may instead be a breakthrough. Ernst Cassirer enshrined the scientific activity in this way:

> The scientist acts upon the principle that even in the most complicated cases he will eventually succeed in finding an adequate symbolism which will allow him to describe his observations in a universal and generally understandable language [...] In face of the immensity of nature many of the greatest scientists may have had that special feeling that was expressed in a famous saying of Newton's. They may have thought that in their own work they were like a child who walks along the shore of an immense ocean and amuses himself occasionally picking up a pebble whose shape or color attracts his eyes. This modest feeling is understandable, but it gives no true and full description of the work of the scientist. The scientist cannot attain his end without strict obedience to the facts of nature. But this obedience is not passive submission. The work of all the great natural scientists —of Galileo and Newton, of Maxwell and Helmholtz, of Planck and Einstein—was not mere fact collecting; it was theoretical, and that means constructive, work. This spontaneity and productivity is the very center of all human activities. It is man's highest power and it designates at the same time the natural boundary of our human world. In language, in religion, in art, in science, man can do no more than to build up his own universe a symbolic universe that enables him to understand and interpret, to articulate and organize, to synthesize and universalize his human experience. Cassirer (2021).

With this in mind, we are going to review classical results, as well as modern attempts to advance research. We will focus on nonequilibrium statistical physics, in which these considerations find their natural home. This subject originated indeed to provide a microscopic understanding of the well-established laws of irreversible thermodynamics, but currently it also faces problems that cannot be framed within thermodynamics.

2 Hydrodynamics and Nonequilibrium Thermodynamics

True ideas are those that we can assimilate, validate, corroborate and verify. False ideas are those that we cannot Pragmatism, a new name for some old ways of thinking,

William James

We should identify "interesting" questions that deserve to be investigated. What makes a question interesting? Trial and error could be one way in which over thousands of years, humans have selected what is interesting from the point of view of survival. Subject to their senses, there was no meaning to infer the shape of the other side of a mountain from its front because for all their existence, if they wanted to know, they could go and look, just as they searched under a rock for food. Therefore, they did not grow an interest in, neither they developed the necessary tools for inferring the shape of the south side of a mount once they had observed its northern side. On the contrary, it was and remains most important to infer what fire can burn, without having to do the test.

The last is an issue concerning our daily life and has to do with Thermodynamics, a subject that received the greatest impulse from the needs of industry and war. Luckily, our senses are engaged all the time with kinds of bombs and turbines: in a sense, such technological devices do not reach too far from cooking food or shielding our dwellings from heat and cold, or from the heat released in the atmosphere by us when we read a book on Statistical Mechanics. Of course, melting metals and powering trains require considerable more energy, but experience taught us that the laws governing these different processes are the same. We know that realms that appear quite different from those just described exist and can be investigated. For instance, on the atomic scale matter does not look like a continuum. In fact, a different

C. Mejía-Monasterio and L. Rondoni, *Nonequilibrium Statistical Mechanics I*, Theoretical and Mathematical Physics,
https://doi.org/10.1007/978-3-031-98249-1_2

description replaces Thermodynamics on that scale. In this chapter, we remain within the realm of Thermodynamics, although allowing departure from global equilibrium. Let us begin by specifying these informal thoughts.

2.1 Continua and Their Properties

To our senses, a macroscopic object looks like a continuum occupying a certain volume $V \subset \mathbb{R}^3$, with a (*average*) mass density $\rho = M/V$, M being its mass. One may then consider the property of which ρ is the average, introducing the local density $\rho(\mathbf{x})$ at each of its points $\mathbf{x} = (x_1, x_2, x_3) \in V$. However, to make sense, ρ cannot be referred to a single point $\mathbf{x}$, which has vanishing mass and occupies a vanishing volume. Rather, one should consider the ration $\rho(\Delta V)$ of the mass Δm contained in a finite volume ΔV to which $\mathbf{x}$ belongs. This quantity may not be stable: it may change if the volume ΔV containing $\mathbf{x}$ changes. In that case, the object is called inhomogeneous. Therefore, one may reduce the size of such a volume, to obtain an intrinsic property of matter around $\mathbf{x}$. Letting the volume be a cube of side $\ell = \sqrt[3]{\Delta V}$, one can investigate the behavior of $\rho(\Delta V)$ as $\ell > 0$ is decreased. There are at least two good reasons to proceed like this:

1. any measurement tool occupies a finite volume, hence it interacts with a finite volume of matter;
2. mathematically, the existence of the $\ell \to 0$ limit,

$$\rho = \lim_{\ell \to 0} \frac{\Delta m}{\Delta V} = \lim_{\ell \to 0} \frac{\Delta m}{\ell^3} \tag{2.1}$$

   does not mean $\ell = 0$ is ever reached; it means that, given any $\epsilon > 0$, there is a $\delta_\epsilon > 0$, such that

$$\left| \frac{\Delta m}{\Delta V} - \rho \right| < \epsilon , \quad \text{if} \quad \ell < \delta_\epsilon \tag{2.2}$$

In other words, considering an exceedingly small ΔV is useless, even from a purely mathematical point of view. The difference between the mathematics and the physics is that there is no mathematical restriction on the tolerance ϵ, while physically the acceptable tolerance is determined by the kind of measurement one wishes to perform. In other words, δ_ϵ can be regarded as the accuracy of volume measurements required to meet the desired accuracy in measurements of the density at $\mathbf{x}$.

In reality, for the density to be an objectively recognizable quantity by different observers, something more is needed: the convergence within ϵ must occur well before scales at which the available measurement tools no longer operate. If matter is a continuum, we cannot and do not need to qualify more precisely ϵ and δ_ϵ, but the fact that matter is made of atoms imposes certain constraints on both. In particular, it makes no sense to measure the density on scales of the atomic size or smaller: they do not pertain to the thermodynamic quantity ρ since, on such a scale, the definition

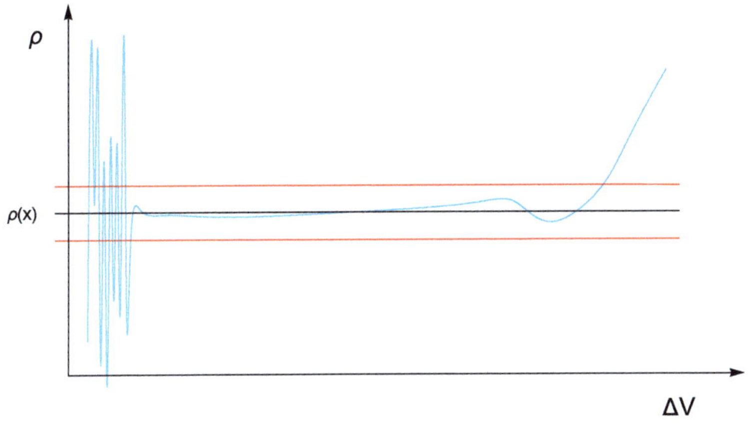

Fig. 2.1 Hypothetical density versus volume graph. For large volumes, the density changes in a macroscopic measurable fashion, due to the inhomogeneities of the mass distribution in space: the notion of density makes sense but the measurement is not accurate, and maybe larger (as in figure) or smaller than $\rho(x)$. When volume decreases below a certain threshold, the density only varies within a narrow band of values: measurement is accurate. If this remains the case for a wide range of scales, density is an intelligible objective quantity. At atomic scales and smaller, the ratio mass to volume wildly fluctuates, depending on whether the volume contains or does not contain atoms or parts of atoms, a condition that changes abruptly and continually. A simple description of the phenomenon, in terms of global constant quantities, like thermodynamic ones, is not possible when this happens

(2.1) yields an impractical quantity. The picture is something like Fig. 2.1. It is not forbidden to consider mass density at atomic scales, but the measurement of such a frantically fluctuating quantity is hard to use; in particular, it does not provide an intelligible quantity that can be described by partial differential equations.

A similar reasoning can be repeated *mutatis mutandis* for all the quantities of thermodynamic interest, which then make sense (we could say: *exist*) only within certain space and time scales and, in particular, these scales exclude the atomic one. Should one conclude that thermodynamics is a *coarse* or *subjective* description? In fact, the answer is not unique, it depends on whether the macroscopic scale is widely separated from the microscopic scale, or not. In the first case, measurements performed within many orders of magnitude of V all yield the same value, i.e. they are indistinguishable and make all observers agree within that level of reality. Figure 2.1 shows a possible scenario: there can be a range of inaccurate measurements (large volumes), within which the density changes substantially when the accuracy is improved (reducing V); that is followed by a wide range of V for which the measurement of the density remains within narrow bounds (red lines). In this case the measurement yields a precise, objective value. Finally, there will be a range within which the notion of density is meaningless (atomic scale). In different situations, such a separation of scales may not be realized, and different operators will not agree. In that case, a kind of subjectivity could indeed creep in.[1]

[1] We say *"could"*, because these situations must be assessed case by case.

This apparently artificial and vague reasoning is, in fact, objective and operationally implementable, because measurement tools and corresponding measurement protocols exist, if stability over a good number of scales, as described above, is realized. In reality, this happens in a vast set of macroscopic phenomena. It is the rule for phenomena occurring in a wide neighborhood of the scale of our daily life[2] cf. Sect. 2.5, in which the notion of Local Thermodynamic Equilibrium (LTE) is illustrated from the microscopic point of view. Macroscopically, LTE is the condition under which the properties of a macroscopic system can be considered continuous functions of space and time. Sections 2.2–2.4 describe some of its consequences. First, LTE rests on the notion of (global) thermodynamic equilibrium, which, in his celebrated Thermodynamics book, Callen (1985), Callen describes as follows:

> Even a cursory consideration of various states of a given system makes it apparent that some states are relatively simple and others are relatively complicated. Thus a fluid at rest clearly is in a simpler state than a fluid in a laminar flow. And fluid in laminar flow, in turn, is in a simpler state than a fluid in a turbulent flow. Moreover, it is experimentally observed that isolated systems generally tend to evolve spontaneously toward simple terminal states.

He then, continues:

> A formal criterion of simplicity is the possibility of description in terms of a small number of variables.

This leads to the following:

POSTULATE: There exist particular states (called equilibrium states) of simple systems that, macroscopically, are characterized completely by their internal energy U, their volume V, and the mole numbers $N_1, N_2, \ldots, N_n$ of their n chemical components. More generally, electric, magnetic, mechanical, etc. properties have to be added to the list of variables characterizing the system. The total will in any case be a limited number of variables.

One consequence of being described by a few parameters, is that, in equilibrium, they are independent of the system's past history. This feature is violated, e.g., in chemically identical pieces of steel, whose properties differ because of different heat treatment. Such systems are not in equilibrium, and are experimentally at variance with thermodynamic predictions. Think also of glasses. M. Zemansky then expresses the equilibrium condition as one in which the following three equilibria are realized together (Zemansky 2011):

[2] Presumably, that is why the thermodynamic formalism is so popular and researchers try to use it in the most diverse situations: from strongly nonequilibrium physics and cosmology, to disciplines quite far from physics, such as neuro-sciences, finance, linguistics, and modern art, to mention just a few. Just do a web search with "entropy" or even "entropy production" together with any other keyword, and be amazed.

1. MECHANICAL EQUILIBRIUM: There are no net external or internal forces acting on the system. In absence of this equilibrium, one observes accelerations, turbulence, waves.
2. CHEMICAL EQUILIBRIUM: There are no variations of the internal structure, meaning of chemical composition and concentrations, and there is no transport of matter.
3. THERMAL EQUILIBRIUM: The state does not change when the system is separated from the environment by adiabatic walls. If not in thermal equilibrium, when separated from the environment, the system evolves toward a thermal equilibrium in which all parts of the system have the same temperature. If in contact with environment, but not in thermal equilibrium, one has heat fluxes.

When all three equilibria hold, the system appears to be the same in all its parts, therefore it can be characterized by just a few parameters, like T, P, and ρ, which remain constant in time. Thermodynamic equilibrium is inherently a homogeneous state.

2.1.1 Basic Thermodynamic Facts

Under thermodynamic equilibrium conditions, many results are obtained, without any recourse to the underlying atomic structure. Here we list a few that will be used in the book. A summary of such a most successful, deep, and wide subject is not intended. Excellent books on thermodynamics, like Ref. Callen (1985), are devoted to that.

Consider a macroscopic system in contact with a heat bath at temperature T. Let λ be an externally controllable parameter, e.g., the position x of a piston moving in a cylinder containing a gas. If λ varies slowly from a value λ_i to another value λ_f, the quantity called entropy changes as follows:

$$\Delta S = S_f - S_i = \int_i^f \frac{dQ}{T} = \frac{Q}{T} \tag{2.3}$$

where Q is the heat exchanged with the bath at temperature T. The fact the process is slow allows the system to adjust to the changing state implied by the variation of λ, so that its temperature and density, for instance, remain uniform in space and the temperature equals T. One also has

$$\Delta F = \Delta U - T\Delta S \tag{2.4}$$

where ΔU is the variation of the quantity known as internal energy , and ΔF of the quantity known as free energy. Then, the work W_r done in such a transformation, known as quasi-static, hence reversible, is given by

$$\Delta S - \frac{1}{T}\Delta U = \frac{1}{T}(Q - \Delta U) = -\frac{1}{T}\Delta F = -\frac{1}{T}W = -\frac{1}{T}W_r$$

The variation of the free energy equals the reversible work done on the system. If the process is not quasi-static, the work W for the same change of state is larger:

$$\Delta S > \frac{Q}{T} \quad \text{implies} \quad -\frac{\Delta F}{T} > -\frac{W}{T} \quad \text{, i.e.,}$$

$$\Delta F < W \quad \text{and} \quad W > W_r$$

because of energy losses.

What can be said when the transformation $\lambda_i \to \lambda_f$ is arbitrarily performed? The work done on the piston is expressed by $W_P = m \int a(x)dx$, where the integral is over the displacement of the piston, m its mass, and $a(x)$ its acceleration. If the process $a(x)$ is fixed, W_P is fixed as well. However, the work done by the piston on the gas may vary from experiment to experiment if the process is not sufficiently slow, because the interactions of the gas molecules with those of the piston will depend on their initial conditions. What the work on the gas might be in such cases lies outside the realm of thermodynamics, it requires an atomistic approach, and will be investigated in Chap. 9.

2.2 Local Thermodynamic Equilibrium

When in equilibrium, the few relevant variables that completely determine the macroscopic state obey relatively simple relations that connect the different properties of the object of interest. For instance, given a system made of n chemical substances of densities $\{\rho_1, \rho_2, \dots, \rho_n\}$, and introducing the mass fractions

$$c_k = \frac{1}{\rho}\rho_k, \quad \text{with} \quad \rho = \sum_{k=1}^{n} \rho_k, \quad \text{so that} \quad \sum_{k=1}^{n} c_k = 1, \quad c_k \geq 0 \tag{2.5}$$

where ρ is the total density, the difference between the properties of two equilibrium states infinitesimally close to each other can be expressed by

$$T ds = du + p\, d\omega - \sum_{k=1}^{n} \mu_k dc_k \tag{2.6}$$

for a system of n components with chemical potentials μ_k, specific (i.e., per unit mass) internal energy u, pressure p, temperature T, specific volume ω (i.e., the inverse of the density) and specific entropy s. Representing equilibrium states, the quantities in Eq. (2.6) depend neither on time nor on space. In practice, a cell is like a small isolated system in thermodynamic equilibrium: at its boundaries only negligible quantities of matter, momentum, energy, etc. are exchanged, so that their amounts in the bulk are practically unaffected.

If the system is not in equilibrium, it may evolve in time. We consider it not too far from equilibrium, provided that it can be decomposed into small cells, each of which is *locally* in equilibrium, in the sense that Eq. (2.6) holds separately for each of these cells, although with different properties from cell to cell. Then, for a mass element followed along its center of mass, one may write

$$T\frac{ds}{dt} = \frac{du}{dt} + p\,\frac{d\omega}{dt} - \sum_{k=1}^{n} \mu_k \frac{dc_k}{dt} \tag{2.7}$$

where the various quantities depend also on space and the substantial derivative is defined by

$$\frac{d}{dt} = \frac{\partial}{\partial t} + v \cdot \nabla$$

and the barycentric velocity v and overall density ρ are defined by

$$v = \sum_{k=1}^{n} \rho_k \frac{v_k}{\rho}; \qquad \rho = \sum_{k=1}^{n} \rho_k \tag{2.8}$$

$v_k = (v_{k_1}, v_{k_2}, v_{k_3})$ being the velocity of substance k.

If relation (2.7) holds at all points of the system, where points stand in fact for small volumes, sufficiently large for the various quantities to make sense, the state of the system is called an LTE state. This nonequilibrium condition is not too far from thermodynamic equilibrium because the thermodynamic quantities still make sense, although they exist as fields that can vary both in space and time.

2.3 Conservation Laws

In this section, we adopt the LTE condition combined with the validity of the first and the second laws of thermodynamics, which are the conservation of energy and the entropy growth in isolated systems. Consequence of LTE is that the properties of interest are continuous functions of space coordinates and time. As we admit multi-component systems, chemical reactions may occur, and external forces may act differently on the different substances. From a microscopic point of view, the conservation of matter, energy, and momentum are direct consequences of the mechanical laws governing the motion of the constituent particles of the system, but at the macroscopic level they have quite a different aspect.

2.3.1 Conservation of Mass

Let $D \subset \mathbb{R}^3$ be a region of space and consider an element of matter about $\mathbf{x} \in D$ at time $t \in R$. Assume there are n distinct chemical components, each of density ρ_k, with $k = 1, \ldots, n$. Let $\mathbf{v}_k \in R^3$ be the velocity of substance k, which may depend on both space and time: $\mathbf{v}_k = \mathbf{v}_k(\mathbf{x}, t)$. For each t, the mass in a generic subregion $W \subset D$ is defined by

$$m(W, t) = \sum_{k=1}^{n} \int_W \rho_k(\mathbf{x}, t)\, \mathrm{d}\mathbf{x} = \sum_{k=1}^{n} m_k(W, t) \tag{2.9}$$

To proceed, it is assumed that the thermodynamic fields $\mathbf{v}_k$ and ρ_k are smooth enough in $D \times R$, so that the standard calculus can be applied. Of course, this assumption is open to criticism not just because LTE is not always guaranteed, but also on purely mathematical grounds: the theory of smooth functions cannot prevent singularities and discontinuities from developing in time, see, e.g., Fig. 2.2. This constitutes a warning for the use of PDEs in general, when dealing with phenomena that are only approximately "continuous." Nevertheless, the continuum assumption leads to so many results, so accurate and general in the treatment of macroscopic phenomena, that makes it one of the most, if not the most used theory of physics. Under smoothness conditions, the rate of change of the mass m_k of substance k in W is then given by

$$\frac{d}{dt} m_k(W, t) = \frac{d}{dt} \int_W \rho_k(x, t) dV = \int_W \frac{\partial \rho_k}{\partial t}(x, t) dV \tag{2.10}$$

This quantity then equals the material flow in and out of ∂W (the boundary of W) plus the production rate of k by chemical reactions:

$$\int_W \frac{\partial \rho_k}{\partial t} dV = -\int_{\partial W} \rho_k \mathbf{v}_k \cdot \hat{n} dA + \int_W \sum_{j=1}^{r} \nu_{kj} R_j dV \tag{2.11}$$

where

- $\mathrm{d}A$ is the surface area element on ∂W;
- $\hat{n}$ is the unit outward normal at all points of ∂W;
- $\mathbf{v}_k \cdot \hat{n}$ is the volume flow rate per unit area;
- $\rho_k \mathbf{v}_k \cdot \hat{n}$ is the mass flow rate per unit area;
- r is the number of chemical reactions;
- R_j is the chemical reaction rate of the jth reaction, whose units are mass per unit volume and unit time;

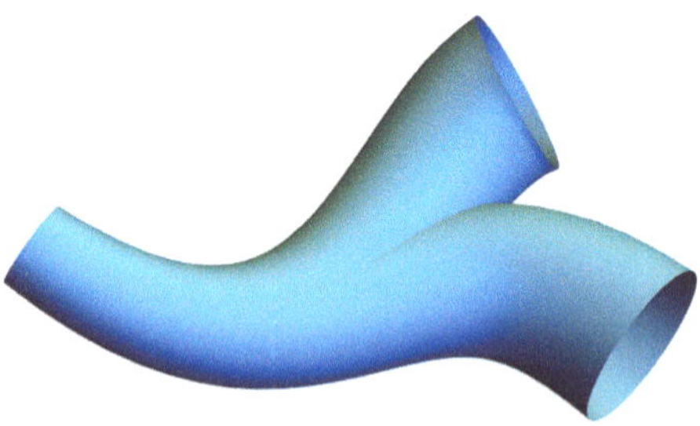

Fig. 2.2 Under the rules imposed by equations of continuum, a single beam can develop singularities, e.g., splitting in two beams

- $\nu_{kj} R_j$ is the production of substance k per unit volume and unit time by reaction j, which is positive if substance k is produced, and negative if it is destroyed, where the coefficient ν_{kj} divided by the molecular mass of substance k is proportional to the stoichiometric coefficient with which substance k appears in the jth reaction.

By the divergence theorem for differentiable vector fields one obtains

$$\int_W \operatorname{div}(\rho_k \mathbf{v}_k)\, dV = \int_{\partial W} \rho_k \mathbf{v}_k \cdot \hat{n}\, dA \tag{2.12}$$

where the dot denotes the scalar product. Because W can be any measurable set, we get

$$\frac{\partial \rho_k}{\partial t} = -\operatorname{div}(\rho_k \mathbf{v}_k) + \sum_{j=1}^{r} \nu_{kj} R_j, \quad k = 1, \ldots, n \tag{2.13}$$

As mass is conserved by chemical reactions, its total variation rate vanishes, $\sum_{k=1}^{n} \nu_{kj} = 0$, hence, summing over all k, the following law is found:

$$\frac{\partial \rho}{\partial t} = -\operatorname{div} \rho \mathbf{v} \tag{2.14}$$

This is the law of conservation of mass, known as *continuity equation*, that gives the local balance of mass in the Eulerian picture, where ρ is the total density and $\mathbf{v}$ the barycentric velocity, (2.8). Introducing the diffusive flow of the substance k, i.e., its motion with respect to the barycentric frame:

$$\mathbf{J}_k = \rho_k(\mathbf{v}_k - \mathbf{v}), \quad \text{which implies} \quad \rho_k \mathbf{v}_k = \rho_k \mathbf{v} + \mathbf{J}_k \quad \text{and} \quad \sum_{k=1}^{n} \mathbf{J}_k = \mathbf{0} \tag{2.15}$$

where $\mathbf{0}$ is the null vector, the following can be written:

$$\begin{aligned}
\frac{\mathrm{d}\rho_k}{\mathrm{d}t} &= -\operatorname{div}(\rho_k \mathbf{v}_k) + \sum_{j=1}^{r} \nu_{kj} R_j + \mathbf{v} \cdot \nabla \rho_k \\
&= -\operatorname{div}(\rho_k \mathbf{v} + \mathbf{J}_k) + \mathbf{v} \cdot \nabla \rho_k + \sum_{j=1}^{r} \nu_{kj} R_j \\
&= -(\nabla \rho_k) \cdot \mathbf{v} - \rho_k \operatorname{div} \mathbf{v} - \operatorname{div} \mathbf{J}_k + \mathbf{v} \cdot \nabla \rho_k + \sum_{j=1}^{r} \nu_{kj} R_j \\
&= -\rho_k \operatorname{div} \mathbf{v} - \operatorname{div} \mathbf{J}_k + \sum_{j=1}^{r} \nu_{kj} R_j
\end{aligned}$$

and

$$\frac{\mathrm{d}\rho}{\mathrm{d}t} = -\mathrm{div}\,(\rho\mathbf{v}) + \mathbf{v}\cdot\nabla\rho = -(\nabla\rho)\cdot\mathbf{v} - \rho\mathrm{div}\,\mathbf{v} + \mathbf{v}\cdot\nabla\rho \tag{2.16}$$

that is to say:

$$\frac{\mathrm{d}\rho}{\mathrm{d}t} = -\rho\,\mathrm{div}\,\mathbf{v} \tag{2.17}$$

which is the continuity equation in the Lagrangian picture, i.e., the expression of the conservation of mass moving along with the fluid. Using the definition of mass fractions, Eq. (2.17) can be rewritten as follows:

$$\rho\frac{\mathrm{d}c_k}{\mathrm{d}t} = -\mathrm{div}\,\mathbf{J}_k + \sum_{j=1}^{r}\nu_{kj}R_j, \quad k = 1, \ldots, n \tag{2.18}$$

The definitions of $\mathbf{J}_k$ and $\mathbf{v}$ imply that only $n-1$ flows $\mathbf{J}_k$ are independent, hence only $n-1$ of the n Eq. (2.18) are independent. Using the specific volume $\omega = 1/\rho$ one also obtains

$$\rho\frac{\mathrm{d}\omega}{\mathrm{d}t} = \mathrm{div}\,\mathbf{v} \tag{2.19}$$

2.3.2 Equations of Motion

The equations of motion of a fluid element contained in a volume element correspond to the local balance of momentum. For an elementary cube of sides Δx, Δy, and Δz, with a corner at $\mathbf{x} = (x, y, z) \in \mathbb{R}^3$, this amounts to an expression for the following equality for a unit of mass:

$$\begin{aligned}&\{\text{rate of change of momentum}\} = \\ &\{\text{rate of momentum input}\} - \{\text{rate of momentum output}\} + \{\text{sum of forces}\}\end{aligned} \tag{2.20}$$

where the input and output contributions correspond to the momentum that enters and exits the volume element because mass enters in and exits from it (convective contribution), while the sum of forces acts as a kind of source or sink of momentum, as determined by the second law of mechanics: the mass in the element is positively or negatively accelerated by forces. Without loss of generality, for simplicity, we consider a single component object, i.e., $k = 1$.

Let us begin with convection in the x-direction. The rate at which the x-component of momentum per unit volume, (ρv_x), flows through the surface element of area $\Delta y\Delta z$, perpendicular to the x-direction in position x, is given by cf. Fig. 2.3:

$$(\rho v_x)\,v_x\Big|_x \Delta y\Delta z \tag{2.21}$$

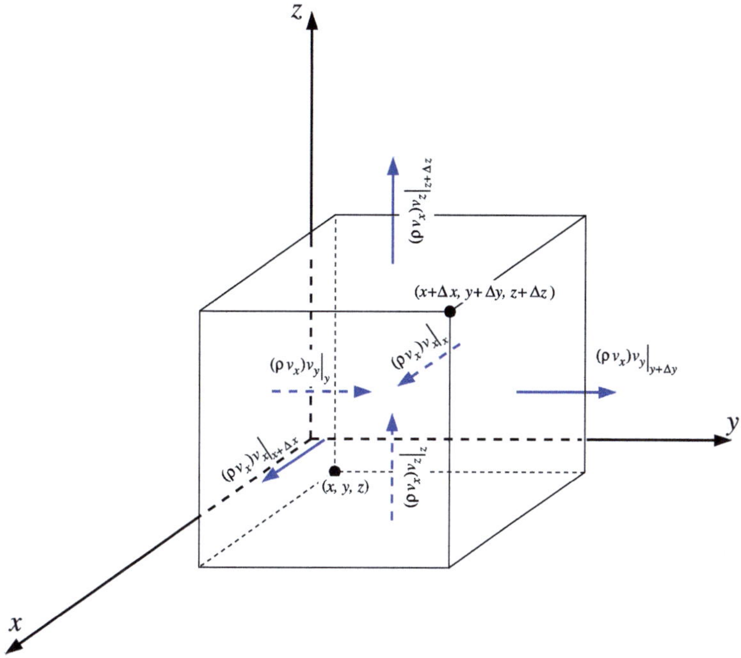

Fig. 2.3 Volume element of sides Δx, Δy, Δz through which a fluid is flowing. The arrows indicate the flux of the x-component of the momentum crossing each of the six faces of the volume

Similarly, at $x + \Delta x$ the x-component of the momentum through the same area flows at the rate

$$(\rho v_x)\, v_x \Big|_{x+\Delta x} \Delta y \Delta z \tag{2.22}$$

The quantity (ρv_x) that flows through the surfaces perpendicular to the y- and z-directions is analogously given by

$$(\rho v_x)\, v_y \Big|_{y} \Delta x \Delta z,\ \ (\rho v_x)\, v_y \Big|_{y+\Delta y} \Delta x \Delta z\ ;\ \text{ and } (\rho v_x)\, v_z \Big|_{z} \Delta x \Delta y,\ \ (\rho v_x)\, v_z \Big|_{z+\Delta z} \Delta x \Delta y \tag{2.23}$$

Therefore, the net flux of (ρv_x) through our elementary cube is given by

$$\left(\rho v_x v_x\Big|_{x} - \rho v_x v_x\Big|_{x+\Delta x}\right) \Delta y \Delta z + \left(\rho v_x v_y\Big|_{y} - \rho v_x v_y\Big|_{y+\Delta y}\right) \Delta x \Delta z + \tag{2.24}$$

$$\left(\rho v_x v_z\Big|_{z} - \rho v_x v_z\Big|_{z+\Delta z}\right) \Delta x \Delta y \cong -\left(\partial_x \rho v_x v_x + \partial_y \rho v_x v_y + \partial_z \rho v_x v_z\right) \Delta x \Delta y \Delta z \tag{2.25}$$

where the approximate equalities

$$\partial_r\, \rho v_x v_r\, \Delta x \Delta y \Delta z \cong \frac{\rho v_x v_r\Big|_{r} - \rho v_x v_r\Big|_{r+\Delta r}}{\Delta r} \Delta x \Delta y \Delta z\,, \quad r = x, y, z \tag{2.26}$$

become exact in the $\Delta r \to 0$ limit. The convective contributions to the variations of the y- and z-components of the momentum in $\Delta x \Delta y \Delta z$ are analogous, hence one eventually gets

$$\{\text{rate of momentum input}\} - \{\text{rate of momentum output}\} \tag{2.27}$$

$$= -\left[\partial_x(\rho v_x \mathbf{v}) + \partial_y(\rho v_y \mathbf{v}) + \partial_z(\rho v_z \mathbf{v})\right] \Delta x \Delta y \Delta z \tag{2.28}$$

Introducing the dyadic product **ab**, which in matrix notation is expressed by

$$\mathbf{ab} = \begin{pmatrix} a_x \\ a_y \\ a_z \end{pmatrix} \begin{pmatrix} b_x & b_y & b_z \end{pmatrix} = \begin{pmatrix} a_x b_x & a_x b_y & a_x b_z \\ a_y b_x & a_y b_y & a_y b_z \\ a_z b_x & a_z b_y & a_z b_z \end{pmatrix}$$

and the divergence of a matrix M, expressed by the following row vector:

$$\text{div} \begin{pmatrix} M_{11} & M_{12} & M_{13} \\ M_{21} & M_{22} & M_{23} \\ M_{31} & M_{32} & M_{33} \end{pmatrix} = \begin{pmatrix} \partial_x & \partial_y & \partial_z \end{pmatrix} \begin{pmatrix} M_{11} & M_{12} & M_{13} \\ M_{21} & M_{22} & M_{23} \\ M_{31} & M_{32} & M_{33} \end{pmatrix} \tag{2.29}$$

$$= \begin{pmatrix} \partial_x M_{11} + \partial y M_{21} + \partial_z M_{31} & \partial_x M_{12} + \partial y M_{22} + \partial_z M_{32} & \partial_x M_{13} + \partial y M_{23} + \partial_z M_{33} \end{pmatrix} \tag{2.30}$$

the contribution (2.28) can be written as

$$-\,\text{div}\, \rho \mathbf{vv}\, \Delta x \Delta y \Delta z \tag{2.31}$$

In turn, the effect of the forces acting on the cube can be subdivided in two contributions: the effect of the contact forces exerted on the surface of the cube $\Delta x \Delta y \Delta z$ by the surrounding matter, and the effect of external forces, acting on the bulk of the cube. The first part is also known as stress, and can be cast in the form:

$$\mathrm{d}F = -P\mathbf{ds}, \quad \mathbf{ds} = \text{surface element} \tag{2.32}$$

where we have introduced the stress tensor P and the vector of magnitude equal to the surface element area, **ds**. This construction is needed because contiguous elements of matter exert on each other forces that may be perpendicular to their contact surfaces that are known as pressure forces. Moreover, contact forces can also be parallel to the surfaces at which they are exerted: they arise when one mass element is dragged by the neighboring one, parallel to their contact surfaces, with or without slip with respect to each other. If there is slip, friction arises cf. Fig. 2.4. Let $P_x = (P_{xx}\ P_{xy}\ P_{xz})$ be the force acting on the x-component of the momentum, ρv_x. It can be exerted through the six different surfaces: normally on $\Delta y \Delta z$ and tangentially on $\Delta x \Delta y$ and $\Delta x \Delta z$. In the first case, i.e., P_{xx}, we have a pressure, while P_{xy} and P_{xz} are shear stresses due to the viscous forces among different layers of matter. The contribution to the variation of ρv_x is given by

$$(P_{xx}|_x - P_{xx}|_{x+\Delta x})\Delta y \Delta z + (P_{xy}|_y - P_{xy}|_{y+\Delta y})\Delta x \Delta z + (P_{xz}|_z - P_{xz}|_{z+\Delta z})\Delta x \Delta y \tag{2.33}$$

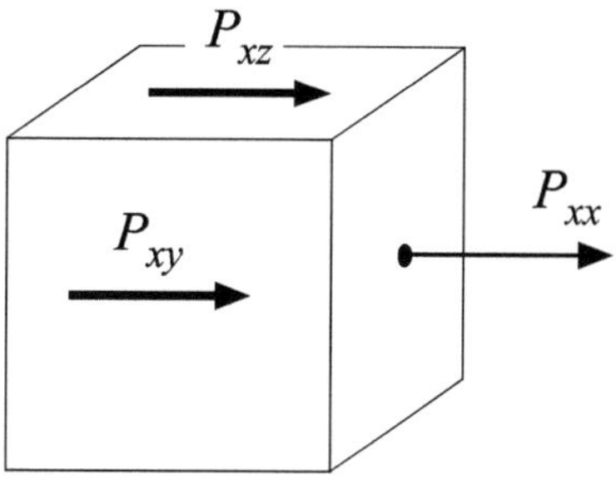

Fig. 2.4 Stress tensor on the surface of a mass element due to contact forces along the x-direction

now dividing by $\Delta x \Delta y \Delta z$ one approximately obtains the divergence of P_x that tends to be exact in the small cube limit:

$$- \text{div } P_x \tag{2.34}$$

which is a scalar. Similarly, the variations of the other components of $\rho \mathbf{v}$ are obtained, making the contribution of the stresses to the variation of $\rho \mathbf{v}$ equal to $-\text{div } P$. Lastly, let $\{\mathbf{F}_k\}_{k=1}^n$ be the external forces per unit mass exerted on chemical k as bulk forces. Their total is given by

$$\sum_{k=1}^{n} \rho_k \mathbf{F}_k \tag{2.35}$$

which directly expresses their contribution to the variations of $\rho \mathbf{v}$. At this point, assembling all contributions in a single equation, the following is obtained:

$$\frac{\partial}{\partial t} \rho \mathbf{v} = -\text{div } (P + \rho \mathbf{v} \mathbf{v}) + \sum_{k=1}^{n} \rho_k \mathbf{F}_k \tag{2.36}$$

in the Eulerian picture; or

$$\rho \frac{\text{d}\mathbf{v}}{\text{d}t} = -\text{div } P + \sum_{k=1}^{n} \rho_k \mathbf{F}_k \tag{2.37}$$

in the Lagrangian picture.

Remark The derivation of the equations of motion has evidenced that the mass in the volume element of interest is considered a block, with no internal structure, consistently with the fact that it is the smallest mass element that can be considered in the present theory. Consequently, there is no energy of interaction within its parts, and its motion is uniquely determined by the action of forces external to it.

This observation is particularly relevant in the energy balance derived in the following subsection.

2.3.3 Kinetic and Potential Energy Balance

The kinetic energy local balance can be derived directly from Eq. (2.37), by scalar multiplication to the right of the vector $\mathbf{v}$:

$$\left(\rho \frac{\mathrm{d}\mathbf{v}}{\mathrm{d}t}\right) \cdot \mathbf{v} = -(\mathrm{div}\ P) \cdot \mathbf{v} + \left(\sum_{k=1}^{n} \rho_k \mathbf{F}_k\right) \cdot \mathbf{v} \tag{2.38}$$

where, in fact, (div P) is a three-dimensional vector, like $\mathbf{v}$ and $\mathbf{F}_k$. The result is

$$\frac{1}{2}\rho \frac{\mathrm{d}v^2}{\mathrm{d}t} = -(\mathrm{div}\ P) \cdot \mathbf{v} + \sum_{k=1}^{n} \rho_k \mathbf{F}_k \cdot \mathbf{v} = -\mathrm{div}\ (P\mathbf{v}) + P : \nabla \mathbf{v} + \sum_{k=1}^{n} \rho_k \mathbf{F}_k \cdot \mathbf{v} \tag{2.39}$$

where v is the magnitude of $\mathbf{v}$, so that $v^2 = \mathbf{v} \cdot \mathbf{v}$, $P : \nabla \mathbf{v}$ denotes the double dot product between P and $\nabla \mathbf{v}$, and $\nabla \mathbf{v}$ is the dyadic product of the vectors $\nabla = (\partial_x\ \partial_y\ \partial_z)$ and $\mathbf{v}$. More explicitly,

$$\mathrm{div}\ (P\mathbf{v}) = \sum_{i,j=1}^{3} \partial_{x_i} \left(P_{ij} v_j\right), \quad P : \nabla \mathbf{v} = \sum_{i,j=1}^{3} P_{ij} \partial_{x_i} v_j \tag{2.40}$$

Similar calculations lead to the equation for the potential energy density

$$\rho \Psi = \sum_{k=1}^{n} \rho_k \Psi_k \tag{2.41}$$

which is assumed to derive from single substance time-independent potentials and to be conserved by chemical reactions. We can thus write

$$\mathbf{F}_k = -\nabla \Psi_k, \quad \sum_{k=1}^{n} \Psi_k \nu_{kj} = 0, \quad \frac{\partial \Psi_k}{\partial t} = 0 \tag{2.42}$$

and

$$\begin{aligned} \frac{\partial \rho \psi}{\partial \mathrm{t}} &= \sum_k \frac{\partial \rho_k}{\partial \mathrm{t}} \psi_k = \sum_{k=1}^{n} \left(-\psi_k \nabla \cdot (\rho_k \mathbf{v}_k) + \sum_j \psi_k \nu_{kj} R_j\right) \\ &= \sum_k \Big(- \nabla \cdot (\rho_k \mathbf{v}_k \psi_k) + \rho_k \mathbf{v}_k \cdot \nabla \psi_k \Big) = \\ &= -\mathrm{div} \left(\sum_k \mathbf{J}_k \psi_k + \rho \psi \mathbf{v}\right) - \sum_k \Big(\rho_k \mathbf{F}_k \cdot \mathbf{v} + \mathbf{J}_k \cdot \mathbf{F}_k\Big) \end{aligned} \tag{2.43}$$

Here, we used $\mathbf{v}_k = \mathbf{v}_k + \mathbf{v} - \mathbf{v}$ in the last line, and the definition $\mathbf{J}_k = \rho_k(\mathbf{v}_k - \mathbf{v})$. Now combining the two expressions for the local balance of kinetic and potential energy we get the local balance equation for the mechanical energy:

$$\frac{\partial}{\partial t}\rho\left(\frac{\mathbf{v}^2}{2} + \psi\right) = -\mathrm{div}\left[æ\left(\frac{\mathbf{v}^2}{2} + \right)\mathbf{v}\right] + \mathrm{P} : \nabla\mathbf{v} - \mathrm{div}\,\mathrm{P}\mathbf{v} - \sum_{\mathrm{k}} \mathbf{J}_{\mathrm{k}} \cdot \mathbf{F}_{\mathrm{k}} - \mathrm{div}\sum_{\mathrm{k}} {}_{\mathrm{k}}\mathbf{J}_{\mathrm{k}} \tag{2.44}$$

where, in particular, the term $P : \nabla\mathbf{v}$ represents a contribution to the kinetic energy due to compression, while the last term of Eq. 2.44 gives the contribution to potential energy due to diffusion processes, since convection is removed in the barycentric fluxes.

Now, the right-hand side of Eq. 2.44 does not merely contain the divergence of the flow of the mechanical energy; source and sink terms appear as well. This means that the mechanical energy, $\rho(\psi + \mathbf{v}^2/2)$, is not locally conserved. Because conservation of energy has resulted in history as a very prolific concept, it is wise to try to save it, assuming that the sum of the kinetic and the potential energy does not represent the whole energy of our element of matter. Thermodynamics follows very successfully this path, introducing the total energy density:

$$e = \frac{\mathbf{v}^2}{2} + \psi + u \tag{2.45}$$

where u is called *internal energy*, since it is not directly evident in observations, and can only be inferred from calculations involving observable quantities. It does not vanish if the kinetic and potential energies do; therefore, it hints at something that is going on, in our little cube, although we do not see it; something that the continuum picture of matter cannot explain. The quantity u is introduced in order to restore conservation of the (total) energy, and to be allowed to write the corresponding continuity equation:

$$\frac{\partial}{\partial t}\rho e = -\mathrm{div}\,\mathbf{J}_e \tag{2.46}$$

Substituting the definition (2.45) of e in Eq. (2.46) provides an explicit expression of $\mathbf{J}_e$ in terms of the other flows and of a new quantity $\mathbf{J}_q$ necessary to complete the balance:

$$\mathbf{J}_e = \rho e\mathbf{v} + P\mathbf{v} + \sum_k \psi_k \mathbf{J}_k + \mathbf{J}_q \tag{2.47}$$

where

- $\rho e\mathbf{v}$ is the convective contribution to the new quantity e;
- $P\mathbf{v}$ is due to the mechanical work;
- $\Psi_k\mathbf{J}_k$ is due to the motion in the field of force;

- $\mathbf{J}_q$ is called *"heat flow"*, i.e., a new form of energy flow introduced to restore energy conservation.

The heat flow constitutes a fundamental and yet subtlest notion of Thermodynamics, because its definition clearly shows that it is not associated to any motion of, or any work done on our matter element; such contributions are accounted for by the other quantities in the energy balance. As observations show that temperature gradients produce some modifications of the state of a given system, excluding phenomena that have no reason to be invoked, such as nuclear reactions, one must admit that $\mathbf{J}_q$ is related to such temperature gradients. In a sense, thermal gradients are sources of heat. With much care, provided $\mathbf{J}_q \neq 0$, we can thus introduce the quantity *"Heat"*, q, which appears to have the units of energy, stating that it obeys

$$\rho \frac{\mathrm{d}q}{\mathrm{d}t} = -\mathrm{div}\, \mathbf{J}_q \tag{2.48}$$

In the end, the local balance equation for the internal energy u may be written as

$$\frac{\partial \rho u}{\partial t} = -\mathrm{div}\,(\rho u \mathbf{v} + \mathbf{J}_q) - P : \nabla \mathbf{v} + \sum_k \mathbf{J}_k \cdot \mathbf{F}_k \tag{2.49}$$

where the heat flow can be interpreted as contributing a certain amount of heat to the internal energy. It should be stressed that it makes no sense to speak of heat in absence of heat flows, which only exist in presence of temperature gradients. Heat refers to one method of energy transfer, it makes no sense without such transfer.

These concepts can be intuitively understood in terms of Callen's pond analogy (Callen 1985) p. 20, in which he draws a parallel between internal energy and heat, on the one hand, and water in a pond and rain, on the other hand. Like rain does not exist when water does not pour from the skies, so heat does not exist when energy is not transferred in response to a temperature gradient.

2.3.4 First Law of Thermodynamics

Having an equation for the internal energy u, we can express the first law of Thermodynamics. To do that, we decompose the stress tensor into two parts: $P = p\mathbb{1} + \Pi$, where p is the pressure, $\mathbb{1}$ the identity matrix, and Π contains the off-diagonal terms of the stress tensor. Taking the time derivative of ρu, and recalling that

$$\omega = \frac{1}{\rho}; \quad \frac{\mathrm{d}\rho}{\mathrm{d}t} = -\rho \mathrm{div}\, \mathbf{v}; \quad \rho \frac{\mathrm{d}q}{\mathrm{d}t} = -\mathrm{div}\, \mathbf{J}_q; \quad \rho \frac{\mathrm{d}\omega}{\mathrm{d}t} = \mathrm{div}\, \mathbf{v} \tag{2.50}$$

we get

$$\begin{aligned}
\frac{\mathrm{d}}{\mathrm{d}t}\rho u &= \frac{\partial \rho u}{\partial t} + \mathbf{v}\cdot\nabla(\rho u)\\
\rho\,\frac{\mathrm{d}u}{\mathrm{d}t} + u\,\frac{\mathrm{d}\rho}{\mathrm{d}t} &= -\mathrm{div}\,(\rho u\mathbf{v} + \mathbf{J}_q) - p\;\mathrm{div}\;\mathbf{v} - \Pi : \nabla\mathbf{v} + \sum_k \mathbf{J}_k\cdot\mathbf{F}_k + \mathbf{v}\cdot\nabla(\rho u)\\
\rho\,\frac{\mathrm{d}u}{\mathrm{d}t} &= \rho\,\frac{\mathrm{d}q}{\mathrm{d}t} - p\,\rho\,\frac{\mathrm{d}\omega}{\mathrm{d}t} - \Pi : \nabla\mathbf{v} + \sum_k \mathbf{J}_k\cdot\mathbf{F}_k\\
\frac{\mathrm{d}u}{\mathrm{d}t} &= \frac{\mathrm{d}q}{\mathrm{d}t} - p\,\frac{\mathrm{d}\omega}{\mathrm{d}t} - \omega\Pi : \nabla\mathbf{v} + \omega\sum_k \mathbf{J}_k\cdot\mathbf{F}_k
\end{aligned} \tag{2.51}$$

which is the First Law of Thermodynamics, expressed in various ways.

2.4 Entropy and the Second Law

Macroscopic systems in thermodynamic equilibrium can be associated with a state function called *"entropy"*, often denoted by the letter S. Its variations can be expressed in full generality as

$$\mathrm{d}S = \mathrm{d}_e S + \mathrm{d}_i S \tag{2.52}$$

where $\mathrm{d}_e S$ is supplied to the system through its surface, while $\mathrm{d}_i S$ is produced inside the system. Calorimetry experiments have led to the formulation of the Second Law of Thermodynamics, which states that:

$$\mathrm{d}_i S \geq 0, \quad \text{with `` = '' only for reversible transformations} \tag{2.53}$$

This means that for isolated systems, i.e., systems that exchange no matter nor energy with their environment, so that $\mathrm{d}_e S = 0$, the following holds:

$$\mathrm{d}S \geq 0 \tag{2.54}$$

Moreover, for closed systems, systems that exchange only energy with their environment, the theorem of Carnot and Clausius implies

$$\mathrm{d}_e S = \frac{\mathrm{d}Q}{T} \tag{2.55}$$

where $\mathrm{d}Q$ is the heat supplied to system, via its surface, at temperature T. Therefore, in general, one can write

$$\mathrm{d}S \geq \frac{\mathrm{d}Q}{T} \tag{2.56}$$

which is another form of the second law.

Introducing the specific entropy s, the total entropy flow per unit area and unit time $J_{s,tot}$, and the entropy production rate per unit volume, one can also write

$$S = \int_D \rho s \, \mathrm{d}V \, ; \quad \frac{d_e S}{dt} = -\int_{\partial D} \mathbf{J}_{s,tot} \cdot \hat{n} \, \mathrm{d}A; \quad \frac{\mathrm{d}_i S}{\mathrm{d}t} = \int_D \sigma \, \mathrm{d}V \tag{2.57}$$

where D is the region of space under consideration, or a part of an object of interest. Then,

$$\int_D \left(\frac{\partial \rho s}{\partial t} + \mathrm{div} \mathbf{J}_{s,tot} - \sigma \right) \mathrm{d}V = 0 \tag{2.58}$$

and because D is any measurable set in $\mathbb{R}^3$, the following holds:

$$\frac{\partial \rho s}{\partial t} = -\mathrm{div} \, \mathbf{J}_{s,tot} + \sigma; \quad \rho \frac{\mathrm{d}s}{\mathrm{d}t} = -\mathrm{div} \, \mathbf{J}_s + \sigma \tag{2.59}$$

where $\mathbf{J}_s = \mathbf{J}_{s,tot} - \rho s \mathbf{v}$, and $\sigma \geq 0$ for $\mathrm{d}S_i$ to always be non-negative.

Now, thermodynamic observations show that entropy is a function of the thermodynamic state, which means that it must depend only on the parameters that characterize such a state, so it can be cast in the form $s = s(u, \omega, c_1, \ldots, c_n)$, where u is the specific internal energy, ω the specific volume, and c_k the mass fraction of substance k. Combining the expressions obtained above and in the previous sections, one obtains (De Groot and Mazur 1984)

$$\rho \frac{ds}{dt} = -\frac{1}{T} \mathrm{div} \, \mathbf{J}_q - \frac{1}{T} \Pi : \nabla \mathbf{v} + \frac{1}{T} \sum_{k=1}^{n} \mathbf{J}_k \cdot \mathbf{F}_k + \frac{1}{T} \sum_{k=1}^{n} \mu_k \mathrm{div} \, \mathbf{J}_k - \frac{1}{T} \sum_{j=1}^{r} R_j \, A_j \tag{2.60}$$

where

$$A_j = \sum_{k=1}^{n} \nu_{kj} \mu_k, \quad j = 1, \ldots, r \tag{2.61}$$

defines the *affinity* A_j corresponding to the jth reaction. Rearranging the terms, one eventually gets

$$\begin{aligned} \rho \frac{\mathrm{d}s}{\mathrm{d}t} = & -\mathrm{div} \left[\frac{1}{T} \left(\mathbf{J}_q - \sum_{k=1}^{n} \mu_k \mathbf{J}_k \right) \right] \\ & - \frac{1}{T^2} \mathbf{J}_q \cdot \nabla T - \frac{1}{T} \sum_{k=1}^{n} \mathbf{J}_k \cdot \left[T \nabla \left(\frac{\mu_k}{T} \right) - \mathbf{F}_k \right] - \frac{1}{T} \Pi : \nabla \mathbf{v} - \frac{1}{T} \sum_{j=1}^{r} R_j \, A_j \end{aligned}$$

Therefore, one may identify the divergence term as the entropy flux:

$$\mathbf{J}_s = \frac{1}{T} \left(\mathbf{J}_q - \sum_{k=1}^{n} \mu_k \mathbf{J}_k \right) \tag{2.62}$$

and the rest as the internal source of entropy:

$$\sigma = -\frac{1}{T^2}\mathbf{J}_q \cdot \nabla T - \frac{1}{T}\sum_{k=1}^{n}\mathbf{J}_k \cdot \left[T\nabla\left(\frac{\mu_k}{T}\right) - \mathbf{F}_k\right] - \frac{1}{T}\Pi : \nabla\mathbf{v} - \frac{1}{T}\sum_{j=1}^{r} R_j\, A_j \tag{2.63}$$

which must verify $\boldsymbol{\sigma} \geq 0$. The separation in flux and production rate is determined by a number of constraints that entropy production must satisfy, in particular:

1. σ must vanish at equilibrium;
2. σ must be invariant under Galilei transformations.

One then realizes that the following holds:

$$\frac{\mathrm{d}s}{\mathrm{d}t} \geq -\int_{\partial D} \frac{1}{T}\mathbf{J}_q \cdot \hat{n}\ \mathrm{d}A \tag{2.64}$$

which is consistent with the Carnot-Clausius theorem, as it needs be. Note that $\mathbf{J}_s$ consists of the reduced heat flow $\mathbf{J}_q/T$ plus a contribution due to the flow of matter. The irreversible entropy production rate σ is thus due to heat conduction, plus diffusion of matter, plus viscous forces, plus chemical reactions.

In particular, the contribution due to diffusion requires the presence of at least two components ($n \geq 2$) that move with respect to each other and effectively interact, giving rise to some friction and hence to Fickian dissipation. In fact $\mathbf{J}_k$ vanishes if only the substance k is present, cf. Eq. (2.15). Moving within walls that do not allow energy exchange, even though they may lead to complicated motions and evolutions, does not lead to any diffusion-related dissipation. In fact, molecules of a single fluid in thermodynamic equilibrium do diffuse through each other, but regardless of the shape of their container, such self-diffusion by the definition of equilibrium does not lead to any energy loss or to any calorimetric effect.

Concerning the non-negative character of $\boldsymbol{\sigma}$, we observe that each of its parts is a product of a flow (either the heat flow $\mathbf{J}_q$ or the diffusion flow, $\mathbf{J}_k$, or the momentum flow represented by Π, or the chemical reaction rate R_j), and of an intensive state variable: gradients of T, μ_k, $\mathbf{v}$, of external potentials, and of differences in chemical affinities A_j, that are collectively called thermodynamic forces or affinities. The result is a bilinear form that turns positive definite under certain conditions. For instance, the most obvious and best understood case is that corresponding to the linear regime, that in which the flows are proportional to the thermodynamic forces.

The above relations are very general. In practice, as long as matter can be regarded as a continuum, with existing first- and second-order derivatives, they hold. At the same time, they are very complicated, hence of little practical use. Under assumptions that turn to be very generally verified at the scales which Thermodynamics addresses, they turn into the standard macroscopic, phenomenological equations, that are simple and of great success. This is the already mentioned linear regime, in which, for instance, the proportionality between the momentum flow and the velocity gradient of a sheared fluid is proportional to each other, via the fluid viscosity. This assumption leads to the Navier-Stokes equations:

$$\rho \frac{d\mathbf{v}}{dt} = -\nabla p + \eta \Delta \mathbf{v} + \left(\frac{1}{3}\eta + \eta_v \right) \nabla \mathrm{div}\, \mathbf{v} \tag{2.65}$$

where η and η_v are known as shear viscosity and bulk viscosity, respectively. Analogously, for one object with $\mathbf{v} = \mathbf{0}$, the energy balance yields the Fourier law of heat conduction:

$$\rho c_v \frac{\partial T}{\partial t} = \lambda \Delta T \tag{2.66}$$

where c_v is the specific heat at constant volume and λ is the heat conductivity. In general, the linear regime is characterized by a set of fluxes $\{J_i\}_{i=1}^n$ that are linear in the thermodynamic forces $\{X_i\}_{i=1}^n$ and that determine the entropy production as

$$\boldsymbol{\sigma} = \sum_{i=1}^{n} J_i X_i \tag{2.67}$$

In a d-dimensional space, the J_i and the X_i are Cartesian components of all fluxes and forces. If there are m such flows and forces, we have $n = dm$. Then, the linear relation between fluxes and thermodynamic forces takes the form:

$$J_i = \sum_{k=1}^{n} L_{ik} X_k \,, \quad i = 1, \ldots, n \tag{2.68}$$

for which the Onsager reciprocal relations hold: i.e., $L_{ik} = L_{ki}$. One then has De Groot and Mazur (1984) Chapter IV:

$$\eta_v = \frac{L_{vv}}{T}, \quad \lambda = \frac{L_{qq}}{T^2} \tag{2.69}$$

and similar relations for all linear transport coefficients. Interestingly, while the terminology may suggest a limited range of applicability of the linear theory, in fact nonlinear phenomena are known, this range is quite wide, in terms of our daily experience and technology.

One notable result concerning the linear regime is Prigogine's minimum entropy production, which states that close to equilibrium and at high temperatures, the following holds:

$$\frac{d\sigma}{dt} \geq 0 \tag{2.70}$$

where equality only holds at a steady state. In other words, steady states in the linear regime are characterized by a minimum of the entropy production allowed by the external constraints.

Clearly, none of the relations derived in this chapter makes any sense if the corresponding thermodynamic fields do not exist, and the condition of their existence is that LTE be verified. In Kubo's words:

> A thermodynamic state in the general sense is a state in which local equilibrium exists. If the system is composed of subsystems, each of which is in thermal equilibrium, it is a natural extension of the definition of the entropy to define it for the whole system as the sum of the entropies of subsystems. Kubo (1968) p. 74.

When this is not the case, we are outside the realm of Thermodynamics. But from within, nothing more can be stated: the thermodynamic fields exist, when they do. A system in which they do exist can then be investigated from the thermodynamic point of view. The good news is that thermodynamics, in fact even its linear regime, widely applies. But how widely?

An answer to this question requires stepping out of the thermodynamic realm and adopting the atomistic picture of matter, in which LTE has a physical interpretation. The next section is devoted to this issue. But first it is interesting to ponder over one general feature of physical theories: they are so successful, and yet cannot justify themselves. In a sense, one may only say that they hold when they do. Quoting Callen's book on the notion of equilibrium on which the whole construction stands:

> In practice, the criterion for equilibrium is circular. Operationally, a system is in an equilibrium state if its properties are consistently described by thermodynamic theory!
>
> It is important to reflect upon the fact that the circular character of thermodynamics is not fundamentally different from that of mechanics. A particle of known mass in a known gravitational field might be expected to move in a specific trajectory; if it does not do so we do not reject the theory of mechanics, but we simply conclude that some additional force acts on the particle. Thus, the existence of an electrical charge on the particle and the associated relevance of an electrical force cannot be known a priori. It is inferred only by circular reasoning in that dynamical predictions are incorrect unless the electric contribution to the force is included. Our model of a mechanical system (including the assignment of its mass, moment of inertia, charge, dipole moment, etc.) is "correct" if it yields successful predictions. Callen (1985), p. 15.

This way of proceeding is in fact very effective, because the apparently vicious circle is broken by experience. That anchors what could merely be a logically consistent speculation to reality and proves that its predictions are consistent with natural phenomena: they can be trusted. Consider, for instance, the discovery of Pluto. The behavior of Uranus would have been consistent with Newtonian mechanics only if some other unobserved celestial body would have been present. Analogously, experience reveals that thermodynamics is much more than a rationally consistent mathematical framework. It applies very widely in the description of macroscopic phenomena and, indeed, it constitutes the starting point for the development of most modern technology, ranging from power production to the bottleneck for the advancement of computer technology, which is constituted by heat generation. Therefore, even the apparently most advanced or appealing theories meant to justify or explain the foundations of Thermodynamics must be looked upon with some concern if they happen to disagree with any point of Thermodynamics. Suppose one day the derivation of the Navier-Stokes equations from the Hamiltonian dynamics of a class of particles is proven impossible, and not just extremely difficult, as is currently believed; shall we throw away the Navier-Stokes equations as invalid description of fluids? Given their success, we will rather conclude that Hamiltonian dynamics or the postulated interactions are not suitable to the task. A different microscopic description will be required. While the success of the Hamiltonian description in a huge variety of equilibrium and nonequilibrium phenomena is not in question, it is good to keep this in mind when new, perhaps exotic or anomalous phenomena are to be investigated from a microscopic point of view. Of course, beyond certain scales, Thermodynamics is not expected to hold, and, in that case, its extensions should instead be taken with a grain of salt.

Let us investigate what our (necessarily limited, but quite deep) understanding of microscopic dynamics may contribute to our understanding of thermodynamics, equilibrium, or nonequilibrium it may be. We conclude this chapter with a glimpse of the atomistic picture of thermodynamic processes.

2.5 Microscopic Conditions for Local Equilibrium

Let us discuss how thermodynamic phenomena might be cast within our picture of the microscopic components of matter. The issue, in fact, is not whether matter is made of atoms and molecules but how our knowledge of these invisible microscopic objects can account for the observable macroscopic phenomena. The fact that atoms could not be seen in the past, and even today they are seen only indirectly through machines such as scanning tunneling microscopes, should not stop us. Indeed, much earlier than sophisticated electronics were devised, atoms have been *counted*, confirming their existence, within a most convincing rational scheme: the theory of the Brownian motion that has been validated a countless times. Feynman then dared to state:

> If, in some cataclysm, all of scientific knowledge had to be destroyed, and only one sentence passed on to the next generation of creatures, what statement would contain the most information in the fewest words? I believe it is the atomic hypothesis (or the atomic fact, or whatever you wish to call it) that all things are made of atoms–little particles that move around in perpetual motion, attracting each other when they are a little distance apart, but repelling upon being squeezed into one another. In that one sentence, you will see, there is an enormous amount of information about the world, if just a little imagination and thinking are applied. Feynman et al. (1964) Vol. 1.

Imagination is certainly needed because even solving the microscopic equations of motion, something so far impossible, would not give us an intelligible picture of the macroscopic world. What would we do with myriads of graphs and numbers with very detailed information on atom trajectories if we were concerned, for instance, about warming up or cooling down a cup of coffee?

Let us think of matter as made of tiny invisible particles that move erratically in space (even in solids, where they remain close to given positions), obeying the laws of mechanics. Consistently with our discussion in Sect. 2.1, a macroscopic (continuum) description appears to require three widely separated scales in space and time, and no more: the microscopic, the mesoscopic, and the macroscopic scales:

$$\ell \ll \delta L \ll L; \quad \tau \ll \delta t \ll t \tag{2.71}$$

within which correspondingly different phenomenologies occur. In the case of gases, ℓ is the mean free path, i.e., the distance that on average particles travel from collision to collision with other particles; τ is the corresponding mean free time; L is the typical system size at the level of observations; and t is the time required by a measurement tool to report a reading. The distance δL and the time δt require more to be characterized. In the first place, δL must be large enough for a cube of volume δL^3 to contain a sufficiently large number N of properly interacting particles, that fluctuations of this number due to particles entering and exiting be negligible. The role of interactions is to randomize the motion of the particles, so that they do not move in straight lines that cross the cell of side δL in an instant, but remain inside for a good amount of time. The randomization of motion also yields, for sufficiently large N, a uniform distribution of mass within the cell centered at x, so that it can be characterized by a single number $\rho(x)$. Then, for sufficiently large N and random motion, the bulk dominates surface effects, making the notion of density sufficiently stable that measuring it is a sensible operation. The same must hold for all other thermodynamic quantities, like the momentum and energy of a cell. These quantities can be exchanged by the particles inside a cell, with particles outside the cell also remaining within the cell, via interparticle interactions. Therefore, to make the exchange with other cells negligible compared to the bulk of the cell, large N does not suffice. It is also required that the range of interactions be short compared to δL.

In the presence of erratic motion, such as that determined by short-range defocusing repulsive interactions, an estimate of the fluctuations of the number of particles inside a cell can be obtained treating N as a random variable, sum of i.i.d. single particles contributions. Then, for sufficiently large N, the fluctuations of N are of order $O(\sqrt{N})$. This provides one criterion to set the size δL, given the density of particles, since $\sqrt{N}$ must be negligible compared to N. At the same time, δL has to be so small with respect to L, that in the equations derived in the previous sections it can be considered as a single point x.

The time δt must be long enough for the particles in δL^3 to realize the above scenario: the density and any other observable of interest must be homogeneous within a cell, so that one value characterizes the whole. At the same time, δt must be short enough to be perceived as an instant at the macroscopic level.

A number of constraints arise on the separation of the three scales mentioned here. Eventually, if all conditions are satisfied, the mass in δL^3 will constitute a small thermodynamic system, with corresponding thermodynamic properties. The condition of LTE holds if all cells of mesoscopic size δL relax to their homogeneous equilibrium in a mesoscopic time, which is infinitesimal with respect to the observation time. Being small and large relative concepts, this reasoning can be scaled. In principle, systems not described by thermodynamic laws within certain (even macroscopic) scales may appear thermodynamic on larger scales.

To be concrete, take the electrons in copper at 273 K, as they collide with the lattice atoms. For them, τ is estimated at about (Ashcroft and Mermin 1976)

$$\tau = 2.7 \cdot 10^{-14}\ \mathrm{s} \tag{2.72}$$

Their Fermi level energy takes the value

$$E_F = \frac{1}{2} m_* v_F^2 = 7\mathrm{eV} \tag{2.73}$$

with an effective mass of[3]

$$m_* = 1.3 m_e \tag{2.74}$$

Substituting the actual values:

$$m_e = 0.511 \frac{M_e V}{c^2} = 9.1095 \cdot 10^{-31}\ \mathrm{kg}; \quad c = 2.998 \cdot 10^8\ \mathrm{m/s} \tag{2.75}$$

one obtains a Fermi level electron speed like:

$$v_F = \sqrt{\frac{2E_F}{m_*}} = \sqrt{\frac{2 \cdot 7\mathrm{eV} \cdot (2.998 \cdot 10^8 \mathrm{m/s})^2}{1.3\dot{0}.511 \cdot 10^6 \mathrm{eV}}} = 13.763 \cdot \frac{10^8}{10^3} \frac{\mathrm{m}}{\mathrm{s}} = 1.376 \cdot 10^6 \mathrm{m/s} \tag{2.76}$$

and a characteristic microscopic distance given by

$$\ell = v_F \tau = 1.376 \cdot 10^6 \frac{\mathrm{m}}{\mathrm{s}} \cdot 10^{-14} 2.7\,\mathrm{s} = 3.716 \cdot 10^{-8} \mathrm{m} \tag{2.77}$$

Therefore, in a cube of side $\delta L = 10\ell = 3.716 \cdot 10^{-7}$ m one expects 10^3 collisions per collision time τ, and in a time interval of $10^2 \tau = 2.7 \cdot 10^{-12}$ s, relaxation to local equilibrium is achieved in that cube. This means that δt can be taken to be of the order of 10^{-12} s in copper at 273 K. Therefore, times of order of 10^{-10} sec and volumes of order $(10^{-6})^3$ are expected to make LTE applicable.

As another example, take an ideal gas at a pressure of 1 atm and a temperature of 298 K. Then in a cube of micrometer sides, which is definitely small to our eyes, there are approximately 10^{13} molecules, making the ratio $\sqrt{N}/N \approx 10^{-7}$, which is certainly negligible. For liquids and solids, the same value corresponds to even smaller volumes. Hence, it is meaningful to assign a mole number density to a volume with a characteristic size of one micrometer.

While these examples show that LTE is a viable condition for experiments on macroscopic objects, current technology can probe situations that violate these restrictions: 10^{-15} sec laser pulses, for instance, can be produced and utilized; likewise 10^{-9}m scales can be engineered. Astrophysical phenomena (e.g., cosmic rays

[3] In a periodic potential, electrons are accelerated by fields as if they had a different mass, which may be larger or smaller than the electron mass; in fact, it may even be negative.

hitting space ships), high vacuum pumps, or transport in micro-porous membranes often violate LTE, although they concern macroscopic objects. The common feature is that the characterizing microscopic and macroscopic scales are not sufficiently separated: in these cases $l \approx L$. Among non-macroscopic cases in which the LTE condition is violated, we have experiments at the nanoscale, in which work is quite rapidly done, and unlike thermodynamic work, it changes from experiment to experiment.

Concerning the internal energy, heat, and dissipation, which are not mechanical quantities, the microscopic picture can be illustrated as follows. Take one object composed by N identical particles of mass m. The velocity $\mathbf{v}_i$ of each particle can be expressed as

$$\mathbf{v}_i = \mathbf{v} + {}^{\circ}{}_i \tag{2.78}$$

where ν_i is the deviation from the velocity $\mathbf{v}$ of the center of mass of the particles. The kinetic energy of the system is then given by:

$$K = \frac{1}{2}\sum_{i=1}^{N} m(\mathbf{v} + \nu_i)^2 = \frac{N}{2} m\mathbf{v}^2 + \frac{1}{2}\sum_{i=1}^{N} m{\nu_i}^2 = K_{\text{macro}} + K_{\text{micro}} \tag{2.79}$$

where the sum of the cross terms vanishes, and K_{macro} is the kinetic energy of the ordered macroscopic motion, while K_{micro} is that of the internal invisible motion. This is the kinetic contribution to the internal energy, due to the disordered molecular motion. In addition to this, the internal energy includes the particles interaction potential energy.

To compare the orders of magnitude of K_{macro} and K_{micro}, let us take water flowing at room temperature $T \approx 300\text{K}$, with speed $|\mathbf{v}| \sim 1\text{cm/s}$. As the mass of an H_2O molecule is $m \approx 3\ 10^{-23}\text{g}$, its kinetic energy corresponding to such a motion is about $1.5\ 10^{-23}$ erg. The average kinetic energy per degree of freedom in the center of mass frame is instead estimated by the energy equipartition principle, discussed in Chap. 4, to be

$$\frac{1}{2} k_B T \approx 3 \cdot 10^{-14}\ \text{erg} \tag{2.80}$$

which leads to $K_{\text{macro}}/K_{\text{micro}}$ quite smaller than 10^{-9}. Of course, the ratio grows if the macroscopic ordered motion increases its speed, but even growing by three or four orders of magnitude, or more, it remains negligible compared to the energy stored in the disordered microscopic motions. Moreover, considering that the internal energy also includes the interparticle interaction energy, one realizes that it is a huge reservoir, although not immediately available to do work. In microscopic terms, it is mechanical energy like that of the ordered motions. However, energy in an ordered form can be used to do work, like pushing the blades of water mills. Thanks to this distinction, convection is a totally different mechanism of macroscopic transport than diffusion or conduction. Such a difference fades if N is small.

One paradigmatic mechanism for increasing the disorder, hence to dissipate energy, is friction between two surfaces in relative motion, or between different layers of a sheared fluid. Microscopically, there is no surface separating the two objects, because the atoms of both fluctuate and do not define a surface. However, if the objects or layers macroscopically move with respect to each other, some atom of the first may invade the space of the second and be scattered by its atoms. As very many such events take place, and because the motions are by and large random, on average the work done to move the two objects mainly increases the disorder. This is then revealed by the heat released to the environment.

However, the degree of disorder and the number of events that justify the above considerations must be very high. Otherwise, no distinction between mechanical and thermodynamical properties can be made. Consider, for instance, the free expansion of a gas through a hole, from a small container of volume V_s to a large adiabatic container of volume V_l. For perfect gases, the initial and final equilibria are easily calculated, and the entropies of the two states, in particular, are proportional to the logarithms of the volumes V_s and V_l. Could one use the entropy balance Eq. (2.59) to follow the variation of the entropy of the system during the process? In general, no. If particles do not interact sufficiently, they would flow through any small region of V_l at incredible speeds, crossing mesoscopic cells in less than the blink of an eye. Before the final equilibrium is nearly approached, LTE does not hold, and none of the thermodynamic relations applies (Fireman 1902). During the free expansion, the disordered motion is not fully developed. In principle, temperature does not exist, since it is precisely a measure of the disordered kinetic energy. In any event, the validity or futility of such a thermodynamic property remains process dependent. And what about the volume or the density of a system made of one or two particles, or, in the case, just a few? Is the volume larger when they are far apart, compared to when they are close? There are various attempts to give a volume to a small number of particles, but whichever one chooses, it will be something that goes beyond Thermodynamics. Only if the volume is filled by a continuum, it safely corresponds to the volume of the system. If particles are only a few, other laws may do better justice to the phenomenon than those of thermodynamics, although the formalism is so robust that violations may be hidden at equilibrium.

The main consequence of violating LTE is that treating matter as a continuum is not justified and that the macroscopic laws may keep a merely statistical meaning. In particular, they may not be robust with respect to parameter's variations, while they are robust if LTE holds. In such circumstances, alternative approaches are, in principle, necessary. Like other theories of physics, thermodynamics has proved to apply more widely than expected. Nevertheless, as far as the theory goes, one cannot overstate the importance of two ingredients for LTE to be established:

1. large N, so that the behavior of single particles and of the whole are well distinguished;
2. interactions among the particles, so that their motion is randomized and, together with large N, makes certain the statistical inferences, for a given single system of interest.

These are required for mass, energy, and momentum to be transferred diffusively, with no matter transport, so that conduction and convection result in substantially different phenomena. When LTE is realized, a coarser than a microscopic eye will not see the granularity of matter, like from the ground we do not see the droplets that make a cloud, and we see a continuum. The form and strength of the interactions, which are intrinsic to the constituents of matter, determine the values of δL and δt, as intrinsic, not subjective, properties of the system.

Much more should be said to characterize the notion of LTE. The interested readers are referred to excellent books, such as Spohn (1991), Kondepudi and Prigogine (1998), Jou et al. (2010), Kreuzer (1981) to mention but a few. In Chap. 6, we will formalize some of the concepts of this section.

2.6 Current Research

Thermodynamics, far from being an old subject, settled and replaced by more advanced atomistic theories, is thriving, being used and developed in the most advanced technological and scientific research areas. Here, it is only possible to mention a few of its countless applications to give an idea of the breadth of its modern uses. Naturally, thermodynamics continues to be used in order to improve the efficiency of all machines, including electronic devices, heating and cooling systems, and insulation including clothing. The experience acquired in centuries of research is now transferred to understanding the behavior of biological matter. This includes microscopic systems such as bacteria, and macroscopic organs and tissues such as the cardiac muscle, cartilage, and tumor structures. Material sciences make ever greater use of thermodynamics, e.g., to develop new materials, such as advanced functional metal oxides for energy applications, and materials to deal with radiation damage and annealing. Atmospheric and environmental thermodynamics is obviously topical, concerning both local human settlements and the global climate.

The Brownian Motion and Its Legacy 3

Interesting questions indeed! One almost regrets having to die long before they are decided.

Uber die Entwicklung der Methoden der theoretischen Physik, Ludwig Boltzmann

How do we know that matter is really made of atoms? Because they are not visible, the question stood wide open from antiquity to recent times. In fact, at the meeting of natural scientists in Munich in 1899, Boltzmann stated:

> Of present day molecular theory, notwithstanding any addition and modifications, will the essential features nevertheless survive, or will the future one day bring an atomic theory that is totally different from today's; or, contrary to my demonstration, will it be found one day that the idea of a pure continuum affords the best picture? [...] Interesting questions indeed! One almost regrets having to die long before they are decided. How immoderate we mortals are! Delight in watching the fluctuations of the contest is our true lot. Boltzmann (1974).

Boltzmann did not know the answer all physicists would have accepted was soon to come. The breakthrough came with Einstein, who set to investigate the Brownian motion with the purpose of demonstrating that atoms do indeed exist (Einstein 1905). Brownian motion was discovered by the groundbreaking observations by the English Botanist Robert Brown in 1928 of the seemingly erratic and incessant motion of tiny particles extracted from pollen grains (Brown 1828). That seminal report marks the birth of diffusive phenomena, and its precise understanding gained by the investigations of Einstein (1905), Sutherland (1905), Smoluchowski (1906), and Langevin (1908), the triumph of the atomistic theory of matter.

C. Mejía-Monasterio and L. Rondoni, *Nonequilibrium Statistical Mechanics I*,
Theoretical and Mathematical Physics,
https://doi.org/10.1007/978-3-031-98249-1_3

3.1 Brownian Motion: Fluctuations Reveal Atoms

Suppose a particle of mass m is dragged by a constant force F in a fluid that is a continuous viscous medium. For simplicity, consider the motion in a single direction, since the reasoning can be repeated separately for the other directions of space. Letting x be the position of the particle and $\dot{x} = v$ its velocity, the motion of the particle is described by the following equations:

$$m\dot{v} + \alpha v = F; \quad v(t) = Ce^{-t/\tau} + v_\infty; \quad v_\infty = \frac{1}{\alpha}F = \frac{1}{m\gamma}F; \quad \mu = \frac{1}{m\gamma} \tag{3.1}$$

Here, $\gamma = \alpha/m$, and $\tau = 1/\gamma$ is the characteristic time scale of the phenomenon, and μ, called *mobility* of the particle in that fluid, is the proportionality constant between the asymptotic velocity v_∞ and the driving force F. The friction force, in turn, is expressed by Stokes' law $F_c = -\alpha v$, where the dimensions of α are $[\alpha] = Kg/s$. Note that Stokes' law is valid in the linear regime, normally attained at sufficiently weak friction forces. If there is no force, $F = 0$, the solution of (3.1) is

$$v(t) = v(0)e^{-t/\tau} \tag{3.2}$$

As an example, let the particle be spherical of radius a, so that the friction coefficient is given by $\alpha = 6\pi a\eta$, where η is the fluid viscosity. Let the sphere represent a pollen grain, for which $a = O(10^{-2})$ cm and $m = O(10^{-7})$ g. Supposing the fluid is water at room temperature and pressure, one has $\eta_{H_2O} = 10^{-2}$g/(cm · s). Hence, $\tau = O(10^{-4})$s, and whatever the initial speed, Eq. (3.2) predicts that in one second $v(0)$ should be reduced by a factor $\exp(-10^4)$!

In contrast to that, the particle continues to move and does not reveal any trend toward immobility; rather the particle moves erratically, in a totally unpredictable fashion. Taking as valid the mechanical description of the particle's motion, one concludes that the Stokes' force is not the only one acting on the particle and because no external action is applied, there must be forces that are generated within the fluid, in addition to the viscous force. Moreover, such forces do not seem to play a role when macroscopic objects are immersed in the fluid. Only very light objects are affected, as if they were hit by invisible objects. What else could these invisible objects be, if not the molecules that constitute the fluid? That would even explain the randomness of the motion of pollen grains.[1] One of the first to propose this picture was Cantoni, who made an impressive series of experiments, in which he also gave evidence of equipartition of energy, and eventually concluded (translated from Italian in Gallavotti (1999)):

[1] In reality, it was much later discovered that water produces short-lived molecular structures, called clusters, which can contribute to the Brownian motion (Gao et al. 2022; Morawietz et al. 2016), but that does not change the substance of Einstein's reasoning.

> I think that the dancing movement of the extremely minute solid particles in a liquid, can be attributed to the different velocities that must be proper at a given temperature of both such solid particles and of the molecules of the liquid that hit them from every side. I do not know whether others did already attempt this way of explaining Brownian motions Cantoni (1867).

Still, it was not clear why molecules should exert a viscous force, that characterizes continua, besides the conservative force arising in collisions with the object of interest. It was not even clear how tiny invisible objects could affect small but visible, hence much larger, objects.[2] Moreover, the effect could have even been contrary to thermodynamics, the most solid theory ever built.

Einstein (1905), and independently von Smoluchowski (1906), took up the task of explaining the Brownian motion, accepting this scenario. In particular, Einstein remarked that particles, even if very large, must obey the laws of statistical mechanics and, in particular, generate osmotic pressure, just as with ordinary solutions (Einstein 1956). The description of the motion of suspended particles could thus be formulated as depicted in Fig. 3.1. There are particles of mass m_i constituting the fluid, and a large particle of mass m suspended in it, which obey the laws of classical mechanics. Their trajectories can be computed, once the initial conditions are given, by integrating the equations of motion:

$$\dot{x} = v, \quad m\dot{v} = -\alpha v + \sum_{i=1}^{N} F_i \qquad \text{and} \qquad \dot{x}_i = v_i. \quad m_i \dot{v}_i = -F_i + \sum_{\substack{j=1 \\ j \neq i}}^{N} F_{ji} \tag{3.3}$$

where F_i represents the interaction of the ith particle with the mass m, and F_{ji} the interaction among the ith and jth particles. Supposing there is a mole of water molecules, the indices i and j run from 1 to Avogadro's number N_A, which was not known when Einstein wrote his paper. This daring picture places the motion of

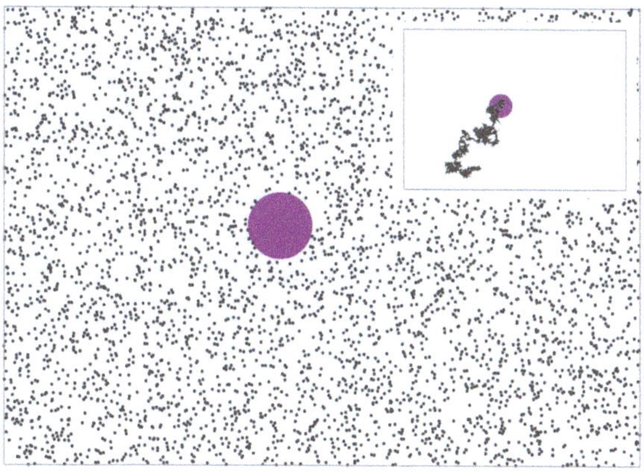

Fig. 3.1 Schematic microscopic representation of a large Brownian particle suspended in a liquid made of much smaller particles in two dimensions. They all follow the laws of classical mechanics, conserving energy and momentum at collisions. Inset: At the macroscopic scales, the Brownian particle follows an erratic trajectory

[2] The mass of a water molecule is about 10^{16} times smaller than that of pollen grains.

pollen in an intermediate situation, between the microscopic and the macroscopic scales, being well distinguished from the two. Indeed, the microscopic motions do not contemplate viscosity, and the macroscopic description does not account for microscopic impacts. Denoting by $m\,\Gamma$ the force on the large particle resulting from such impacts, one may nevertheless write

$$\dot{v} = -\gamma v + \Gamma \tag{3.4}$$

with initial conditions

$$x(0) = x_0,\ v(0) = v_0,\ x_1(0) = x_{10},\ v_1(0) = v_{10}, \ldots,\ x_N(0) = x_{N0},\ v_N(0) = v_{N0} \tag{3.5}$$

Integration yields

$$v(t) = e^{-\gamma t}\left(v_0 + \int_0^t e^{\gamma s}\, \Gamma(s; x_{10}, \ldots, v_{N0})\, \mathrm{d}s\right) \tag{3.6}$$

where we stress that the force $m\,\Gamma$ depends on the initial conditions and on time.

This direct approach meets, however, insurmountable problems, because one cannot solve 10^{23} equations of motion in practice, and also because the initial conditions of the water molecules are not known. Furthermore, even if all necessary information were available and equations could actually be solved, the approach would be useless. Every time the process is repeated, the initial condition is different, the calculations would have to be repeated, and no general prediction would ever be made. In practice, knowledge of previous calculations/experiments would be irrelevant for future experiments and no real understanding of the phenomenon would be secured.

A different approach is necessary. The idea was to pass from the exact description of the phenomenon to a statistical description, which is less detailed, but more meaningful, cf. Chap. 1. In practice, one may consider a large number of independent suspended particles or, equivalently, a sequence of experiments performed with a single particle, so that the possible initial conditions are adequately sampled, and then one may be satisfied with the corresponding averages.[3] This way, one does not need precise knowledge of the force $m\,\Gamma$; its statistical properties, that can be inferred from observation, will do. The result will not allow us to make predictions on the single suspended particle, unless some special condition is verified, but that may indeed be of no interest. This was the program proposed by Paul Langevin in 1908, in which Eq. 3.4, nowadays known as Langevin equation, considers Γ as random (Langevin 1908). Later in 1930 Leonard Ornstein and George E. Uhlenbeck formalized the Langevin equation in the modern language of stochastic processes

[3] For instance, it could happen that all the different values of a given microscopic quantity are experienced by a single particle in time. Then, averaging over the ensemble of independent particles yields the average of the values experienced in time by the single particle. Whether this actually happens or not depends on the system and on the quantity at hand, and only experience can tell. Also, this requires observation scales substantially longer than the scales of the microscopic events.

(Uhlenbeck and Ornstein 1930). In the rest of this subsection we follow the Ornstein and Uhlenbeck account.

Let us then do as if Γ was not known, apart from its average and its two times autocorrelation function. Observation suggests the following:

- In an isolated system, the forces F_i do zero net work on a mass m, otherwise they would use internal energy of the fluid, that would be eventually exhausted, and the process would halt. In a closed system,[4] heat could flow to keep the system going, but no drift of the center of mass of the suspended particles is observed in experiments. Therefore, no net work is done by $m\Gamma$, and one may write

$$\mathbb{E}[\Gamma(t)] = 0 \tag{3.7}$$

  where $\mathbb{E}[\Gamma(t)]$ stands for the expectation (mean) value of Γ at any time t, which is computed averaging over the collection of independent particles or, equivalently, over repetitions of single particle experiments.
- The motion of a suspended particle looks like uncorrelated in time if observed at times t and t'. Therefore, observing that $\Gamma(t)$ is naturally correlated to itself, the accelerations due to the unknown forces should satisfy

$$\mathbb{E}[\Gamma(t)\Gamma(t')] = 0 \qquad \text{for } |t - t'| \geq \tau_0 > 0 \tag{3.8}$$

  where the correlation time τ_0 is empirically determined for the fluid and the suspended particles at hand.
- In the Brownian motion, τ_0 appears much shorter than $\tau = 1/\gamma$, therefore for the sake of simplicity, one may write

$$\mathbb{E}[\Gamma(t)\Gamma(t')] = q\delta(t - t') \tag{3.9}$$

  where δ is the Dirac delta function and q a constant to be determined.

Because, the force $m\,\Gamma$ is known only on average, we cannot limit its impact on the velocity of m. Therefore, we cannot guarantee that the process is differentiable and the equation of motion is better expressed by something like:

$$\mathrm{d}v + \gamma v \mathrm{d}t = \mathrm{d}\tilde{\Gamma} \tag{3.10}$$

where the differentials $\mathrm{d}v$ and $\mathrm{d}\tilde{\Gamma}$ represent the variations of the functions v and $\tilde{\Gamma}$ in an elementary time interval $\mathrm{d}t$. This way one indicates that such functions are not bound to be differentiable in time. Only when it makes sense to define $\Gamma = d\tilde{\Gamma}/dt$ also the derivative of v with respect to time exists, and one may use (3.4).

[4] A closed system exchanges energy with the environment, but no mass.

Equation (3.10) is called a *stochastic differential equation*. One may formally integrate it, but it yields concrete results only on the statistical properties of the quantities occurring in it. Therefore, its terms should not be interpreted literally as velocities or forces; in particular, one should not expect $m\Gamma$ to obey the action-reaction principle,[5] especially in nonequilibrium conditions, see, e.g., Buenzli and Soto (2008); Pinheiro (2011).

With the intention to compute only statistical quantities, we may then first formally integrate Eqs. (3.4), (3.10), with initial condition $v(0) = v_0$ and average afterward. In principle, one should first check that the operations of integration and averaging commute. Physically, however, averaging means summing the contributions of an arbitrarily large, but finite, number of particles, N, and dividing by this number. That does commute with the integration with respect to time.[6] Therefore, we first have

$$v(t) = e^{-\gamma t}\left(v_0 + \int_0^t e^{\gamma t'}\Gamma(t')\,\mathrm{d}t'\right) = v_0 e^{-\gamma t} + \int_0^t e^{-\gamma(t-t')}\Gamma(t')\,\mathrm{d}t' \tag{3.11}$$

and then

$$\mathbb{E}[v(t)] = v_0 e^{-\gamma t} + \int_0^t e^{-\gamma(t-t')}\mathbb{E}[\Gamma(t')]\,\mathrm{d}t' = v_0 e^{-\gamma t} \tag{3.12}$$

because $\mathbb{E}[\Gamma(t')] = 0$. This result means that *on average* the velocity of suspended (Brownian) particles follows the classical hydrodynamic law: although no particle really slows down, *on average* their speed decreases exponentially in time. Analogously, the correlation function can be computed, starting from the following formal expression:

$$\begin{aligned} v(t_1)v(t_2) &= \left(v_0 e^{-\gamma t_1} + \int_0^{t_1} e^{-\gamma(t_1-t_1')}\Gamma(t_1')\,\mathrm{d}t_1'\right)\left(v_0 e^{-\gamma t_2} + \int_0^{t_2} e^{-\gamma(t_2-t_2')}\Gamma(t_2')\,\mathrm{d}t_2'\right) \\ &= v_0^2 e^{-\gamma(t_1+t_2)} + v_0\int_0^{t_2} e^{-\gamma(t_1+t_2-t_2')}\Gamma(t_2')\,\mathrm{d}t_2' + v_0\int_0^{t_1} e^{-\gamma(t_1+t_2-t_1')}\Gamma(t_1')\,\mathrm{d}t_1' \\ &+ \int_0^{t_1}\mathrm{d}t_1'\int_0^{t_2}\mathrm{d}t_2'\, e^{-\gamma(t_1+t_2-t_1'-t_2')}\Gamma(t_1')\Gamma(t_2') \end{aligned} \tag{3.13}$$

[5] Observe that $m\Gamma$ expresses an average behavior. Therefore, unless the dynamics takes place in a straight line, in general there is no reason to consider $m\Gamma$ instantaneously directed along the straight line joining two interacting points, which is when the action-reaction principle of mechanics is expected to hold. One may still assume action-reaction and judge it from its consequences, knowing it may fail.

[6] Averaging here means: $\mathbb{E}[v] = \sum_{i=1}^{N} v_i/N$, where v_i denotes the velocity of the ith Brownian particle. More sophisticated mathematical points of view can be adopted. However, the success of the model presented here in describing the Brownian motion validates our interpretation.

Averaging again over all realizations of the process, we obtain

$$\begin{aligned}
\mathbb{E}[v(t_1)v(t_2)] &= v_0^2 e^{-\gamma(t_1+t_2)} + v_0 \int_0^{t_1} e^{-\gamma(t_1+t_2-t_1')}\mathbb{E}[\Gamma(t_1')]\,dt_1' + v_0 \int_0^{t_2} e^{-\gamma(t_1+t_2-t_2')}\mathbb{E}[\Gamma(t_2')]\,dt_2' \\
&+ \int_0^{t_1} dt_1' \int_0^{t_2} dt_2'\, e^{-\gamma(t_1+t_2-t_1'-t_2')}\mathbb{E}[\Gamma(t_1')\Gamma(t_2')] \\
&= v_0^2 e^{-\gamma(t_1+t_2)} + \int_0^{t_1} dt_1' \int_0^{t_2} dt_2'\, e^{-\gamma(t_1+t_2-t_1'-t_2')} q\delta(t_1'-t_2')
\end{aligned} \tag{3.14}$$

To compute the double integral, we consider two cases: $t_1 > t_2$ and $t_1 < t_2$. For $t_1 > t_2$, we can write

$$\begin{aligned}
\int_0^{t_2} dt_2' \int_0^{t_1} dt_1'\, e^{-\gamma(t_1+t_2-t_1'-t_2')} q\delta(t_1'-t_2') &= q e^{-\gamma(t_1+t_2)} \int_0^{t_2} e^{2\gamma t_2'}\, dt_2' \\
&= \frac{q}{2\gamma}\left[e^{-\gamma(t_1-t_2)} - e^{-\gamma(t_1+t_2)}\right] = \frac{q}{2\gamma}\left[e^{-\gamma|t_1-t_2|} - e^{-\gamma(t_1+t_2)}\right]
\end{aligned} \tag{3.15}$$

and swapping t_1 and t_2, for $t_1 < t_2$ we have

$$\frac{q}{2\gamma}\left[e^{-\gamma(t_2-t_1)} - e^{-\gamma(t_1+t_2)}\right] = \frac{q}{2\gamma}\left[e^{-\gamma|t_1-t_2|} - e^{-\gamma(t_1+t_2)}\right] \tag{3.16}$$

Finally, combining the two cases yields

$$\mathbb{E}[v(t_1)v(t_2)] = \frac{q}{2\gamma}\left[e^{-\gamma|t_1-t_2|} - e^{-\gamma(t_1+t_2)}\right] + v_0^2 e^{-\gamma(t_1+t_2)} \tag{3.17}$$

which holds for $t_1 = t_2 = t$ as well.[7] Then, for large t_1 and t_2, $\mathbb{E}[v(t_1)v(t_2)]$ does not depend on the initial velocity v_0 and takes the form:

$$\mathbb{E}[v(t_1)v(t_2)] \approx \frac{q}{2\gamma} e^{-\gamma|t_1-t_2|} \tag{3.18}$$

The asymptotic average of the kinetic energy E then writes

$$\mathbb{E}[E] = \lim_{t\to\infty} \frac{m}{2}\,\mathbb{E}[v(t)v(t)] = \frac{mq}{4\gamma} \tag{3.19}$$

If we assume that the system is in equilibrium (no net external force, no net torque, no dissipation) and energy follows the equipartition principle, $\mathbb{E}[E] = k_B T/2$, we obtain

$$\frac{1}{2}k_B T = \frac{mq}{4\gamma} \quad \text{which implies} \quad q = \frac{2k_B T\gamma}{m} \tag{3.20}$$

[7] Recall the physical meaning of the Dirac δ function: a very peaked and correspondingly narrow function.

Consider now the squared distance traveled by a particle in a time t:

$$(x(t)-x_0)^2 = \int_0^t v(t_1)\,\mathrm{d}t_1 \int_0^t v(t_2)\,\mathrm{d}t_2 = \int_0^t \int_0^t v(t_1)v(t_2)\,\mathrm{d}t_1\,\mathrm{d}t_2 \tag{3.21}$$

Again, averaging over all realizations, one obtains

$$\begin{aligned}\mathbb{E}[(x(t)-x_0)^2] &= \int_0^t \int_0^t \mathbb{E}[v(t_1)v(t_2)]\,\mathrm{d}t_1\,\mathrm{d}t_2 \\ &= \int_0^t \mathrm{d}t_1 \int_0^t \mathrm{d}t_2 \left[\frac{q}{2\gamma}\left(e^{-\gamma|t_1-t_2|} - e^{-\gamma(t_1+t_2)}\right) + v_0^2 e^{-\gamma(t_1+t_2)}\right] \\ &= \int_0^t \mathrm{d}t_1 \int_0^t \mathrm{d}t_2 \left[\frac{q}{2\gamma} e^{-\gamma|t_1-t_2|} + \left(v_0^2 - \frac{q}{2\gamma}\right) e^{-\gamma(t_1+t_2)}\right]\end{aligned} \tag{3.22}$$

where

$$\int_0^t \int_0^t e^{-\gamma(t_1+t_2)}\,\mathrm{d}t_1\,\mathrm{d}t_2 = \int_0^t e^{-\gamma t_1}\,\mathrm{d}t_1 \int_0^t e^{-\gamma t_2}\,\mathrm{d}t_2 = \frac{1}{\gamma^2}(e^{-\gamma t}-1)^2 \tag{3.23}$$

and

$$\begin{aligned}\int_0^t \int_0^t e^{-\gamma|t_1-t_2|}\,\mathrm{d}t_1\,\mathrm{d}t_2 &= \int_0^t \left[\int_0^{t_1} e^{-\gamma(t_1-t_2)}\,\mathrm{d}t_2 + \int_{t_1}^t e^{\gamma(t_1-t_2)}\,\mathrm{d}t_2\right]\mathrm{d}t_1 \\ &= \int_0^t \left[e^{-\gamma t_1}\frac{e^{\gamma t_1}-1}{\gamma} + e^{\gamma t_1}\frac{e^{-\gamma t_1}-e^{-\gamma t}}{\gamma}\right]\mathrm{d}t_1 \\ &= \int_0^t \left[\frac{2}{\gamma} - \frac{1}{\gamma}e^{-\gamma t_1} - \frac{1}{\gamma}e^{\gamma t_1}e^{-\gamma t}\right]\mathrm{d}t_1 \\ &= \frac{2}{\gamma}t + \frac{1}{\gamma^2}(e^{-\gamma t}-1) - \frac{1}{\gamma^2}e^{-\gamma t}(e^{\gamma t}-1) \\ &= \frac{2}{\gamma}t + \frac{2}{\gamma^2}(e^{-\gamma t}-1)\end{aligned} \tag{3.24}$$

so that

$$\mathbb{E}[(x(t)-x_0)^2] = \left(v_0^2 - \frac{q}{2\gamma}\right)\frac{(1-e^{-\gamma t})^2}{\gamma^2} + \frac{q}{2\gamma}\left[\frac{2}{\gamma}t - \frac{2(1-e^{-\gamma t})}{\gamma^2}\right] \tag{3.25}$$

which grows linearly in time. Introducing the (one-dimensional) diffusion coefficient as

$$D := \lim_{t\to\infty} \frac{\mathbb{E}[(x(t)-x_0)^2]}{2t} \tag{3.26}$$

one obtains

$$D := \frac{1}{2}\lim_{t\to\infty}\left[\left(v_0^2 - \frac{q}{2\gamma}\right)\frac{(1-e^{-\gamma t})^2}{\gamma^2 t} + \frac{q}{\gamma^2} - \frac{q}{\gamma^3 t}\left(1-e^{-\gamma t}\right)\right] = \frac{q}{2\gamma^2} = \frac{k_B T}{m\gamma} \tag{3.27}$$

where the last equality holds if equipartition does. Equation (3.27) yields the celebrated Einstein relation:

$$D = \mu k_B T, \quad \text{where} \quad \mu = \frac{1}{m\gamma} \tag{3.28}$$

that links diffusion coefficient and temperature, via the mobility μ, Eq. (3.1).[8] The presence of fluctuations, i.e., the mean square displacement $\mathbb{E}[(x(t) - x_0)^2]$, and of dissipation, i.e., μ, makes Eq. (3.27) the first fluctuation-dissipation relation, an amazing result, that has been experimentally validated first by Jean Perrin (1908), and that irrevocably directed the course of science toward atomism, since it allowed the calculation of Avogadro's number N_A.

Indeed, a measurement of the variance of the positions of the Brownian particles yields D; the temperature T, the friction coefficient γ, and the mass m are known, and the Boltzmann constant, which can be expressed as the ratio $k_B = R/N_A$, where R is the universal constant of gases, is also known. Then one obtains

$$D = \frac{k_B T}{m\gamma} = \frac{RT}{6\pi\eta a}\frac{1}{N_A} \tag{3.29}$$

from which one gets $N_A = 6.022 \cdot 10^{23}$ mol^{-1}. The discussion on the existence of atoms, up to that point considered at best a useful mathematical trick even by prominent scientists, such as Wilhelm Ostwald and Ernst Mach, came to halt. Within the rational framework of the Brownian motion and kinetic theory, atoms can be counted!

A generalization to d-dimensional systems of the one-dimensional theory illustrated above takes the following form:

$$\dot{v}_i = -\gamma v_i + \Gamma_i(t), \qquad i = 1, \ldots, d \tag{3.30}$$

In absence of correlations among the different coordinates, one may assume:

$$\mathbb{E}[\Gamma_i(t)] = 0; \qquad \mathbb{E}[\Gamma_i(t)\Gamma_j(t')] = q\delta_{ij}\delta(t - t') \tag{3.31}$$

so that each variable can be treated separately. Moreover, when equipartition applies, we can write

$$\mathbb{E}[E] = \sum_{i=1}^{d} \frac{1}{2} m\mathbb{E}[v_i^2] = \frac{d}{2} k_B T, \quad \text{and} \quad \frac{\mathbb{E}[\overrightarrow{\Delta x}(t)^2]}{t} \to \frac{2dk_B T}{m\gamma}, \quad \text{for } t \to \infty \tag{3.32}$$

where $\overrightarrow{\Delta x}(t) = (\vec{x}(t) - \vec{x}(0))$ is a d-dimensional vector.

[8] This equation was derived independently by William Sutherland in 1905 to estimate the molecular weight of different molecules. Sometimes, Eq. 3.28 is referred to as the Einstein-Sutherland relation.

3.1.1 Teachings of Brownian Motion

The relation (3.29) contains an incredible amount of information. In the first place, it links the microscopic realm with the macroscopic one through $k_B = 1.380649\, 10^{-23} J/K$, which marks the huge "distance" between the two. Then, once the value of N_A has been established, Eq. (3.29) allows us to compute a macroscopic nonequilibrium property, the viscosity η, from an equilibrium experiment in which no work is done: one merely observes fluctuations. Note that viscosity does not exert any force when the fluid is at rest with respect to its container or to the walls of an object in it. In a direct experimental measurement of the viscosity, the object in the fluid is dragged by a force F; then, viscosity imposes a limit velocity v_∞, and the mobility is defined by the linear response relation $F = \mu v_\infty$. Equation (3.29) bypasses all that work, and relates nonequilibrium properties to equilibrium fluctuations, which are made visible by objects belonging to an intermediate, mesoscopic, realm.

In this respect, it is important to stress what does the scale separation mathematically assumed above physically mean. In the case of Brownian particles, the relevant time scales, for instance, are 10^{-12}, 10^{-4}, 10^{1} s, i.e., eight to six orders of magnitude separation. Perhaps at variance with intuition, but in accord with the picture illustrated in Sect. 1 and Fig. 1.1, this is enough to move from one representation to an "orthogonal" one for the very same object of interest. For differences below four orders of magnitude, the separation of scales turns shaky and the phenomenology uncertain. In such situations, the applicability of formalisms has to be checked case by case.

This theory has been confirmed in countless instances; it effectively allows us to count atoms. Therefore, atoms must definitely exist! Hard to beat the beauty and importance of such a result. Einstein's theory of Brownian motion, experimentally demonstrated by Perrin (1908, 1909), and with later refinements by Ivar Nordlund (1914), dispelled all doubts about atoms, and several decades later Feynman could tell that atomism contains the most information one can get from a few words, with *just a little imagination and thinking*.

Imagination is needed because solving the corresponding equations of motion, something so far impossible, does not amount to understanding. What would myriads of graphs and numbers describing atomic trajectories mean for warming up or cooling down a cup of coffee? The thermodynamic description, not counting at all on microscopic data, is much more natural for that kind of phenomena. However, the macroscopic laws must be supplemented with the values of thermophysical constants, such as the transport coefficients, that require experiments or must derived from microscopic approaches. One of the goals of statistical mechanics is to predict these parameters. The Brownian motion theory is a fundamental step in that direction.

3.1.2 Gaussian Processes

The solutions of the Langevin equation (3.4), or (3.10), define a stochastic process: a process that is intrinsically probabilistic, in which only the statistical properties of the variables of interest are relevant (see Appendix C). Among the large variety of stochastic processes, Gaussian processes stand out. A stochastic process $Z(t)$ indexed by a parameter t, e.g., time, is called Gaussian if the joint probability density of its values $z_1, z_2, \ldots, z_n$ at times $t_1, t_2, \ldots, t_n$ is given by

$$W_n(z_1, t_1; \ldots; z_n, t_n) = ce^{-\frac{1}{2}\sum_{i,j=1}^{n} a_{ij}(z_i - m_i)(z_j - m_j)} \tag{3.33}$$

where c is the normalization constant, $m_k = \mathbb{E}[Z(t_k)]$ and $A = (a_{ij})$ is a positive definite $n \times n$ matrix, whose inverse has entries defined by $(A^{-1})_{ij} = \mathbb{E}[(Z(t_i) - m_i)(Z(t_j) - m_j)]$, for $i, j = 1, \ldots, n$. The first three *cumulants* of the probability distribution of a random variable Z, $\mathbb{E}[\cdot]_c$, are defined by

$$\begin{aligned}\mathbb{E}[Z]_c &= \mathbb{E}[Z]\\ \mathbb{E}[Z^2]_c &= \mathbb{E}[Z^2] - \mathbb{E}[Z]^2\\ \mathbb{E}[Z^3]_c &= \mathbb{E}[Z^3] - 3\mathbb{E}[Z^2]\mathbb{E}[Z] + 2\mathbb{E}[Z]^3\end{aligned} \tag{3.34}$$

A normal distribution $N(m, \sigma)$ implies

$$\mathbb{E}[Z]_c = m, \qquad \mathbb{E}[Z^2]_c = \sigma^2, \qquad \mathbb{E}[Z^n]_c = 0 \ \text{ for } \ n \geq 3 \tag{3.35}$$

and the distribution is normal if all cumulants of order $n \geq 3$, $\mathbb{E}[Z_1 Z_2 \cdots Z_n]_c$, vanish. This also holds for multivariate Gaussian distributions such as Eq. 3.33.

Defining a linear transformation of a stochastic process $Z(t)$ as

$$Y(t) = \int_a^b c(t, t')Z(t')\,\mathrm{d}t' \tag{3.36}$$

the cumulants of $Y(t)$, for any order n, are given by

$$\begin{aligned}&\mathbb{E}[Y(t_1) \cdots Y(t_n)]_c = \\ &\int_a^b \mathrm{d}t_1' \int_a^b \mathrm{d}t_2' \cdots \int_a^b \mathrm{d}t_n'\; c(t_1, t_1')c(t_2, t_2') \cdots c(t_n, t_n')\mathbb{E}[Z(t_1') \cdots Z(t_n')]_c\end{aligned} \tag{3.37}$$

Hence, if $Z(t)$ is a zero mean Gaussian random variable at all t, $Y(t)$ is as well, because all cumulants of order $n \geq 3$ vanish. Then, the velocity V of the Brownian motion is a Gaussian process if the stochastic term $\Gamma(t)$ is a Gaussian process, because it results from an integral like (3.36). Considering that in the $t \to \infty$ limit one has $\mathbb{E}[V]_c = 0$ and $\sigma^2 = \mathbb{E}[V^2]_c = q/2\gamma$, the asymptotic probability density of V is given by

$$f_V(v) = ce^{-\frac{v^2}{(2q/2\gamma)}} = ce^{-\frac{\gamma v^2}{q}} = \sqrt{\frac{m}{2\pi k_B T}}e^{-\frac{mv^2}{2k_B T}} = f_{MB}(v) \tag{3.38}$$

where, accepting the equipartition principle , $q/\gamma = 2k_BT/m$, we have obtained the celebrated *Maxwell-Boltzmann* distribution, f_{MB}. Because its variance takes the form $\sigma^2 = k_BT/m$, fixing the temperature of the fluid T and placing a large mass m in it yield small uncertainty on the velocity: the velocity is practically deterministic. On the contrary, a small mass m implies large uncertainty: in the small m limit, the velocity is totally random. This can already be seen in the Langevin equation, when the intensity of Γ is expressed by q: large q/γ (relatively small mass) means that random forces dominate the deterministic viscous force, while small q/γ is the opposite situation in which hydrodynamics holds. For a test particle of the size of water molecules, the motion is random and there is no viscosity, while for a boat the motion is deterministic, viscous, and not affected by atomic impacts. The Brownian particle lays at the border between the micro- and the macro-worlds that coexist, even though *very* different, in a single reality, and reveals both. While microscopic dynamics know no viscosity and no energy dissipation, which is why the motion does not stop, macroscopic dynamics knows no molecules, it dissipates energy, hence the motion stops. These two pictures are not contradictory: the ordered motion of the macroscopic object passes its energy to the disordered motion of the fluid molecules, contributing to the internal energy of the fluid; no energy is lost, it is simply randomized, which means made unavailable for work.

3.2 Evolution of Probabilities in Stochastic Processes

Let us introduce the notion of transition probability density P, to express the time evolution of a probability density W:

$$W(z, t + \tau) = \int P(z, t + \tau|z', t)\, W(z', t)\, \mathrm{d}z' \tag{3.39}$$

which means that the probability density at time $t + \tau$ for the random variable Z to take values around z cumulates the contributions coming from all possible values z' at time t, weighted with the probability density of transiting from z' at time t, to z at time $t + \tau$. By definition, P obeys

$$P(z, t + \tau|z', t) \geq 0; \quad \int P(z, t + \tau|z', t)\, \mathrm{d}z = 1 \tag{3.40}$$

because transition probabilities are probabilities, and any event z' at a time t surely becomes some other event z (including $z = z'$) within the total set of events, whatever later time τ one considers. We also require that the process be sufficiently regular that

$$\lim_{\tau \to 0} P(z, t + \tau|z', t) = \delta(z - z') \tag{3.41}$$

and that the moments of P are known:

$$M_n(z', t, \tau) = \int_{-\infty}^{\infty} (z - z')^n P(z, t + \tau|z', t)\, \mathrm{d}z \tag{3.42}$$

Then, one can construct the characteristic function, cf. Appendix C, defined by

$$c(u, z', t, \tau) = \int_{-\infty}^{\infty} e^{iu(z-z')} P(z, t+\tau|z', t)\, \mathrm{d}z = 1 + \sum_{n=1}^{\infty} \frac{(iu)^n}{n!} M_n(z', t, \tau) \tag{3.43}$$

which can be inverted to give

$$\begin{aligned} P(z, t+\tau|z', t) &= \frac{1}{2\pi} \int_{-\infty}^{\infty} e^{-iu(z-z')} c(u, z', t, \tau)\, \mathrm{d}u \\ &= \frac{1}{2\pi} \int_{-\infty}^{\infty} e^{-iu(z-z')} \left[1 + \sum_{n=1}^{\infty} \frac{(iu)^n}{n!} M_n(z', t, \tau) \right] \mathrm{d}u \end{aligned} \tag{3.44}$$

hence

$$P(z, t+\tau|z', t) W(z', t) \tag{3.45}$$

$$= \frac{1}{2\pi} \int_{-\infty}^{\infty} e^{-iu(z-z')} \left[W(z', t) + \sum_{n=1}^{\infty} \frac{(iu)^n}{n!} M_n(z', t, \tau) W(z', t) \right] \mathrm{d}u \tag{3.46}$$

where each product $M_n(z', t, \tau)W(z', t)$ constitutes a single function of z', like the left-hand side (3.45) of the equality. This is then inserted in the integral in Eq. (3.39). It is convenient now to introduce the following representation of the Dirac δ function:

$$\delta(x) = \frac{1}{2\pi} \int_{-\infty}^{\infty} e^{ikx}\, \mathrm{d}k \tag{3.47}$$

which, for f continuous in a neighborhood of $x = 0$, yields

$$\begin{aligned} \int_{-\infty}^{\infty} \mathrm{d}x\, f(x) \frac{1}{2\pi} \int_{-\infty}^{\infty} \mathrm{d}k e^{ikx} &= \int_{-\infty}^{\infty} \mathrm{d}k\, \frac{1}{2\pi} \int_{-\infty}^{\infty} \mathrm{d}x\, f(x) e^{-i(-k)x} \\ &= \int_{-\infty}^{\infty} \tilde{f}(-k)\, \mathrm{d}k = \int_{-\infty}^{\infty} \tilde{f}(-k)\, 1\, \mathrm{d}k = \int_{-\infty}^{\infty} \tilde{f}(-k) e^{i(-k)0}\, \mathrm{d}k = f(0) \end{aligned} \tag{3.48}$$

where $\tilde{f}$ is the Fourier transform of f and the last integral performs the inverse transform, evaluated at $x = 0$. We also have

$$\frac{1}{2\pi} \int_{-\infty}^{\infty} \mathrm{d}k (ik)^n e^{i(-k)x} = \frac{1}{2\pi} \int_{-\infty}^{\infty} \mathrm{d}k (-1)^n \frac{\mathrm{d}^n}{\mathrm{d}x^n} e^{i(-k)x} = (-1)^n \frac{\mathrm{d}^n}{\mathrm{d}x^n} \delta(x) \tag{3.49}$$

which implies

$$\begin{aligned} W(z, t+\tau) &= \int W(z', t) \delta(z - z')\, \mathrm{d}z' - \frac{\partial}{\partial z} \int M_1(z', t, \tau) W(z', t) \delta(z - z')\, \mathrm{d}z' + \cdots \\ &= W(z, t) - \frac{\partial}{\partial z} M_1(z, t, \tau) W(z, t) + \frac{1}{2} \frac{\partial^2}{\partial z^2} M_2(z, t, \tau) W(z, t) + \cdots \end{aligned} \tag{3.50}$$

This expression is useful to derive an evolution equation for W. Moving to the left-hand side the term $W(z,t)$, dividing by τ and taking the $\tau \to 0$ limit, one obtains

$$\lim_{\tau\to 0}\frac{W(z,t+\tau)-W(z,t)}{\tau}=\frac{\partial W}{\partial t}=\lim_{\tau\to 0}\sum_{n=1}^{\infty}\frac{1}{n!}\left(-\frac{\partial}{\partial z}\right)^n\left[\frac{M_n(z,t,\tau)}{\tau}W(z,t)\right] \tag{3.51}$$

which is the Kramers-Moyal expansion. A famous theorem due to Pawula (1967) states that the Kramers-Moyal expansion is either truncated at $n=1$, which corresponds to a deterministic process, or it stops at $n=2$, which corresponds to the diffusion of probability, or it does not stop at any finite n. Any truncation of the expansion at some finite $n>2$ may yield functions that take negative values, which cannot be accepted as probability densities. In particular, given a process $Z(t)$ obeying the Langevin equation with Gaussian δ-correlated noise,

$$\dot{z}=h(z,t)+g(z,t)\Gamma(t),\ \ \mathbb{E}[\Gamma(t)]=0;\ \ \mathbb{E}[\Gamma(t)\Gamma(t')]=2\delta(t-t') \tag{3.52}$$

the moments of order higher than 2 vanish: $M_n=0$ if $n\geq 3$, while M_1 and M_2 take the general form Risken (1989):

$$\begin{cases}D^{(1)}(z,t)=h(z,t)+g'(z,t)g(z,t)\\ D^{(2)}(z,t)=g^2(z,t)\end{cases}\quad \text{where } D^{(k)}(z,t)=\lim_{\tau\to 0}\frac{1}{k!}\frac{M_k(z,t,\tau)}{\tau},\ \ g'=\frac{\partial g}{\partial x} \tag{3.53}$$

The term $D^{(1)}$ is called drift coefficient, and $D^{(2)}$ is called diffusion coefficient because they appear in the Fokker-Planck equation:

$$\frac{\partial W}{\partial t}=-\frac{\partial}{\partial z}D^{(1)}(z,t)W(z,t)+\frac{\partial^2}{\partial z^2}D^{(2)}(z,t)W(z,t)=-\frac{\partial}{\partial z}J(z,t) \tag{3.54}$$

formally as contributions from a probability drift and from the diffusion of probability, with

$$J(z,t)=D^{(1)}(z,t)W(z,t)-\frac{\partial}{\partial z}D^{(2)}(z,t)W(z,t) \tag{3.55}$$

representing a probability current. The Fokker-Planck equation is thus a balance equation, analogous to those of conserved quantities in thermodynamics, since probability is conserved. If there is no noise, it becomes Liouville's equation for the deterministic evolution.

For more general cases, in which the number of stochastic variables is N, we can merge the equations of motion in the following system:

$$\dot{\vec{z}}=\vec{h}+G\Gamma \tag{3.56}$$

where

$$\vec{\dot{z}} = \begin{pmatrix} \dot{z}_1 \\ \dot{z}_2 \\ \vdots \\ \dot{z}_N \end{pmatrix}, \quad \vec{h} = \begin{pmatrix} h_1 \\ h_2 \\ \vdots \\ h_N \end{pmatrix}, \quad G = \begin{pmatrix} g_{1,1} & g_{1,2} & \cdots & g_{1,N} \\ g_{2,1} & g_{2,2} & \cdots & g_{2,N} \\ \vdots & \vdots & \ddots & \vdots \\ g_{N,1} & g_{N,2} & \cdots & g_{N,N} \end{pmatrix}, \quad \Gamma = \begin{pmatrix} \Gamma_1 \\ \Gamma_2 \\ \vdots \\ \Gamma_N \end{pmatrix}$$

Therefore, we will have N different $D^{(1)}$'s and N^2 different $D^{(2)}$'s, which is an immediate extension of the one-dimensional case. On the contrary, even simple operations, such as the change of variables or of observables, are not so immediate, because of certain properties of stochastic processes, cf. Appendix D.

3.3 Fluctuation-Dissipation Theorem

Let Z_t be a stationary process, which is observed exclusively within the finite-time interval $t \in [0, \tau]$. Leaving τ as large as one wants, larger in particular than the physically relevant time scales, this limitation poses no difficulty. Moreover, it allows us to assume that Z_t is periodic of period τ, justifying the Fourier analysis of the signal $z(t)$ produced by the random variable Z_t. One may thus write

$$z(t) = \sum_{n=-\infty}^{+\infty} a_n e^{i\omega_n t}; \quad \text{where } a_n = \frac{1}{\tau}\int_0^\tau z(t) e^{-i\omega_n t}\,\mathrm{d}t; \quad \omega_n = \frac{2\pi n}{\tau} \tag{3.57}$$

Setting $a_n = \alpha_n + i\beta_n$ with $\alpha, \beta \in \mathbb{R}$, one needs $a_{-n} = a_n^* = \alpha_n - i\beta_n$ for Z_t to be real, and because Z_t is a random variable, so are the Fourier coefficients a_n. Their averages over the realizations of the process can then be expressed as

$$\mathbb{E}[a_0] = \frac{1}{\tau}\int_0^\tau \mathbb{E}[Z_t]\,\mathrm{d}t = \mathbb{E}[Z_t] = \mathbb{E}[Z]$$
$$\mathbb{E}[a_n] = \frac{1}{\tau}\int_0^\tau \mathbb{E}[Z_t] e^{-i\omega_n t}\,\mathrm{d}t = \mathbb{E}[Z]\frac{1}{\tau}\int_0^\tau e^{-i\omega_n t}\,\mathrm{d}t = 0 \quad \text{if} \quad n \neq 0$$

where we have stressed that the process is stationary, namely, its statistical properties do not change in time and one can write $\mathbb{E}[Z_t] = \mathbb{E}[Z]$. For a single realization of the process, i.e., for one given set of coefficients $\{a_n\}_{-\infty}^{\infty}$, the time average is expressed by

$$\overline{z}^\tau = \frac{1}{\tau}\int_0^\tau z(t)\,\mathrm{d}t = a_0 \tag{3.58}$$

which in general differs from $\mathbb{E}[a_0]$. The process is called *ergodic* if

$$\lim_{\tau\to\infty} \overline{z}^\tau = \mathbb{E}[Z] \tag{3.59}$$

The ergodicity condition is commonly satisfied by stochastic processes so that $a_0 = \mathbb{E}[a_0] = \mathbb{E}[Z]$, even if conceptually, time averages and expectation values of a given random variable are totally different quantities.[9]

The average strength of the Fourier component a_n of a given process Z_t is defined by

$$\mathbb{E}[|a_n|^2] = \mathbb{E}[\alpha_n^2] + \mathbb{E}[\beta_n^2] \tag{3.60}$$

and the average intensity I_T around the frequency ω is defined by

$$I_\tau(\omega;\, \Delta\omega)\Delta\omega = \sum_{\omega_n \in [\omega, \omega+\Delta\omega)} \mathbb{E}\left[|a_n|^2\right] \tag{3.61}$$

where a positive amount is always obtained if $\Delta\omega$ is larger than the distance $2\pi/\tau$ between two consecutive angular frequencies. Moreover, if τ is large, and $z(t)$ sufficiently regular, a_n can be considered practically continuous in n, and (relatively) small intervals $[\omega, \omega + \Delta\omega)$ contain Fourier components which are approximately equal to each other. One may then write

$$I_\tau(\omega;\, \Delta\omega)\Delta\omega = \mathbb{E}\left[|a_{\widehat{\omega}}|^2\right] \frac{\tau}{2\pi} \Delta\omega + o(\Delta\omega) \tag{3.62}$$

where $\widehat{\omega}$ is an angular frequency in $[\omega, \omega + \Delta\omega)$, $a_{\widehat{\omega}}$ is defined by Eq. (3.57) with ω_n replaced by $\widehat{\omega}$, and $\tau/2\pi$ is the number of frequencies per frequency unit, so that $\tau\Delta\omega/2\pi$ approximately equals the number of frequencies in an interval of width $\Delta\omega$. As $\widehat{\omega} \to \omega$ for $\Delta\omega \to 0$, the continuity of the process yields $a_{\widehat{\omega}} \to a_\omega$ for $\Delta\omega \to 0$. Now, given the process Z_t, the correction $o(\Delta\omega)$ in Eq. (3.62) can be made negligible by taking the small $\Delta\omega$ limit together with the large τ limit in such a way that $\tau\Delta\omega = \text{const} \gg 2\pi$, i.e., that $2\pi/\tau$ is negligible with respect to $\Delta\omega$. In fact, one may also take

$$\tau\Delta\omega \sim f(\tau) \quad \text{with} \quad \lim_{\tau\to\infty} f(\tau) = \infty \quad \text{and} \quad \lim_{\tau\to\infty} \frac{f(\tau)}{\tau} = 0 \tag{3.63}$$

to produce an ever denser set of frequencies, which better approximates a continuum. This is to ensure many frequencies ω_n in $[\omega, \omega + \Delta\omega)$ while $\Delta\omega$ is very small, so that $\widehat{\omega} \approx \omega$. Then, for the power of the signal in a fixed frequency interval to reach a finite value while τ grows, the power associated with each frequency in the interval

[9] In fact they refer to two kinds of experiments whose connection is far from obvious. Time averages of Z characterize the evolution of a single object observed for a long time, as thermodynamic measurements are supposed to be. Expectations of Z and its moments do not involve time: they characterize the spectrum of possible values that the random variable can intrinsically take at any instant of time, regardless of the measurement processes. In the classical notion of probability, such expectation values are interpreted as resulting from an ensemble of identical objects in different phases of their existence. These facts will be thoroughly examined in Chap. 6.

must tend to 0. If this is the case, as it is common in physical applications, one may divide by $\Delta\omega$ and define the intensity spectrum, or power spectrum of Z_t as follows:

$$I(\omega) = \lim_{\substack{\tau\to\infty;\ \tau\Delta\omega=\text{ const}\\ \widehat{\omega}\in[\omega,\omega+\Delta\omega)}} I_\tau(\widehat{\omega}, \Delta\omega) = \lim_{\substack{\tau\to\infty;\tau\Delta\omega=\text{ const}\\ \widehat{\omega}\in[\omega,\omega+\Delta\omega)}} \frac{\tau}{2\pi}\mathbb{E}\left[|a_{\widehat{\omega}}|^2\right] \tag{3.64}$$

The terminology derives from the theory of time-dependent currents in electrical circuits. Letting I be the current, R the resistance, and V the electric potential, the dissipated power P under Ohm's law is given by

$$P(t) = I(t)V(t) = R\,|I(t)|^2 \tag{3.65}$$

Expressing I in terms of its Fourier components, $I(t) = \sum_n a_n \exp(i\omega_n t)$, the average dissipated power then becomes

$$\overline{P} = \frac{R}{\tau}\int_0^\tau I(t)I^*(t)\,\mathrm{d}t = \frac{R}{\tau}\sum_{n=-\infty}^{\infty}\sum_{m=-\infty}^{\infty} a_n a_m^* \int_0^\tau e^{i(\omega_n-\omega_m)t}\,\mathrm{d}t = R\sum_{n=-\infty}^{\infty}|a_n|^2 \tag{3.66}$$

This shows that the square modulus of the nth Fourier component of I is proportional to the average power dissipated at angular frequency ω_n.

The Fourier coefficients b_n of the process $Y_t = Z_t - \mathbb{E}[Z]$ obey

$$\mathbb{E}[b_n] = 0 \quad \forall n \in \mathbb{Z} \tag{3.67}$$

Because the subtraction of a constant from a stochastic process does not alter its significance, let us now assume that the average of the signal Z_t vanishes. Its autocorrelation function is defined by

$$\Phi_Z(t_1, t_2) = \mathbb{E}[Z_{t_1} Z_{t_2}] = \mathbb{E}[Z_{t_1} Z_{t_1+t}] \tag{3.68}$$

where the time $t = t_2 - t_1$ has been introduced to stress the fact that Φ_Z depends on the initial time, and on the difference between initial and final observation time. This turns useful in a stationary state, which may be obtained in the $\tau \to \infty$ limit and considering such late times t_1 and t_2 that initial transients have decayed. In this case, time translation invariance holds, and the initial time makes no difference, and one may simply write

$$\Phi_Z(t) = \mathbb{E}[Z_{t_1} Z_{t_1+t}], \quad \forall t_1 \in \mathbb{R} \tag{3.69}$$

The Wiener-Khinchin theorem asserts that the spectral power density of the stationary process Z_t, I_Z say, and its autocorrelation Φ_Z are each other's Fourier transforms:

$$I_Z(\omega) = \frac{1}{2\pi}\int_{-\infty}^{\infty} \Phi_Z(t)e^{-i\omega t}\,\mathrm{d}t, \quad \Phi_Z(t) = \int_{-\infty}^{\infty} I_Z(\omega)e^{i\omega t}\,\mathrm{d}\omega, \tag{3.70}$$

i.e., knowledge of the power spectrum is equivalent to knowledge of the autocorrelation. Note, here the observation period τ is taken to be infinitely long, and the Fourier series has become an integral. This makes the factor $2\pi/\tau$ in Eq. (3.57), corresponding to the spacing of Fourier components, and the factor $1/\tau$ in the definition of the Fourier coefficients, simplify each other, leaving only 2π to distinguish Fourier transform and anti-transform. Therefore, there is an inconsequential 2π factor difference between the notation for finite τ, and that for infinite τ.[10]

Reconsidering the Brownian motion in its stationary regime, we may now proceed as follows. Let the equation of motion be

$$m\dot{v} = -m\gamma v + m\Gamma \quad \text{or} \quad \dot{v} = -\gamma v + \Gamma \tag{3.71}$$

whose terms can be expanded as

$$\Gamma(t) = \sum_{n=-\infty}^{+\infty} \Gamma_n e^{i\omega_n t}, \quad v(t) = \sum_{n=-\infty}^{+\infty} v_n e^{i\omega_n t} \tag{3.72}$$

where v_n is the Fourier coefficient of v and Γ_n that of Γ, at frequency ω_n. As Γ is a stochastic process, the coefficients of its expansion are random variables. Then, substituting (3.72) in (3.71), one obtains

$$\sum_{n=-\infty}^{\infty} i\omega_n v_n e^{i\omega_n t} = -\sum_{n=-\infty}^{\infty} \gamma v_n e^{i\omega_n t} + \sum_{n=-\infty}^{\infty} \Gamma_n e^{i\omega_n t} \tag{3.73}$$

which holds for all times $t \in [0, T]$ if and only if

$$i\omega_n v_n = -\gamma v_n + \Gamma_n \quad \text{i.e.} \quad v_n = \frac{\Gamma_n}{i\omega_n + \gamma} \tag{3.74}$$

[10] This can be seen as follows. Given a periodic function f, let f_n be its nth Fourier component and introduce $\widehat{f}(\omega_n) = \tau f_n/2\pi = \int_{-\tau/2}^{\tau/2} f(t)\exp(-i\omega_n t)\,\mathrm{d}t/2\pi$ Denoting by the spacing between frequencies by $\delta\omega = 2\pi/\tau$, we can write

$$f(t) = \sum_{n=-\infty}^{\infty} f_n\, e^{i\omega_n t} = \frac{2\pi}{\tau}\sum_{n=-\infty}^{\infty} \widehat{f}(\omega_n)e^{i\omega_n t} = \frac{2\pi}{\tau}\sum_{n=-\infty}^{\infty} \widehat{f}(\omega_n)e^{i\omega_n t}\frac{\tau}{2\pi}\delta\omega = \sum_{n=-\infty}^{\infty} \widehat{f}(\omega_n)e^{i\omega_n t}\delta\omega$$

In the $\tau \to \infty$ limit, which is also the $\delta\omega \to 0$ limit, the sum on the right becomes an integral from $-\infty$ to ∞, and one has

$$f(t) = \int_{-\infty}^{\infty} \widehat{f}(\omega)\exp(i\omega t)\,\mathrm{d}\omega, \quad \text{where} \quad \widehat{f}(\omega) = \frac{1}{2\pi}\int_{-\infty}^{\infty} f(t)\exp(-i\omega t)\,\mathrm{d}t$$

The power spectra of v and Γ are thus related as

$$I_V(\omega) = \lim_{\substack{\tau\to\infty;\, \tau\Delta\omega=\text{const}\\ \hat{\omega}\in[\omega,\omega+\Delta\omega)}} \frac{\tau}{2\pi}\mathbb{E}\left[\frac{|\Gamma_{\hat{\omega}}|^2}{\hat{\omega}^2+\gamma^2}\right] \tag{3.75}$$

$$= \lim_{\substack{\tau\to\infty;\, \tau\Delta\omega=\text{const}\\ \hat{\omega}\in[\omega,\omega+\Delta\omega)}} \frac{1}{\hat{\omega}^2+\gamma^2}\mathbb{E}\left[\frac{\tau|\Gamma_{\hat{\omega}}|^2}{2\pi}\right] = \frac{I_\Gamma(\omega)}{\omega^2+\gamma^2} \tag{3.76}$$

In the simplest case, $I_\Gamma(\omega) = I_\Gamma =$ constant, all Fourier components contribute the same power, thus the spectrum of Γ is called white, and that of v

$$I_V(\omega) = \frac{I_\Gamma}{\omega^2+\gamma^2} \tag{3.77}$$

is called Lorentzian. In a stationary state, taking $t_1 > t_2$, one can write

$$\Phi_\Gamma(t_1-t_2) = \mathbb{E}[\Gamma_{t_1}\Gamma_{t_2}] = \int_{-\infty}^{\infty} I_\Gamma e^{i\omega(t_1-t_2)}\,\mathrm{d}\omega \tag{3.78}$$

and, using the representation (3.47) of the Dirac delta function, one gets

$$\Phi_\Gamma(t_1-t_2) = 2\pi I_\Gamma\,\delta(t_1-t_2) \tag{3.79}$$

which shows that $q = 2\pi I_\Gamma$ in the treatment of Sect. 3.1.2. Analogously, the auto-correlation of the velocity V reads

$$\Phi_V(t) = \int_{-\infty}^{+\infty} \frac{I_\Gamma}{\omega^2+\gamma^2} e^{i\omega t}\,\mathrm{d}\omega \tag{3.80}$$

Then, for $t = t_1 - t_2$, one may write

$$\Phi_V(t_1-t_2) = \mathbb{E}[V_{t_1}V_{t_2}] = I_\Gamma \int_{-\infty}^{+\infty} \frac{e^{i\omega(t_1-t_2)}}{\omega^2+\gamma^2}\,\mathrm{d}\omega = I_\Gamma \frac{\pi}{\gamma} e^{-\gamma(t_1-t_2)} \quad \text{for} \quad t_1 > t_2 \tag{3.81}$$

or

$$\mathbb{E}[V_{t_1}V_{t_2}] = \frac{\pi I_\Gamma}{\gamma} e^{-\gamma|t_1-t_2|}, \quad \forall t_1, t_2 \in \mathbb{R} \tag{3.82}$$

The velocity correlation decays exponentially in time. Then, taking $t_1 = t_2 = t$, and t large enough that the steady state has been reached, and averages do not depend on time anymore, one obtains

$$\mathbb{E}[V^2] = \frac{\pi I_\Gamma}{\gamma} \qquad I_\Gamma = \frac{\gamma}{\pi}\mathbb{E}[V^2] \tag{3.83}$$

If the Brownian particle has equilibrated with the fluid, equipartition of energy holds, $m\mathbb{E}[V^2] = k_B T$, hence

$$I_\Gamma = \frac{k_B T}{\pi m}\gamma = \frac{q}{2\pi} \tag{3.84}$$

in accord with Eq. (3.20). Equation (3.84) states that the power of the random force is proportional to the friction coefficient and to the thermal energy. In turn, Einstein relation (3.27) represents an inverse proportionality between the diffusion coefficient and the friction coefficient. Both results demonstrate that the dissipation of energy caused by friction is intimately related to the equilibrium fluctuations, due to the incessant randomizing molecular motions. The name "Fluctuation-Dissipation relation" given to Eq. (3.27) is thus vindicated.

3.3.1 Derivation of Eq. (3.82)

Consider the inverse Fourier transform of a given function f:

$$\int_{-\infty}^{+\infty} f(\omega)e^{i\omega t}\,\mathrm{d}\omega \tag{3.85}$$

where $t \in \mathbb{R}_+$ and f is analytic in the upper half plane, except for a finite number of poles, and assume $\lim_{|\omega|\to\infty} f(\omega) = 0$, with $\arg(\omega) \in [0, \pi]$ (Fig. 3.2).
Recall that

$$\oint_C F(\omega)\,\mathrm{d}w = \int_{-R}^{R} F(x)\,\mathrm{d}x + \int_0^{\pi} F(Re^{i\theta})iRe^{i\theta}\,\mathrm{d}\theta \tag{3.86}$$

$$= 2\pi i \sum \text{residues in upper half plane} \tag{3.87}$$

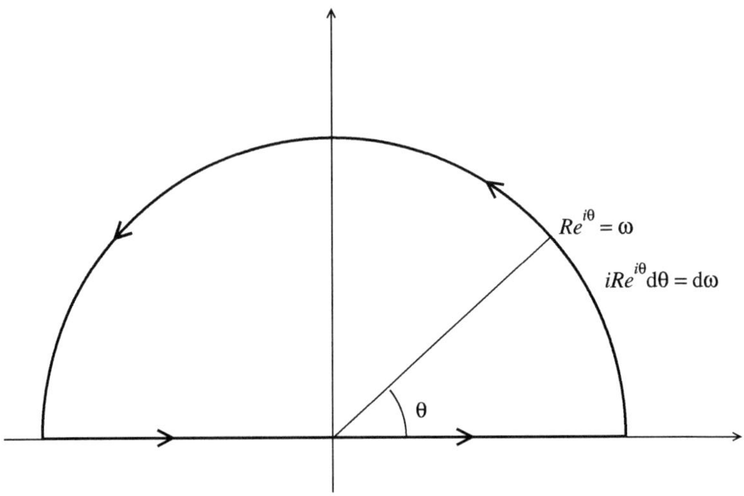

Fig. 3.2 Circuit for integration in the complex plane of Eq. 3.86

and that Cauchy's formula yields

$$\oint_c \frac{F(\omega)}{\omega - \omega_0}\, \mathrm{d}\omega = 2\pi i F(\omega_0) \tag{3.88}$$

if ω_0 is within the circuit c. Introduce the finite R approximation of the inverse Fourier transform of f, then consider

$$I_R = \int_0^{\pi} f(Re^{i\theta}) e^{it(R\cos\theta + iR\sin\theta)} iRe^{i\theta}\, \mathrm{d}\theta \tag{3.89}$$

Let R be large enough that $|f(Re^{i\theta})| < \varepsilon$. Therefore

$$|I_R| \le \ \varepsilon R \int_0^{\pi} \left| e^{itR\cos\theta} e^{-tR\sin\theta)} e^{i\frac{\pi}{2}} e^{i\theta} \right| \mathrm{d}\theta = \varepsilon R \int_0^{\pi} e^{-tR\sin\theta}\, \mathrm{d}\theta \tag{3.90}$$

$$= \ 2\varepsilon R \int_0^{\pi/2} e^{-tR\sin\theta}\, \mathrm{d}\theta \tag{3.91}$$

Now, $\frac{2}{\pi}\theta \le \sin\theta$ for $\theta \in [0, \pi/2]$, hence

$$|I_R| \le 2\varepsilon R \int_0^{\pi/2} e^{-tR2\frac{\theta}{\pi}}\, \mathrm{d}\theta = 2\varepsilon R \frac{e^{-tR2\theta/\pi}}{-tR2/\pi}\Bigg|_0^{\frac{\pi}{2}} = \varepsilon \frac{1 - e^{-tR}}{t\pi} = \frac{\pi\varepsilon}{t}(1 - e^{-tR}) \tag{3.92}$$

Because ε is arbitrarily small (it suffices to take R large), we have

$$\lim_{R\to\infty} \int_0^{\pi} f(Re^{i\theta}) e^{itR(\cos\theta + i\sin\theta)} Re^{i\theta} i\, \mathrm{d}\theta = 0 \tag{3.93}$$

Hence

$$\begin{aligned}\int_{-\infty}^{+\infty} f(\omega) e^{i\omega t}\, \mathrm{d}\omega &= \lim_{R\to\infty} \left[\int_{-R}^{R} f(\omega) e^{i\omega t}\, \mathrm{d}\omega + \underbrace{\int_0^{\pi} f(Re^{i\theta}) e^{itRe^{i\theta}} iRe^{i\theta}\, \mathrm{d}\theta}_{\to 0} \right] \\ &= 2\pi i \sum \text{upper half plane residues}\end{aligned} \tag{3.94}$$

if $t > 0$ (c oriented counterclockwise). Now, the residue of a pole of order m of a given F is defined by

$$a_{-m} = \frac{1}{(m-1)!} \frac{\mathrm{d}^{m-1}}{\mathrm{d}z^{m-1}} [(z - z_0)^m F(z)] \Big|_{z=z_0} \tag{3.95}$$

For example

$$a_{-1} = (z - z_0) F(z) \Big|_{z_0} \tag{3.96}$$

Now, $\frac{e^{i\omega t}}{\omega^2+\gamma^2} = \frac{e^{i\omega t}}{(\omega+i\gamma)(\omega-i\gamma)}$ has only one pole in the upper half plane: $\omega_0 = i\gamma$ and $f(\omega) \xrightarrow[|\omega|\to\infty]{} 0$. Therefore, we can apply the above calculations:

$$a_{-1} = (\omega - i\gamma)\frac{e^{i\omega t}}{\omega^2+\gamma^2}\bigg|_{i\gamma} = \frac{e^{i\omega t}}{\omega+i\gamma}\bigg|_{i\gamma} = \frac{e^{-\gamma t}}{2i\gamma} \tag{3.97}$$

Hence

$$\int_{-\infty}^{+\infty} \frac{e^{i\omega t}}{\omega^2+\gamma^2}\,\mathrm{d}\omega = \pi\frac{e^{-\gamma t}}{\gamma} = \frac{\pi}{\gamma}e^{-\gamma t} \tag{3.98}$$

For negative t, use the clockwise contour and obtain $\frac{\pi}{\gamma}e^{\gamma t} = \frac{\pi}{\gamma}e^{-\gamma|t|}$. The negative sign that emerges in the denominator is canceled by the clockwise rotation.

3.3.2 Application: Johnson-Nyquist Noise

Consider an RC circuit in equilibrium at temperature T, with a time-dependent potential difference $U(t)$ at the ends of its resistance (Fig. 3.3). This describes an experiment performed by Johnson, and explained by Nyquist in 1928. It is indeed possible to observe one electron at a time that is emitted by a hot filament (Schottky). Because of their temperature, the electrons in the circuit move up and down in it, but with equal probabilities, so there is no net current, if averaged in time. However, on short time scales, one may find more electrons going up than down and vice versa. Thus, $U(t)$ fluctuates with zero mean. Moreover, the variance of these fluctuations appears to be related only to the value of T. Letting $Q(t) = CU(t)$ be the charge in the capacitor, one may try this description:

$$R\frac{\mathrm{d}Q}{\mathrm{d}t} = -\frac{1}{C}Q + \eta(t);\quad \dot{Q} + \frac{1}{RC}Q = \frac{1}{R}\eta;\quad \dot{U} + \frac{1}{RC}U = \frac{1}{RC}\eta \tag{3.99}$$

where $\mathrm{d}Q/\mathrm{d}t = I$ is a rapidly fluctuating current. Using $\mathbb{E}[\eta(t)\eta(t')] = 2Rk_BT\,\delta(t-t')$ leads to a Langevin equation that is formally identical to the one for the Brownian motion, with $\mathbb{E}[\Gamma(t)\Gamma(t')] = q\delta(t-t')$. In particular, comparing the two, the viscosity of the Brownian motion γ is replaced by $1/RC$, and one may write

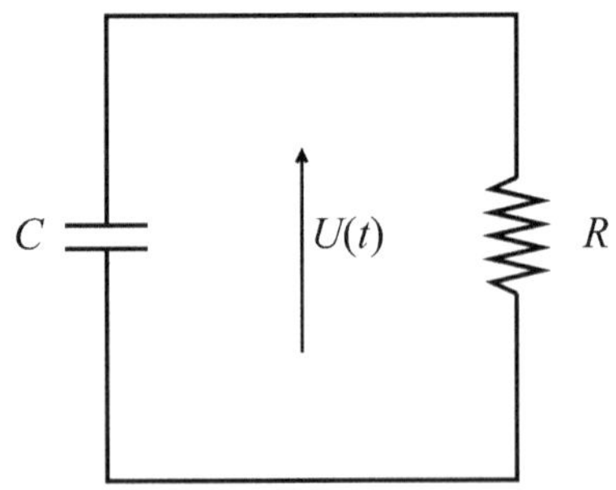

Fig. 3.3 RC circuit with time-dependent current due to a time-dependent potential $U(t)$

$$
\begin{aligned}
\mathbb{E}\left[\frac{\eta(t)}{R}\frac{\eta(t')}{R}\right] &= \frac{2k_BT}{R}\delta(t-t') \quad \Rightarrow \quad q_Q = \frac{2k_BT}{R} \\
\mathbb{E}\left[\frac{\eta(t)}{RC}\frac{\eta(t')}{RC}\right] &= \frac{2k_BT}{RC^2}\delta(t-t') \quad \Rightarrow \quad q_U = \frac{2k_BT}{RC^2}
\end{aligned}
\tag{3.100}
$$

as well as

$$
\begin{aligned}
&\mathbb{E}[V] = 0 \;\Rightarrow\; \mathbb{E}[Q] = 0 \;\Rightarrow\; \mathbb{E}[U] = 0 \\
&\mathbb{E}[V^2] = \frac{q}{2\gamma} = \frac{k_BT}{m} \;\Rightarrow\; \mathbb{E}[Q^2] = \frac{q_Q RC}{2} = Ck_BT \;\Rightarrow\; \mathbb{E}[U^2] = \frac{q_U RC}{2} = \frac{k_BT}{C}
\end{aligned}
\tag{3.101}
$$

We also have $\mathbb{E}[Q(t)Q(t')] = k_BTCe^{-|t-t'|/RC}$, and from Eqs. (3.20), (3.82), and (3.83) which yield

$$
I_V(\omega) = \frac{q/2\pi}{\gamma^2+\omega^2} = \frac{\gamma k_BT}{\pi m(\omega^2+\gamma^2)} \tag{3.102}
$$

we obtain

$$
I_U(\omega) = \frac{q_U/2\pi}{\gamma^2+\omega^2} = \frac{Rk_BT}{\pi[1+(RC\omega)^2]} \tag{3.103}
$$

that for low frequencies, $\omega \ll (RC)^{-1}$, implies that I_U does not depend on C: it only depends on temperature: $I_U(\omega) \approx 2Rk_BT$. This has been experimentally confirmed, giving a glimpse of the immenseness of the range of applicability of the theory of Brownian motion, for which it suffices to interpret its mathematical symbols in different fashions.

3.3.3 Fokker-Planck Equation When Inertia Matters

Let us obtain the probability distribution for a Brownian particle of mass $m = 1$, in a potential ϕ, whose dynamics are expressed by

$$
\begin{cases}
\ddot{x} + \gamma\dot{x} + \phi'(x) = \sigma\Gamma(t) \\
\mathbb{E}[\Gamma(t)] = 0 \\
\mathbb{E}[\Gamma(t)\Gamma(t')] = \delta(t-t')
\end{cases}
\tag{3.104}
$$

To use the Fokker-Planck equation, we introduce $v = \dot{x}$, so that the process be written as

$$
\begin{cases}
\dot{x}(t) = v(t) \\
\dot{v}(t) = -\gamma v(t) - \phi'(x(t)) + \sigma\Gamma(t)
\end{cases}
\tag{3.105}
$$

The coefficients of Eq. (3.52) take the form:

$$
\begin{aligned}
&h_x = v(t) \quad g_{xx} = 0 \quad g_{xv} = 0 \\
&h_v = -\gamma v(t) - \phi'(x(t)) \quad g_{vx} = 0 \quad g_{vv} = \sigma
\end{aligned}
$$

and those of Eq. (3.53) read

$$\begin{aligned} &D_x^{(1)} = v(t) \quad D_v^{(1)} = -\gamma v(t) - \phi'(x(t)) \\ &D_{xx}^{(2)} = 0 \quad D_{xv}^{(2)} = 0 \quad D_{vx}^{(2)} = 0 \quad D_{vv}^{(2)} = \sigma^2 \end{aligned}$$

The Fokker-Planck equation (3.54) is thus the following:

$$\begin{aligned} \frac{\partial W}{\partial t} &= \left[-\frac{\partial}{\partial x} D_x^{(1)} - \frac{\partial}{\partial v} D_v^{(1)} + \frac{\partial^2}{\partial x^2} D_{xx}^{(2)} + \frac{\partial^2}{\partial x \partial v} D_{xv}^{(2)} + \frac{\partial^2}{\partial v \partial x} D_{vx}^{(2)} + \frac{\partial^2}{\partial v^2} D_{vv}^{(2)} \right] W \\ &= -\frac{\partial}{\partial x} v W - \frac{\partial}{\partial v} \Big[-\gamma v(t) - \phi'(x(t)) \Big] W + \frac{\partial^2}{\partial v^2} \sigma^2 W \end{aligned} \tag{3.106}$$

Setting $\sigma^2 = \gamma k_B T/m = \gamma v_{th}^2$ and $-\phi'(x) = -\omega^2 x$, which is a harmonic force, we obtain

$$\frac{\partial W}{\partial t} = -v \frac{\partial W}{\partial x} + \gamma \left(W + v \frac{\partial W}{\partial v} \right) + \omega x \frac{\partial W}{\partial v} + \gamma v_{th}^2 \frac{\partial^2 W}{\partial v^2} \tag{3.107}$$

The solution in the stationary state, which is obtained separating the variables as

$$W(x, v) = \Phi(x) \Psi(v)$$

is expressed by

$$W_{st}(x, v) = \frac{\omega}{2\pi v_{th}^2} e^{-\frac{1}{2}\frac{v^2}{v_{th}^2}} e^{-\frac{1}{2}\frac{\omega^2 x^2}{v_{th}^2}} = \frac{m\omega}{2\pi k_B T} e^{-\frac{mv^2 + m\omega^2 x^2}{2k_B T}} = \frac{m\omega}{2\pi k_B T} e^{-\frac{E}{k_B T}} \tag{3.108}$$

Marginalizing, we obtain the stationary densities for the two observables:

$$W_{st}(x) = \frac{m\omega}{2\pi k_B T} e^{-\frac{m\omega^2 x^2}{2k_B T}} \int_{-\infty}^{+\infty} e^{-\frac{mv^2}{2k_B T}} \mathrm{d}v \tag{3.109}$$

Substituting coordinates:

$$\frac{mv^2}{k_B T} = \frac{u^2}{s^2} \quad \Rightarrow \quad du = s \sqrt{\frac{m}{k_B T}} \mathrm{d}v \quad \Rightarrow \quad \mathrm{d}v = \sqrt{\frac{k_B T}{m s^2}} du \tag{3.110}$$

Therefore,

$$W_{st}(x) = \frac{m\omega}{2\pi k_B T} e^{-\frac{m\omega^2 x^2}{2k_B T}} \int_{-\infty}^{+\infty} e^{-\frac{u^2}{2s^2}} \sqrt{\frac{k_B T}{m s^2}} du \tag{3.111}$$

$$= \frac{m\omega}{2\pi k_B T} e^{-\frac{m\omega^2 x^2}{2k_B T}} \sqrt{\frac{2\pi k_B T}{m}} = \sqrt{\frac{m\omega^2}{2\pi k_B T}} e^{-\frac{m\omega^2 x^2}{2k_B T}} \tag{3.112}$$

Similarly

$$W_{st}(v) = \frac{m\omega}{2\pi k_B T} e^{-\frac{mv^2}{2k_B T}} \int_{-\infty}^{+\infty} e^{-\frac{m\omega^2 x^2}{2k_B T}} \,\mathrm{d}x = \sqrt{\frac{m}{2\pi k_B T}} e^{-\frac{mv^2}{2k_B T}} \tag{3.113}$$

This implies that

$$\mathbb{E}[x] = 0; \quad \mathbb{E}[v] = 0$$
$$\mathbb{E}[x^2] = \frac{k_B T}{m\omega^2}; \quad \mathbb{E}[v^2] = \frac{k_B T}{m}$$

3.3.4 Variations of Brownian Motion: Detection of Gravitational Waves

Here, we consider one variation of the Brownian motion, showing that it also applies to sophisticated nonequilibrium phenomena, such as the noise in gravitational waves detectors (Bonaldi et al. 2009). Among these, the resonant bar detectors are massive objects meant to resonate with gravitational waves. In the absence of gravitational waves, these bars vibrate and, like any solid in an equilibrium state at a given temperature T, the variance of such vibrations is proportional to T, as predicted by the fluctuation-dissipation relation. This is the thermal noise which a classical macroscopic object cannot avoid, and interferes with the vibrations induced by the gravitational waves. However, it can be reduced by reducing T. In particular, the bar known as AURIGA is an aluminum bar of mass 2.2×10^3 Kg and length of 3 m, which is cooled to liquid helium temperature $T_0 = (4.6 \pm 0.2)$ K. Then, in order to improve the stability of the device, its modes of vibration are further "cooled" by a feedback mechanism, which acts as a kind of viscosity. Such an experimental device can be described through separate normal modes, each of which behaves like an RLC series electrical circuit. Denoting the effective inductance of a mode by L, the capacitance

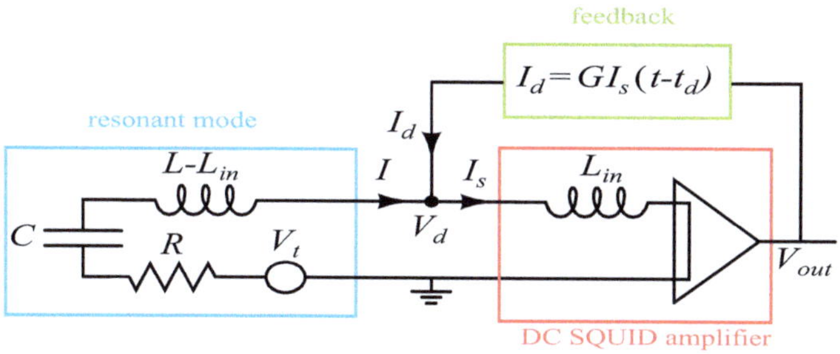

Fig. 3.4 RLC circuit describing one mode of vibration of a feedback cooled resonant bar. A dc SQUID is represented as current amplifier. The observable is the current I_s, and the feedback corresponds to a current I_d which is a delayed copy of I_s reduced by a factor $G \ll 1$. The SQUID output voltage is $V_{out} = AI_s$

by C and the resistance by R, cf. Fig. 3.4, the corresponding evolution equations take the form:

$$(L - L_{in})\frac{d^2Q(t)}{dt^2} + R\frac{dQ(t)}{dt} + \frac{Q(t)}{C} = V_T(t) - V_d(t) \tag{3.114a}$$

$$V_d(t) = L_{in}\frac{dI_s(t)}{dt}, \quad \text{where } I(t) + I_d(t) = I_s(t) \tag{3.114b}$$

Here, Q is the charge in the capacitor, $I = dQ(t)/dt$ is the current through the inductance L, V_d is the voltage at the node where the feedback takes place, and L_{in} is the input inductance of the SQUID amplifier. The feedback form chosen in Ref. Bonaldi et al. (2009) is

$$I_d(t) = GI_s(t - t_d), \quad \text{with } t_d = \frac{\pi}{2\omega_r} \text{ and } G \ll 1 \tag{3.115}$$

It is assumed that each oscillator is driven by a stochastic voltage of the form $V_T(t) = \sqrt{2k_BT_0R}\,\Gamma(t)$, where Γ is a Gaussian noise. Moreover, the very high quality factor of the oscillators implies that the currents I_s, I_d, and I have a frequency quite close to $\omega_r = 1/\sqrt{LC}$, while their amplitudes and phases change very slowly. Therefore, the following approximation

$$I(t) = \hat{I}(t)\sin[\omega_r t + \hat{\phi}(t)]$$

and the analogous ones for the other currents are quite accurate. Although Eq. (3.115) implies memory effects, due to contributions to time t coming from earlier times $t - t_d$, the above quasi-harmonic approximation implies $I_s(t - t_d) \simeq \omega_r Q_s(t)$, and $I_s(t) = dQ_s(t)/dt$ obeys

$$L\frac{dI_s(t)}{dt} + I_s(t)\,[R + R_d] + \frac{Q_s(t)}{C} = \sqrt{2k_BT_0R}\,\Gamma(t) \tag{3.116}$$

with $R_d = G\omega_r L_{in}$ plays the role of the viscous damping. Introducing $g = R_d/R$, as the efficiency of the feedback mechanism, Eq. (3.116) turns identical to the Langevin equation with damping $R + R_d$, in an equilibrium state at the fictitious temperature $T_{\text{eff}} = T_0/(1 + g)$.

Although the fact that T_{eff} differs from the bath temperature T_0 reveals the nonequilibrium nature of the phenomenon, one may formally solve Eq. (3.116) as usual in the equilibrium case. This was done in Ref. Bonaldi et al. (2009), comparing with experimental data collected in several years time. The above representation of the gravitational waves detector can be written as

$$\begin{cases} \dot{Q}(t) = I(t) \\ \dot{I}(t) = -\dfrac{R_k + R_d}{L_k}I - \dfrac{1}{L_kC_k}Q + \sqrt{2k_BT_{eff}(R_k + R_d)}\Gamma \end{cases} \tag{3.117}$$

which corresponds to the following identifications with the Brownian motion equation:

$$\gamma = \frac{R_k + R_d}{L_k}; \quad \omega^2 = \frac{1}{L_k C_k}; \quad \sigma = \sqrt{2k_B T_{eff}(R_d + R_k)}$$

As we took $\sigma^2 = \frac{\gamma k_B T}{m}$, the above averages can be written as

$$\mathbb{E}[x] = 0; \quad \mathbb{E}[v] = 0$$
$$\mathbb{E}[x^2] = \frac{\sigma^2}{\gamma\omega^2}; \quad \mathbb{E}[v^2] = \frac{\sigma^2}{\gamma}$$

Consequently, this specific model has Gaussian distributions with

$$\mathbb{E}[Q] = 0; \quad \mathbb{E}[I] = 0 \tag{3.118}$$

and

$$\mathbb{E}[Q^2] = \mathrm{var}(Q) = \frac{\sigma^2}{\gamma\omega^2} = \frac{2k_B T_{eff}(R_k + R_d)}{\frac{(R_k+R_d)}{L_k}\frac{1}{L_k C_k}} = 2k_B T_{eff} L_k^2 C_k$$

$$\mathbb{E}[I^2] = \mathrm{var}(I) = \frac{\sigma^2}{\gamma} = \frac{2k_B T_{eff}(R_k + R_d)}{\frac{(R_k+R_d)}{L_k}} = 2k_B T_{eff} L_k$$

To obtain the probability distribution functions of

$$Q_\tau(t) = \frac{1}{\tau}\int_t^{t+\tau} Q(s)\,\mathrm{d}s \quad \text{and of} \quad I_\tau(t) = \frac{1}{\tau}\int_t^{t+\tau} I(s)\,\mathrm{d}s \tag{3.119}$$

it is enough to recall that these are a Gaussian process; hence, the distributions of its variables observed at different times results in a joint Gaussian distribution.

3.3.5 Variations of Brownian Motion: Active Matter

Unlike molecules or inert objects, in general, *active particles* propel themselves by harvesting energy from their environment, such as, e.g., certain types of bacteria. If immersed in a fluid, one possible representation of their dynamics is a variation of the Brownian Motion, with a supplementary force. In fact, bacteria can even be smaller than pollen and should be similarly affected by molecular impacts. A two-dimensional active fluid may be represented by persistent Brownian particles that receive a push in the same direction as their orientation:

$$\dot{x} = -\mu_t \partial_x U(x) + v\cos\vartheta + \sqrt{2D_t}\Gamma_x \tag{3.120}$$
$$\dot{y} = v\sin\vartheta + \sqrt{2D_t}\Gamma_y \tag{3.121}$$
$$\dot{\vartheta} = \mu_r G(\vartheta) + \sqrt{2D_r}\Gamma_\theta \tag{3.122}$$

where (x, y) is the position (e.g., center of mass) of an active particle, ϑ is its orientation with respect to x-axis, $U(x)$ is a potential representing a wall, G is the torque this wall produces, D_t, D_r are diffusivities, μ_t, μ_r are two coefficients related to friction, Γ_x, Γ_y, Γ_θ are independent Gaussian noises, and ν is the activity coefficient, which models the effect of the engine of the particle.

This model assumes that inertia is negligible. In case it is not, another model can be considered. For instance:

$$\dot{x} = v_x, \tag{3.123}$$

$$\dot{y} = v_y, \tag{3.124}$$

$$\dot{\theta} = v_\theta, \tag{3.125}$$

$$\dot{v}_x = -\frac{1}{m}\frac{\partial U}{\partial x} + \frac{\nu \cos\frac{\theta}{\sqrt{J}}}{m\mu_t} - \frac{v_x}{m\mu_t} + \frac{\sqrt{2D_t}}{m\mu_t}\Gamma_x, \tag{3.126}$$

$$\dot{v}_y = -\frac{1}{m}\frac{\partial U}{\partial y} + \frac{\nu \sin\frac{\theta}{\sqrt{J}}}{m\mu_t} - \frac{v_y}{m\mu_t} + \frac{\sqrt{2D_t}}{m\mu_t}\Gamma_y, \tag{3.127}$$

$$\dot{v}_\theta = \frac{G[\frac{\theta}{\sqrt{J}}]}{mJ^{1/2}} - \frac{v_\theta}{\mu_r m J} + \frac{\sqrt{2D_r}}{\mu_r m J^{1/2}}\Gamma_\theta \tag{3.128}$$

where $J = (a^2 + b^2)/5$ is the inertial moment per unit mass of the ellipse, and the orientation angle is rescaled as $\theta = \vartheta J^{1/2}$. In principle, the underdamped picture should converge to the overdamped picture, when inertia, i.e., the mass of particles becomes small. However, this limit is known to be singular, in general, and the activity of particles may make it even more troublesome. This second approach leads to different equations for the probability distribution of active particles, and to different expressions for the thermodynamic quantities, pressure included (De Groot and Mazur 1984). The investigation of such and many other tremendously interesting questions has just begun.

The examples reported above indicate that the legacy of the Brownian motion extends to nonequilibrium systems. Much research is still required to fully grasp the meaning and extent of this statement. Stochastic thermodynamics is one framework in which this program is actively pursued (Peliti and Pigolotti 2021).

3.4 Generalization of Fluctuation-Dissipation Relation

The study of Brownian motion is important because its range of applicability includes an incredibly wide set of fluctuating phenomena. At the same time, its validity is strictly demonstrated within the realm of equilibrium, which is characterized by fast decay of correlations and by equipartition. This excludes, for instance, the gravitational antenna described above, which, in fact, violates equipartition and relies on effective temperatures, whose physical interpretation depends on the case at hand.[11]

[11] In the case of vibrating solids, it has been related to an excess of energy stored in low-frequency oscillations (Conti et al. 2013a).

To move away from equilibrium, a celebrated approach introduces retarded frictions (Kubo et al. 1998).

Consider the following model:

$$\dot{v}(t) = -\int_{-\infty}^{t} \gamma(t-t')v(t')\,\mathrm{d}t' + \Gamma(t), \quad \mathbb{E}[\Gamma(t)] = 0 \tag{3.129}$$

which is linear and can be treated by harmonic analysis using the Fourier integrals:

$$v(t) = \int_{-\infty}^{\infty} \tilde{v}(\omega)e^{iwt}\,\mathrm{d}\omega; \quad \Gamma(t) = \int_{-\infty}^{\infty} \tilde{\Gamma}(\omega)e^{iwt}\,\mathrm{d}\omega; \quad \gamma(t) = \int_{-\infty}^{\infty} \tilde{\gamma}(\omega)e^{iwt}\,\mathrm{d}\omega \tag{3.130}$$

where

$$\tilde{v}(\omega) = \frac{1}{2\pi}\int_{-\infty}^{\infty} v(t)e^{-i\omega t}\,\mathrm{d}t; \; \tilde{\Gamma}(\omega) = \frac{1}{2\pi}\int_{-\infty}^{\infty} \Gamma(t)e^{-i\omega t}\,\mathrm{d}t; \; \tilde{\gamma}(\omega) = \frac{1}{2\pi}\int_{-\infty}^{\infty} \gamma(t)e^{-i\omega t}\,\mathrm{d}t \tag{3.131}$$

In this model, causality, meaning that events occurring at times $t' > t$ do not affect the process at time t, implies that friction obeys

$$\gamma(t) = 0 \quad \text{if } t < 0 \tag{3.132}$$

Therefore, the integral in Eq. (3.129) can be written as a convolution integral:

$$\int_{-\infty}^{t} \gamma(t-t')v(t')\,\mathrm{d}t' = \int_{-\infty}^{\infty} \gamma(t-t')v(t')\,\mathrm{d}t' \tag{3.133}$$

Substituting Eq. (3.130) in Eq. (3.129) leads to

$$i\omega\tilde{v}(\omega) = -\tilde{\gamma}(\omega)\tilde{v}(\omega) + \tilde{\Gamma}(\omega), \tag{3.134}$$

hence to this:

$$\tilde{v}(\omega) = \frac{\tilde{\Gamma}(\omega)}{i\omega + \tilde{\gamma}(\omega)} \quad \text{and} \quad I_V(\omega) = \frac{I_\Gamma(\omega)}{|i\omega + \tilde{\gamma}(\omega)|^2} \tag{3.135}$$

Many different possible routes can be taken, now, corresponding to different relations between γ and Γ. However, given the stunning success of the Brownian motion theory, it is useful to develop a framework that reduces to that theory, when conditions return to those of the Brownian motion. In particular, the autocorrelation function of V, that is derived from the spectrum of Γ, must be recovered when memory vanishes. That imposes conditions on I_Γ. For instance, one may require

$$I_\Gamma(\omega) = \frac{1}{2\pi}\int_{-\infty}^{\infty} \mathbb{E}\left[\Gamma(0)\Gamma(t)\right] e^{-i\omega t}\,\mathrm{d}t = \frac{k_B T}{\pi m}\,\mathrm{Re}[\tilde{\gamma}(\omega)], \quad \text{and} \quad I_\Gamma(\omega) \geq 0 \tag{3.136}$$

to mimic Eq. (3.84), where

$$\tilde{\gamma}(\omega) = \frac{1}{2\pi}\int_{-\infty}^{\infty} \mathrm{d}t\, \gamma(t)e^{-i\omega t} = \frac{1}{2\pi}\int_{0}^{\infty} \mathrm{d}t\, \gamma(t)e^{-i\omega t} = \mathrm{Re}[\tilde{\gamma}(\omega)] + i\,\mathrm{Im}[\tilde{\gamma}(\omega)] \tag{3.137}$$

is the Fourier transform of the real function γ. Equation (3.136) is obtained if the autocorrelation of the random term obeys

$$\mathbb{E}\left[\Gamma(t_1)\Gamma(t_2)\right] = \frac{k_B T}{2\pi m}\Big[\gamma(t_1 - t_2) + \gamma(t_2 - t_1)\Big] \tag{3.138}$$

and the retarded friction γ is such that the following holds: $\mathrm{Re}[\tilde{\gamma}(\omega)] \geq 0$. Using time translation invariance for a steady state, and taking $t_1 = t$ and $t_2 = 0$, one obtains

$$\begin{aligned}\frac{1}{2\pi}\int_{-\infty}^{\infty} \mathbb{E}\left[\Gamma(0)\Gamma(t)\right]e^{-i\omega t}\,\mathrm{d}t &= \frac{1}{2\pi}\int_{-\infty}^{\infty} \frac{k_B T}{2\pi m}\Big[\gamma(t) + \gamma(-t)\Big]e^{-i\omega t}\,\mathrm{d}t \\ &= \frac{k_B T}{2\pi m}\Big[\tilde{\gamma}(\omega) + \tilde{\gamma}(-\omega)\Big] = \frac{k_B T}{2\pi m}\, 2\mathrm{Re}\left[\tilde{\gamma}(\omega)\right] = \frac{k_B T}{\pi m}\,\mathrm{Re}\left[\tilde{\gamma}(\omega)\right] \geq 0\end{aligned} \tag{3.139}$$

Furthermore, taking $\gamma(t) = 2\pi\gamma\delta(t)$ with $\gamma > 0$, one falls back in the original situation with δ-correlated noise and obtains

$$I_\Gamma(\omega) = \frac{1}{2\pi}\int_{-\infty}^{\infty} \frac{k_B T}{2\pi m}\, 4\pi\gamma\,\delta(t)\,e^{-i\omega t}\,\mathrm{d}t = \frac{k_B T}{\pi m}\gamma \tag{3.140}$$

Note that in the Langevin equation (3.71), γ and Γ are independent, while in Eq. (3.138), they are not. Their relation expressed by Eq. (3.84) within the theory of Brownian motion is not necessary, as it follows from the energy equipartition principle, which, however, is not necessarily verified, especially away from equilibrium. Moreover, the viscosity of a fluid and the temperature can be changed independently, for instance, by taking different fluids with different viscosities at the same temperature. The condition (3.138) is the cost required by the new theory to be consistent with Brownian motion.

Correspondingly, to generalize the mobility note first that:

$$\mu = \frac{D}{k_B T} = \lim_{t\to\infty} \frac{\mathbb{E}\left[(x(t) - x(0))^2\right]}{2tk_B T} \tag{3.141}$$

$$= \lim_{t\to\infty} \frac{1}{2tk_B T}\int_0^t \mathrm{d}t_1 \int_0^t \mathrm{d}t_2 \mathbb{E}\left[v(t_1)v(t_2)\right] \tag{3.142}$$

$$= \lim_{t\to\infty} \frac{1}{2tk_B T}\int_0^t \mathrm{d}t_1 \int_0^t \mathrm{d}t_2 \mathbb{E}\left[v(0)v(t_2 - t_1)\right] \tag{3.143}$$

where (3.143) follows from (3.142) because in a steady state the origin of time plays no role, and we have

$$\mathbb{E}\left[v(t_1)v(t_2)\right] = \phi_V(t_2 - t_1) = \mathbb{E}\left[v(0)v(t_2 - t_1)\right] \tag{3.144}$$

For the same reason,

$$\phi_V(t) = \mathbb{E}[v(0)v(t)] = \mathbb{E}[v(-t)v(0)] = \phi_V(-t) \quad \text{hence} \quad \phi_V(t_2 - t_1) = \phi_V(t_1 - t_2) \tag{3.145}$$

Let us now change integration variables from (t_1, t_2) to $(s = t_1 - t_2, r = t_2)$, if $t_1 \geq t_2$, and to $(s = t_2 - t_1, r = t_1)$ if $t_2 > t_1$, which implies unit Jacobian determinant in both cases. This amounts to consider the square $[0, t] \times [0, t]$ as the union of two triangles. In the first case, integrating over r at constant s means that $\phi_V(s)$ is constant, while r ranges from 0 (the smallest value of t_2) to $t - s$ (the largest value of $t_2 = t_1 - s$). In the second case, integrating over r at fixed s yields again constant ϕ_V, and r ranging from 0 (the smallest value of t_1) to $t - s$ (the largest value of $t_1 = t_2 - s$). The following integrations in s range from 0 (the smallest values s can take because of its definition) to t (the largest difference between t_1 and t_2). Therefore, one gets

$$\int_0^{t-s} \mathrm{d}r\, \phi_V(s) = (t - s)\phi_V(s) \tag{3.146}$$

and

$$\int_0^t \mathrm{d}t_1 \int_0^t \mathrm{d}t_2 \mathbb{E}[v(0)v(t_2 - t_1)] = \int_0^t \mathrm{d}t_1 \left\{ \int_0^{t_1} \mathrm{d}t_2 \phi_V(t_2 - t_1) + \int_{t_1}^t \mathrm{d}t_2 \phi_V(t_2 - t_1) \right\} \tag{3.147}$$

$$= 2\int_0^t \mathrm{d}s\, (t - s)\phi_V(s) \tag{3.148}$$

which implies

$$\mu = \lim_{t\to\infty} \frac{1}{k_B T t} \int_0^t (t - s)\mathbb{E}[v(0)v(s)]\, \mathrm{d}s \tag{3.149}$$

Then, the condition

$$\int_0^t s\, \mathbb{E}[v(0)v(s)]\, \mathrm{d}s = o(t) \quad \text{for } t \to \infty \tag{3.150}$$

yields

$$\mu = \frac{1}{k_B T} \lim_{t\to\infty} \int_0^t \mathbb{E}[v(0)v(s)]\, \mathrm{d}s \tag{3.151}$$

which gives a finite mobility coefficient if the following condition is also satisfied:

$$\left| \int_0^\infty \phi_V(s)\mathrm{d}s \right| < \infty \tag{3.152}$$

The validity of conditions (3.152) and (3.150) must be separately required for Eq. (3.151) to hold, but both show that a transport coefficient like μ exists only when correlations decay sufficiently fast.

Expression (3.151) is one example of Green-Kubo integral. For every linear transport coefficient, there is one such Green-Kubo integral, whose integrand is the autocorrelation of the current of interest. A transport process is called *normal* if such autocorrelation decays sufficiently fast in time that the integral, hence the transport coefficient, exists.

As the integral (3.151) is a kind of Fourier transform at zero frequency, one may adopt the following expression for a frequency dependent mobility:

$$\mu(\omega) = \frac{1}{k_B T} \int_0^\infty \mathbb{E}\left[v(0)v(t)\right] e^{-i\omega t}\, \mathrm{d}t = \frac{1}{k_B T} \int_0^\infty \phi_V(t) e^{-i\omega t}\, \mathrm{d}t, \qquad (3.153)$$

and corresponding frequency dependent diffusion coefficient: $D(\omega) = k_B T \mu(\omega)$. We can thus write

$$\begin{aligned} I_V(\omega) &= \frac{1}{2\pi} \int_{-\infty}^\infty \phi_V(t) e^{-i\omega t} \mathrm{d}t = \frac{1}{2\pi} \int_{-\infty}^0 \phi_V(t) e^{-i\omega t} \mathrm{d}t + \frac{1}{2\pi} \int_0^\infty \phi_V(t) e^{-i\omega t} \mathrm{d}t \\ &= \frac{k_B T}{2\pi} \Big[\mu(\omega) + \mu(-\omega)\Big] = \frac{k_B T}{\pi} \mathrm{Re}\left[\mu(\omega)\right] \end{aligned} \qquad (3.154)$$

where the last equality holds because for real ϕ_V one has

$$\mu(-\omega) = \frac{1}{k_B T} \int_0^\infty \phi_V(t) \cos \omega t\ \mathrm{d}t - \frac{i}{k_B T} \int_0^\infty \phi_V(t) \sin \omega t\ \mathrm{d}t = \mu(\omega)^* \qquad (3.155)$$

Equation (3.154) is consistent with (3.28), for which $\mu = 1/m\gamma$. Indeed, Eqs. (3.135), (3.136) and (3.154) imply

$$\mathrm{Re}\left[\mu(\omega)\right] = \frac{\pi}{k_B T} \frac{I_\Gamma(\omega)}{|i\omega + \tilde{\gamma}(\omega)|^2} = \frac{\pi}{k_B T} \frac{k_B T}{\pi m} \frac{\mathrm{Re}[\tilde{\gamma}(\omega)]}{|i\omega + \tilde{\gamma}(\omega)|^2} = \frac{\mathrm{Re}[\tilde{\gamma}(\omega)]}{m\, |i\omega + \tilde{\gamma}(\omega)|^2} \qquad (3.156)$$

and $\mu(0)$ is real, because $\mu(0) = \mu(0)^*$, like $\tilde{\gamma}(0)$ is, because γ is real. Then, recalling that we have assumed $\mathrm{Re}[\tilde{\gamma}(\omega)] \geq 0$, and that $\tilde{\gamma}(0)$ is real since it equals $\tilde{\gamma}(0)^*$, one gets

$$\mu(0) = \frac{\tilde{\gamma}(0)}{m\, |\tilde{\gamma}(0)|^2} = \frac{1}{m\tilde{\gamma}(0)} \qquad (3.157)$$

A second way to compute I_V is derived from Eqs. (3.135) and (3.136):

$$I_V(\omega) = \frac{I_\Gamma(\omega)}{|i\omega + \tilde{\gamma}(\omega)|^2} = \frac{k_B T}{m\pi} \frac{\mathrm{Re}[\tilde{\gamma}(\omega)]}{|i\omega + \tilde{\gamma}(\omega)|^2} = \frac{k_B T}{m\pi} \frac{1}{2} \left[\frac{1}{i\omega + \tilde{\gamma}(\omega)} + \frac{1}{-i\omega + \tilde{\gamma}(\omega)^*}\right] \qquad (3.158)$$

where two complex conjugate terms have been separated. Again, comparing Eq. (3.154) with Eq. (3.158), and using Eq. (3.153), the mobility may be written as

$$\mu(\omega) = \frac{1}{m\left[i\omega + \tilde{\gamma}(\omega)\right]} = \frac{1}{k_B T} \int_0^\infty \mathrm{d}t\ \mathbb{E}\left[v(0)v(t)\right] e^{-i\omega t} \qquad (3.159)$$

Analogously, given that $\phi_\Gamma(t) = \mathbb{E}[\Gamma(0)\Gamma(t)]$ is real and a symmetric function, cf. Eqs. (3.138), (3.139) solved for the friction coefficient yields

$$\begin{aligned} \mathrm{Re}[\tilde{\gamma}(\omega)] &= \frac{\pi m}{k_B T}\frac{1}{2\pi}\int_{-\infty}^{\infty} \mathrm{d}t\, \phi_\Gamma(t) e^{-i\omega t} \\ &= \frac{m}{2k_B T}\left[\int_{-\infty}^{0} \mathrm{d}t\, \phi_\Gamma(t) e^{-i\omega t} + \int_{0}^{\infty} \mathrm{d}t\, \phi_\Gamma(t) e^{-i\omega t}\right] \\ &= \frac{m}{2k_B T}\left[\int_{0}^{\infty} \mathrm{d}t\, \phi_\Gamma(t) e^{-i(-\omega) t} + \int_{0}^{\infty} \mathrm{d}t\, \phi_\Gamma(t) e^{-i\omega t}\right] \end{aligned} \tag{3.160}$$

where the last two integrals are complex conjugate. Therefore, we can write

$$\tilde{\gamma}(\omega) = \frac{m}{k_B T}\int_0^\infty \mathrm{d}t e^{-i\omega t}\mathbb{E}[\Gamma(0)\Gamma(t)] \tag{3.161}$$

Equations (3.159) and (3.161) are called *Fluctuation-Dissipation Relation* (FDR) of first kind and of second kind, respectively. The first kind gives the complex mobility (admittance, in general) in terms of the autocorrelation of the velocity (flow, in general). The second kind gives the complex viscosity (impedance, in general) in terms of the autocorrelation of the random force. These two kinds of FDR imply that the response of a system to external actions which perturb its equilibrium is linked to the spontaneous thermal fluctuations in absence of perturbing forces. The FDR of the first kind has to be considered more fundamental than the second, since it refers to experimentally accessible quantities (the flows), while the second kind relies on the distinction between frictional and random forces, which is made problematic by Eq. (3.138). Other popular Green-Kubo integrals are the following:

$$D_i = \frac{1}{3}\int_0^\infty \mathrm{d}t\, \mathbb{E}[v_i(0)v_i(t)], \quad \text{self-diffusion in 3D} \tag{3.162}$$

$$\lambda = \frac{V}{3k_B T^2}\int_0^\infty \mathrm{d}t\, \mathbb{E}\left[J_q(0)J_q(t)\right], \quad \text{thermal conductivity} \tag{3.163}$$

$$\eta = \frac{V}{3k_B T}\int_0^\infty \mathrm{d}t\, \mathbb{E}\left[P_{xy}(0)P_{xy}(t)\right], \quad \text{shear viscosity} \tag{3.164}$$

where V is the volume occupied by the system of interest and T is its temperature. It should be noted that these formulae are obtained equally the same from stochastic as well as deterministic processes.

To conclude this section, it is interesting to reflect on the procedure we have followed: given the result we wanted to obtain—a consistent generalization of the theory of Brownian motion—we have searched for the conditions that produce it. Consequently, the model we have constructed is forced to yield the desired result. This is standard practice in physics, often more useful than the straight logical deductions from general principles, which are frequently cumbersome or even impossible. The question is whether there exist systems of physical interest satisfying the imposed conditions. Speaking of fluctuation-dissipation relations and Green-Kubo formulae, we are lucky: there are countless successful applications of the theory outlined above. As a matter of fact, one often relies on macroscopic observations to deduce microscopic properties and conditions theories should obey. For instance, the attractive part of the Lennard-Jones potential is justified by quantum mechanical calculations, while the repulsive part is only conjectured and its parameters are fitted comparing empirical data and the theoretical equations of state that such potential implies. When the microscopic structure is determined, it can however be used to obtain results that the macroscopic theory does not yield. The theory of Brownian motion is remarkable also from this point of view, since general ideas on the microscopic dynamics, leading to given predictions on the macroscopic behavior, have been subsequently experimentally verified.

3.5 Current Research

As partly indicated in the main body of this chapter, the legacy of the Brownian motion is far from extinguished. Current applications of the Brownian motion, diffusive processes, the Langevin equation and their variations are countless. Among those concerning nonequilibrium phenomena and most familiar to us, we mention the first passage time problem developed from Brownian motion as the underlying dynamics (Redner 2001), its fluctuations under additional constraints (Mejía-Monasterio et al. 2011), and countless biological applications, to cell kinetics, drug delivery, etc. Belousov et al. (2024). A particular area of current active research is that of target search problems (Grebenkov et al. 2024). Brownian motion is also the departure to the vast amount of models to describe anomalous transport, namely, processes that are not diffusive (Metzler and Klafter 2000). Quantum effects expected to appear at low temperatures have raised great interest in the extensions of Brownian motion to the quantum realm (Lampo et al. 2019). Quantum Brownian motion is the basis of the more general research of quantum noise (Gardiner and Zoller 2004), as well as extensions to non-Markovian processes in open quantum systems (de Vega and Alonso 2017). Stochastic control in classical and quantum dynamics (Guerra and Morato 1983) stays an active area which has recently found interesting connections with Stochastic Thermodynamics and Optimal Transport (Aurell et al. 2011). The theory known as Stochastic Thermodynamics (Sekimoto 2010; Seifert 2012) is based on the Langevin equation, and applied to many mesoscopic systems (Peliti and Pigolotti 2021). This theory moves along a narrow ridge that borders macroscopic and microscopic physics, modifying definitions of macroscopic observables to apply them to fluctuating small-scale quantities. Nevertheless, the border appears to be sufficiently wide to allow verification of its results in certain experimental situations, involving low-dimensional dynamics (Ciliberto 2017). That includes molecular machines, proteins, and bacteria. Indeed, the theory of the Brownian motion itself lives on that narrow border, although limited to the safer grounds of equilibrium phenomena. A more general theory of fluctuating thermodynamic quantities that is consistent with the equilibrium macroscopic theory and does not require macroscopic observables to be defined specifically for small-scale systems was developed, e.g., in Ref. Greene and Callen (1951), and has been developed to apply to nonequilibrium systems in, e.g., Ref. Porporato et al. (2024), where the notion of thermodynamic Langevin equation has been introduced. Notable are the applications of relativistic Brownian motion that concern, for instance, the physics of particles accelerators and astrophysical gases (Dunkel and Hänggi 2009). Brownian motion is the essential tool in modeling various economic and finance phenomena like stock prices and market volatility due to their inner random unpredictable behavior (Hull 2021). Last, but definitely not least, a voluminous literature is developing about applications of diffusion models and Langevin equations in machine learning and artificial intelligence (Yang et al. 2023).

Energy Equipartition, Virial Theorem, Boyle's Law

4

If the world were clear, art would not exist.

The Myth of Sisyphus, Albert Camus

4.1 Equipartition of Energy

In this section, we scrutinize the energy equipartition principle, a pillar of statistical physics, that can be easily derived under the assumption that the canonical ensemble applies to the phenomenon of interest. This principle constitutes the basis of countless results of both equilibrium and nonequilibrium physics, since it is needed to connect the microscopic molecular motions with the thermodynamic macroscopic quantities. However, its simplicity, particularly within the realm of the canonical ensemble, may lead into error about the thermodynamic descriptions. In Gallavotti's words:

> this principle, after some shining initial successes, reveals itself as the sign of statistical mechanics inadequacy to solve important problems that fall into its domain. Likewise other well-known important consequences of statistical mechanics are affected by serious paradoxes and theoretical problems. Gallavotti (1999).

Avoiding errors and paradoxes is of course possible, but the applicability of the principle has to be thoroughly understood, which is a bit of an art, although it can be explained. Therefore, the derivation of the equipartition principle will be followed, here, by the derivation of the virial theorem, and by a critical analysis of how Boyle's law of gases and its generalizations result from them. The conclusion will be that, to the very least, very many molecules are required for this derivation of Boyle's law to make sense, although a straight application of the equipartition principle, or of the canonical ensemble, seems to proceed smoothly for any number N of molecules.

C. Mejía-Monasterio and L. Rondoni, *Nonequilibrium Statistical Mechanics I*,
Theoretical and Mathematical Physics,
https://doi.org/10.1007/978-3-031-98249-1_4

This suffices to our purposes; the interested readers are referred to Ref. Gallavotti (1999) for a thorough and detailed analysis of the subject.

The principle of equipartition of energy in classical statistical mechanics states that, given an object made of N particles in d space dimensions in equilibrium at temperature T, the mean energy associated with each of its variables contributing a quadratic term to its Hamiltonian H equals $k_B T/2$. Assuming that mean is computed with respect to the canonical ensemble, the proof proceeds as follows. Arrange the canonical coordinates $(p_1, ..., p_N, q_1, ..., q_N)$, where each q_i and p_j is d-dimensional, as a vector $(x_1, ..., x_n)$ whose first r entries appear only quadratically in H, with $n = 2dN$. Decompose H as a sum of two pieces:

$$H(x_1, ..., x_n) = H_0(x_1, ..., x_r) + H_1(x_{r+1}, ..., x_n) \tag{4.1}$$

where H_0 depends on part or all the variables that appear only quadratically in H, and H_1 depends on all variables that do not appear only quadratically in H, and on those that appear only quadratically in H but are not included in H_0. Adopting the following definitions:

Positive definite function Folland (1999): *A function $f : \mathbb{R}^n \to \mathbb{C}$ is called positive definite if*

$$\sum_{i,j=1}^{m} f(x_i - x_j)\rho_i \rho_j^* \geq 0 \tag{4.2}$$

for any $m \in \mathbb{N}$, for all $x_1, ..., x_m \in \mathbb{R}^n$ and all $\rho_1, ..., \rho_m \in \mathbb{C}$, where * denotes complex conjugation.

and

Homogeneous function Apostol (1981): *A function f defined on an open set $S \subset \mathbb{R}^n$ is called homogeneous of degree p in S, if*

$$f(\lambda x) = \lambda^p f(x) \tag{4.3}$$

for every $\lambda \in \mathbb{R}$ and every $x \in S$ such that $\lambda x \in S$.

One concludes that H_0 is positive definite and homogeneous of degree 2. In addition, H_0 must be differentiable for the N particles to obey Hamilton's equations of motion. Therefore Euler's theorem implies that:

$$\sum_{i=1}^{r} x_i \frac{\partial H_0}{\partial x_i} = 2H_0(x_1, ..., x_r) \tag{4.4}$$

Setting $\beta = 1/k_B T$ and defining the canonical partition function as

$$Z \equiv \int_{\mathbb{R}^n} e^{-\beta(H_0+H_1)}\, \mathrm{d}x_1 \cdots \mathrm{d}x_n \tag{4.5}$$

the canonical ensemble average of H_0 is expressed by

$$\begin{aligned} \mathbb{E}_c(H_0) \equiv & \frac{1}{Z}\int_{\mathbb{R}^n} H_0\, e^{-\beta(H_0+H_1)}\, dx_1\cdots dx_n \\ = & \frac{1}{Z}\int_{\mathbb{R}^n} \frac{1}{2}\sum_{i=1}^{r} x_i \frac{\partial H_0}{\partial x_i}\, e^{-\beta(H_0+H_1)}\, dx_1\cdots dx_n \\ = & -\frac{1}{2\beta}\sum_{i=1}^{r}\frac{1}{Z}\int_{\mathbb{R}^r} dx_1\cdots dx_r \int_{\mathbb{R}^{n-r}} dx_{r+1}\cdots dx_n \left[x_i \frac{\partial}{\partial x_i} e^{-\beta(H_0+H_1)}\right] = r\frac{k_B T}{2} \end{aligned} \tag{4.6}$$

where the second line derives from Euler's theorem, and the third line from the fact that x_i appears only in H_0. Assuming that x_i^2 appears with a positive coefficient in H_0, as usually is the case, one can write

$$\int_{-\infty}^{\infty} \frac{x_i}{2}\frac{\partial H_0}{\partial x_i}\, e^{-\beta(H_0+H_1)}\, dx_i = -\frac{1}{2\beta}\int_{-\infty}^{\infty} x_i \frac{\partial}{\partial x_i} e^{-\beta(H_0+H_1)} dx_i \tag{4.7}$$

$$= -\left.\frac{x_i e^{-\beta(H_0+H_1)}}{2\beta}\right|_{x_i=-\infty}^{\infty} + \frac{1}{2\beta}\int_{-\infty}^{\infty} e^{-\beta(H_0+H_1)} dx_i = \frac{1}{2\beta}\int_{-\infty}^{\infty} e^{-\beta(H_0+H_1)} dx_i \tag{4.8}$$

because the divergence of the factor x_i is killed by the decay of the exponential, in which x_i appears as a negative quadratic term. In other words, each of the r variables on which H_0 depends contributes the amount $k_B T/2$ to the internal energy. Then, the kinetic energy K of unconstrained systems or of systems subjected to time-independent constraints, which is only quadratic in the momenta, appears to be proportional to the temperature:

$$K = dN\,\frac{k_B T}{2} \tag{4.9}$$

Note, this connection between microscopic degrees of freedom and the macroscopically measurable quantity known as temperature arises from the assumption that the statistics of the microscopic phases are described by the canonical ensemble, in which the temperature is assumed to exist and to appear in a specific form.[1] The applicability of the above derivations is quite wide since the kinetic energy is typically only quadratic in the momenta, and often even the potential energy can be approximated by quadratic expressions in the configuration coordinates.

Let us investigate the consequences of energy equipartition in a few simple examples. Suppose the temperature T of a system in a rigid container of volume V is high.

[1] Equipartition is also obtained under the other equilibrium ensembles, by properly adjusting the necessary conditions and the relevant framework. For instance, in the microcanonical ensemble, which pertains to isolated systems, equipartition is obtained in the large N limit, using the Boltzmann entropy S_B, and relating it to temperature *via* the corresponding thermodynamic relation:

$$\frac{1}{T} = \frac{\partial S_B}{\partial U} \tag{4.10}$$

where U is the internal energy of the system (Huang, 1987).

Then the kinetic energy is also high, and may dominate the potential energy Ψ. In fact, because Ψ depends on the distances between particles, it is reasonable to assume that it does not change sensibly with T on average, if the walls of the container are not affected by variations of T. Therefore, sufficiently high T allows us to write $U = \mathbb{E}(K + \Psi) \approx \mathbb{E}(K) \propto T$ for the internal energy of the system. An analogous situation is generated by weakly interacting systems, such as rarefied gases, whose average kinetic energy dominates the average potential energy, because atomic interactions last very short times, although they can be very energetic. For instance, given a monatomic gas in three dimensions constituted by N atoms, recalling that each particle has three degrees of freedom, one may write

$$U = 3N\frac{k_BT}{2} = \frac{3}{2}Nk_BT \tag{4.11}$$

and the heat capacity at constant volume takes the form:

$$C_v = \left.\frac{\partial U}{\partial T}\right|_V = \frac{3}{2}Nk_B \tag{4.12}$$

If one has instead rigid diatomic molecules, whose particles have five degrees of freedom, one obtains

$$C_v = \frac{5}{2}Nk_B \tag{4.13}$$

For a crystal made of N point particles which oscillate with small amplitudes around their equilibrium positions, one may assume that the dynamics is determined by Hooke's forces, hence not only the momenta but also the positions appear quadratically in the Hamiltonian H. Therefore, one obtains

$$U = \frac{3}{2}Nk_BT + \frac{3}{2}Nk_BT = 3\,Nk_BT \quad \text{and} \quad C_v = 3\,Nk_B \tag{4.14}$$

For the oscillations to be small, so that the forces are guaranteed to be harmonic, the kinetic energy has to be small, hence T should be small, unlike the monatomic case, which requires high T. In case Ψ is not quadratic in the particles positions, equipartition only concerns the kinetic energy, and the calculation of C_v is not immediate.

These results must be modified if relativistic or quantum mechanical effects are not negligible, because the canonical ensemble does not apply in such cases (Gallavotti, 1999). Indeed, it turns out that the heat capacities we have computed agree with the experimental observations on rarefied monatomic gases, while deviations are evident in the case of diatomic molecules, even in rarefied conditions, and in the case of crystals. Quantum mechanical effects, in particular, call for different probability distributions, as well as for different enumeration of the *relevant* degrees of freedom. For instance, take N one-dimensional independent quantum harmonic oscillators, whose energies are given by the eigenvalues $(n + 1/2)\hbar\omega$ of the Hamiltonian H, where ω is the angular frequency and $n = 0, 1, 2, \ldots$. Restricting our analysis to this single observable, thus avoiding the contributions of non-commuting operators, the calculations are analogous to the classical case, with partition function for N independent oscillators in equilibrium with a bath at temperature T given by Peliti (2011)

$$Z = \left[\sum_{n=0}^{\infty} \exp \left(-\frac{\left(n+\frac{1}{2}\right)\hbar\omega}{k_B T} \right) \right]^N = \left[\frac{\exp\left(-\hbar\omega/2k_B T\right)}{1-\exp\left(-\hbar\omega/k_B T\right)} \right]^N \tag{4.15}$$

The internal energy and the heat capacity then take the form:

$$U = -\frac{\partial \ln Z}{\partial \beta} = N\frac{\hbar\omega}{2} + \frac{N\hbar\omega}{\exp\left(\frac{\hbar\omega}{k_B T}\right)-1}, \quad C_v = \frac{N k_B \left(\frac{\hbar\omega}{k_B T}\right)^2 \exp\left(\frac{\hbar\omega}{k_B T}\right)}{\left[\exp\left(\frac{\hbar\omega}{k_B T}\right)-1\right]^2} \tag{4.16}$$

If we have three-dimensional oscillators, C_v is multiplied by 3, because the energy of one oscillator is given by $(n_x + n_y + n_z + 3/2)\hbar\omega$, that sensibly differs from the classical expressions, since it vanishes, rather than being constant, in the $T \to 0$ limit. However, C_v restores its classical behavior for a single space dimension when T gets large, showing that the classical calculation is not a waste, but is useful in its proper realm.

Despite these differences and other possible considerations,[2] which indicate that the principle of equipartition must be carefully adapted to the different situations, the above calculations are conceptually important. They clearly identify the ingredients underlying many results of statistical mechanics, and set a stage for more accurate calculations, where necessary. In this regard, Tolman stated that the above are: *"among the results of the classical statistical mechanics which have proved most illuminating in the development of physical chemical ideas"* (Tolman, 1955). Although the enumeration of relevant degrees of freedom and the proper statistics may change from case to case, the procedure to bridge the gap separating microscopic dynamics and thermodynamic equilibrium properties holds in general.

4.2 The Virial Theorem

Consider a system made of N classical point masses, subjected to generic forces $F_i = \dot{p}_i = m_i a_i$, where p_i is the momentum and a_i the acceleration of particle i. Introduce the quantity:

[2] For instance, insisting with the enumeration of degrees of freedom, and supposing that atoms are made of more elementary blocks, which are in turn made of still more elementary blocks, and continuing like that, one may reach the conclusion that heat capacities exceed all bounds. This kind of paradox neglects that certain degrees of freedom are practically frozen (Gallavotti, 1999), and that exciting them requires too much energy or too long times to do. However, it also shows that equipartition cannot be blindly embraced.

$$G = \sum_{i=1}^{N} p_i q_i \tag{4.17}$$

and calculate its time derivative:

$$\frac{dG}{dt} = \sum_{i=1}^{N} \left(\dot{p}_i q_i + p_i \dot{q}_i \right) \tag{4.18}$$

Observe that:

$$\sum_{i=1}^{N} p_i \dot{q}_i = \sum_{i=1}^{N} \frac{p_i}{m_i} p_i = \sum_{i=1}^{N} \frac{p_i^2}{m_i} = 2K \; ; \qquad \sum_{i=1}^{N} \dot{p}_i q_i = \sum_{i=1}^{N} F_i q_i \tag{4.19}$$

from which it follows that:

$$\frac{dG}{dt} = 2K + \sum_{i=1}^{N} F_i q_i \tag{4.20}$$

Denoting by Φ^t the evolution operator of points Γ in phase space, so that $\Gamma_t = \Phi^t \Gamma_0$ is the solution of the equations of motion $\dot{\Gamma} = G(\Gamma)$ with initial condition $\Gamma_0 = (q_{10}, p_{10}, ..., q_{N0}, p_{N0})$, and averaging Eq. (4.20) over a time interval $[0, \tau]$ along the trajectory starting at the phase space point , one obtains

$$\frac{1}{\tau} \int_0^\tau \frac{dG}{dt} (\Phi^t \Gamma_0) \mathrm{d}t \tag{4.21}$$
$$= \frac{2}{\tau} \int_0^\tau K(\Phi^t \Gamma_0) \mathrm{d}t + \frac{1}{\tau} \int_0^\tau \sum_{i=1}^{N} F_i(\Phi^t \Gamma_0) q_i(\Phi^t \Gamma_0) \mathrm{d}t$$

where $q_i(\Phi^t \Gamma_0)$ is the i-th space coordinate of particle i at time t. Denoting by "$\overline{\cdot}^{\tau}$" a time average from time 0 up to time τ, Eq. (4.21) can be rewritten as

$$2\overline{K}^{\tau}(\Gamma_0) = -\sum_{i=1}^{N} \overline{F_i \cdot q_i}^{\tau}(\Gamma_0) + \frac{G(\Phi^\tau \Gamma_0) - G(\Gamma_0)}{\tau} \tag{4.22}$$

where the role of the initial condition Γ_0 cannot be neglected, in general: whatever dynamics Φ^t is considered, each initial condition gives rise to a distinct trajectory, and to a correspondingly distinct finite-time average.[3] Assume now that positions q and momenta p are bounded, as for a system with finite energy confined within walls

[3] Whether letting τ grow makes different time averages converge to the same limit or not has to be investigated on a case-by-case basis.

delimiting a finite three-dimensional volume. Then G is finite, and $[G(\Phi^\tau \Gamma_0) - G(\Gamma_0)]$ is likewise finite. Consequently, the $\tau \to \infty$ limit yields

$$\lim_{\tau \to \infty} \overline{K}^\tau(\Gamma_0) = \lim_{\tau \to \infty} \left[\frac{G(\Phi^\tau \Gamma_0) - G(\Gamma_0)}{2\tau} - \frac{1}{2} \sum_{i=1}^{N} \overline{F_i \cdot q_i}^\tau(\Gamma_0) \right] \quad (4.23)$$

$$= -\frac{1}{2} \lim_{\tau \to \infty} \sum_{i=1}^{N} \overline{F_i \cdot q_i}^\tau(\Gamma_0) \quad (4.24)$$

which remains dependent on Γ_0, in general. Denoting by "$\overline{\cdot}$" an infinite time average, this result may be rewritten as

$$\overline{K}(\Gamma_0) = -\frac{1}{2} \sum_{i=1}^{N} \overline{F_i \cdot q_i}(\Gamma_0) \quad (4.25)$$

which is known as the *Virial Theorem*. Its right hand is called the virial of Clausius.

In case $F_i = c_i + f_i$ is a sum of conservative forces, c_i, and friction forces (proportional to velocities) f_i, only the conservative forces matter. Indeed, letting a friction force be expressed by $f_i = -\gamma_i \dot{q}_i$, we obtain

$$\frac{1}{\tau} \int_0^\tau (-\gamma_i \dot{q}_i) \cdot q_i dt = -\frac{\gamma_i}{2\tau} \int_0^\tau \frac{d}{dt} q_i^2 dt$$
$$= -\frac{\gamma_i}{2\tau} \left[q_i^2(\Phi^\tau \Gamma_0) - q_i^2(\Gamma_0) \right] \longrightarrow 0 \text{ as } \tau \to \infty \quad (4.26)$$

if positions are bounded, as assumed here.[4] Naturally, friction has to be balanced by an energy source, otherwise the motion stops, when all the available energy has been dissipated. Therefore, one may focus only on conservative forces that are derived from a potential Ψ, assuming that $F_i = -\partial \Psi / \partial q_i$, thus obtaining:

$$\overline{K}(\Gamma_0) = \frac{1}{2} \sum_{i=1}^{N} \overline{\frac{\partial \Psi}{\partial q_i} \cdot q_i}(\Gamma_0) \quad (4.27)$$

In the case of central potentials acting on a single particle, e.g., $\Psi = a r^{m+1}$ with a a real constant and r the distance of the particle from the center of the potential, Eq. (4.27) takes the form:

$$\overline{K}(\Gamma_0) = \frac{1}{2}(m+1)\overline{\Psi}(\Gamma_0) \quad (4.28)$$

[4] The same actually holds even if positions grow more slowly than diffusively.

Then, for harmonic potentials, $m = 1$, one obtains the particularly suggestive and well-known virial expression:

$$\overline{K}(\Gamma_0) = \overline{\Psi}(\Gamma_0) \tag{4.29}$$

which says that the time averages of kinetic and potential energy are equal, along any possible trajectory.

The boundedness of the positions and momenta of particles has played a crucial role here. Were they not bounded, Eq. (4.25) would have had one extra term:

$$\overline{K}(\Gamma_0) = -\frac{1}{2}\sum_{i=1}^{N} \overline{F_i \cdot q_i}(\Gamma_0) + \lim_{\tau\to\infty} \frac{G(\Phi^\tau \Gamma_0)}{2\tau} \tag{4.30}$$

On the other hand, there is no equilibrium for a finite system whose components expand in time without bounds, like for a freely expanding gas in infinite space.

4.3 The Equation of Perfect Gases

Consider N particles constituting a monatomic gas at absolute temperature T, contained in a three-dimensional volume V delimited by rigid walls. If energy equipartition and ergodicity hold, every atom attains a mean kinetic energy of $3k_B T/2$, and the whole system enjoys $\overline{K}(\Gamma_0) = 3Nk_B T/2$, independently of the initial conditions.[5]

For a perfect gas, the time averaged interatomic interaction energy $\overline{\Psi}$ is negligible with respect to $\overline{K}$, but the forces exerted by the walls of V are important. If we assume that these boundary forces are localized and do not extend to a positive fraction of V, we may say that they act only when a particle impacts a wall. The corresponding potential energy is ideally infinitely large and repulsive at the wall, and vanishes in the interior of V. In this picture, Boyle's law describes the pressure that arises from the collisions of the gas molecules with the walls delimiting V, cf. Fig. 4.1. Consider a surface element of area $\mathrm{d}A$ centered about a point q of the wall. The pressure corresponds to the momentum exchange between the colliding particles and the surface element, which takes place in a sufficiently long time for thermodynamic notions to make sense, and it represents the normal average force per unit area produced by the collisions. Assuming that the wall is smooth, so that it exerts no force in the directions parallel to itself, the pressure P on $\mathrm{d}A$ can therefore be defined by

$$-P\,\hat{n}\,\mathrm{d}A = \overline{\mathrm{d}F} = \sum_{i \text{ hitting } \mathrm{d}A} \overline{F_i} \tag{4.31}$$

[5] As discussed in greater detail in Chap. 6, ergodicity, which makes the initial conditions irrelevant, is also possible for single particle systems, i.e., for $N = 1$.

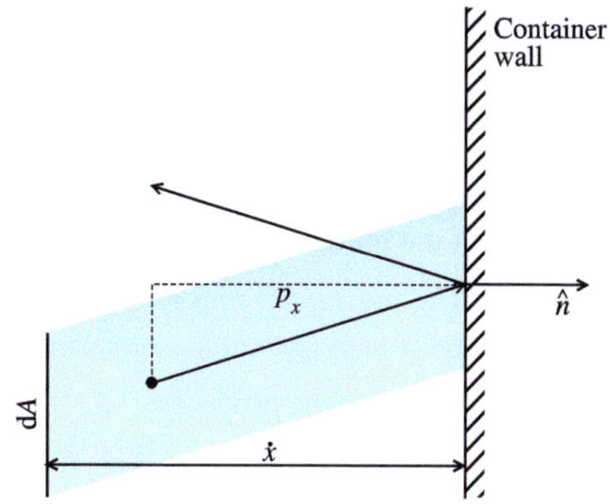

Fig. 4.1 Calculation of pressure

where $\hat{n}$ is the outward unit normal to $\mathrm{d}A$; $\overline{\mathrm{d}\,F}$ is the sum of the time averages of the instantaneous forces $F_i = (F_{ix}, F_{iy}, F_{iz})$ exerted by the wall on i-th particle hitting $\mathrm{d}A$; and the minus sign is due to the fact that the force exerted by the wall on the atoms is directed inward, while the action of the atoms on the wall is directed outward.

Because $\mathrm{d}A$, hence $\hat{n}$, do not depend on time, pressure can be seen as the time average of a fluctuating quantity $\mathscr{P}$ defined by

$$-\,\mathscr{P}\,\hat{n}\,\mathrm{d}\,A = \sum_{i \text{ hitting d} A} F_i \tag{4.32}$$

so that one can write $P = \overline{\mathscr{P}}$. To be more explicit, take $\mathrm{d}A$ orthogonal to the x direction, and let F_{ix} be the x component of the instantaneous force exerted by the wall on the approaching atom i. If this wall is on the right of the bulk of the gas, the atom has a positive velocity $\dot{x}_{in} > 0$ in the x-direction. Due to the wall and in the absence of other forces, the post-collision x-component of the velocity is given by $\dot{x}_{out} = -\dot{x}_{in}$. Therefore, the time average of F_{ix} about a collision occurring at a time t_ℓ is given by

$$\overline{F}^{2\varepsilon}_{ix} = \frac{1}{2\varepsilon}\int_{t_\ell-\varepsilon}^{t_\ell+\varepsilon} m\ddot{x}_i\,\mathrm{d}\,s = \frac{m}{2\varepsilon}(\dot{x}_{out} - \dot{x}_{in}) = -\frac{2}{2\varepsilon}m\dot{x}_{in} = \frac{1}{2\varepsilon}\Delta p_{ix}\,, \quad \text{and} \quad \int_{t_\ell-\varepsilon}^{t_\ell+\varepsilon} F_{ix}\,\mathrm{d}\,s = \Delta p_{ix} \tag{4.33}$$

where $\varepsilon > 0$ can be small (ideally infinitesimal), because the interaction does not extend to the bulk of V, and Δp_{ix} is the variation of the momentum of the atom due to the collision. The quantity $\overline{F}^{2\varepsilon}_{ix}$ is the average force acting on the particle i during a single collision; the shorter the integration time 2ε the greater this quantity, if the force is a real hard-core repulsion. But the time average defining the pressure remains limited and converges to a positive number, because each collision contributes only the finite amount Δp_{ix}:

$$\overline{\mathrm{d}\,F} = \sum_{i \text{ hitting d} A} \overline{F}_i = \sum_{i \text{ hitting d} A} \lim_{t\to\infty} \frac{1}{t} \sum_{\ell=1}^{c_i(t)} \Delta p_{ix}(\ell) \tag{4.34}$$

where $c_i(t)$ is the number of collisions of particle i with $\mathrm{d}A$ up to time t, and $\Delta p_{ix}(\ell)$ is the momentum exchanged during the ℓ-th collision. Taking longer and longer averaging times, one sums over all collisions of atom i with $\mathrm{d}A$. Adding the analogous averages obtained for all the atoms of the gas eventually leads to the total force averaged in time that is exerted on the element of the wall surface, or the total variation of the momentum per unit time of the atoms, determined by $\mathrm{d}A$.

Remark: *This reasoning makes use of the fact that particles only rarely or very weakly interact with each other, so that Δp_{ix} can be fully attributed to the wall.*

Consider now the scalar product $F_i \cdot q_i$ of the instantaneous wall force and position q_i of atom i, in $\mathrm{d}A$, when the wall exerts the force F_i on it. Because $\mathrm{d}A$ is infinitesimal, q_i can be identified with the center q of $\mathrm{d}A$, hence its dependence on the initial condition Γ_0 is insignificant, and one may write

$$\sum_{i \text{ hitting } \mathrm{d}A} (F_i \cdot q_i)(\Gamma_0) = q \cdot \sum_{i \text{ hitting } \mathrm{d}A} F_i(\Gamma_0) \tag{4.35}$$

Therefore, neglecting other forces, the time average of $F_i \cdot q_i$ along the trajectory of atom i determined by the initial condition Γ_0 constitutes the contribution to the virial associated with the interaction of atom i with the wall element $\mathrm{d}A$. Equation (4.32) then yields

$$\sum_{i \text{ hitting } \mathrm{d}A} (F_i \cdot q_i)(\Gamma_0) = -\mathscr{P}(\Gamma_0)\, q \cdot \hat{n}\, \mathrm{d}A \tag{4.36}$$

which can be averaged in time, leading to the expression:

$$\sum_{i \text{ hitting } \mathrm{d}A} \overline{F_i \cdot q_i}(\Gamma_0) = -\overline{\mathscr{P}}(\Gamma_0)\, q \cdot \hat{n}\, \mathrm{d}A = -P(\Gamma_0)\, q \cdot \hat{n}\, \mathrm{d}A \tag{4.37}$$

Here, we have stressed the dependence on the initial conditions of the trajectories of the atoms colliding with the wall in $\mathrm{d}A$. If one changes the surface element, hence its center, one may find that P changes as well. However, this is excluded by the definition of thermodynamic equilibrium, provided that P is indeed a thermodynamic property. Therefore, assuming the system is in thermodynamic equilibrium and the mechanical theory developed above applies, the integral over the whole surface S of the container yields

$$P(\Gamma_0) \int_S \hat{n} \cdot q\, \mathrm{d}A = -\sum_{i=1}^{N} \overline{F_i \cdot q_i}(\Gamma_0) \tag{4.38}$$

where the sum runs over all the collisions of all the atoms with the walls. Using Gauss' theorem to pass from integration over S to integration over the volume V delimited by S, one obtains

$$P(\Gamma_0) \int_S \hat{n} \cdot q\, \mathrm{d}A = P(\Gamma_0) \int_V \mathrm{div}\, q\, dV = 3P(\Gamma_0) V \tag{4.39}$$

because the divergence of q with respect to q is 3.

In principle, P depends on the initial conditions, even if it is uniform in S. However, assuming energy equipartition holds, and ensemble averages equal time averages, we can write

$$\frac{3}{2}P(\Gamma_0)V = -\frac{1}{2}\sum_{i=1}^{N}\overline{F_i \cdot q_i}(\Gamma_0) = \overline{K}(\Gamma_0) = \frac{3}{2}Nk_BT \tag{4.40}$$

where the dependence on Γ_0 has been eliminated, because the average of K has been fixed by equipartition. One finally obtains the celebrated law of perfect gases:

$$PV = Nk_BT \tag{4.41}$$

In summary, the following has been either explicitly or implicitly used in the derivation of Eq. (4.41):

- Newtonian dynamics for the atoms.
- Very large number of particles N which, thanks to mutual interactions, enjoy disordered motions and rapidly enable a uniform distribution of their properties in V. This makes their collisions with S a continuous process both in space and time.
- $\overline{\Psi} \ll \overline{K}$: interactions between particles are on average energetically negligible in calculations, as appropriate for rarefied gases, and only particle-wall interactions contribute to the virial.[6]
- The validity of ergodicity and energy equipartition.[7]

Because of the last point, our derivation of Eq. (4.41) is not merely based on the mechanics of particles; the statistical assumptions, although justified by observations, are not evident in the dynamics. Ergodic theory was born to provide these justifications, but the problem remains open, as it will be discussed in Chapter 6. For the time being, one has to accept that thermodynamic properties do not directly derive from the dynamics so far attributed to atoms. They must emerge in some kind of limit that may apparently require impossible conditions, such as number of atoms

[6] The intermolecular forces do not need to be weak. They can be strong, but rare in time, so that their time averages contribute little to the virial. Also, one thing is small Ψ, another is $\Psi = 0$. In absence of interactions, the necessary randomization of the distributions of mass, momentum, energy, etc. cannot be established: K hardly forgets Γ_0, there can be no equipartition, nor identification of the discrete (particle) system with a continuous one. In general, however, interactions must be repulsive at short range and vanish shortly beyond. Long-range forces usually do not randomize, violate the condition of locality, and give large contributions to the virial.

[7] Although equipartition is commonly derived applying one of the classical ensembles, thermodynamics only requires its validity for time averages concerning a single system. Independence on initial conditions is also required, which is tantamount to ergodicity, but without specification of the ensemble.

$N \to \infty$, their cross section $\sigma \to 0$, and time $t \to \infty$. This is the case because theories arise from mathematical idealizations of the phenomenon of interest, cf. Chap. 1. In reality, scales such as those of Sect. 2.5 are perfectly satisfactory.

In the case in which particle interactions cannot be neglected, the expression of the virial changes, cf. e.g. Ref. Toda et al. (1998); Tuckerman (2023). Then, one must rewrite Eq. (4.25) as

$$\overline{K}(\Gamma_0) = -\frac{1}{2}\sum_{i=1}^{N}\left[\overline{F_i^{\text{ext}} \cdot q_i}(\Gamma_0) + \overline{F_i^{\text{int}} \cdot q_i}(\Gamma_0)\right] \tag{4.42}$$

where the forces acting on atom i due to external agents, F_i^{ext}, have been separated from those due to the interactions with the other atoms, F_i^{int}. When F_i^{ext} is only due to the walls of the container, whose action can be considered instantaneous, hence dominant over F_i^{int} when sufficiently close to the surface S,[8] Eq. (4.40) turns into:

$$\begin{aligned}\frac{3}{2}Nk_BT = \overline{K}(\Gamma_0) &= -\frac{1}{2}\sum_{i=1}^{N}\left[\overline{F_i^{\text{ext}} \cdot q_i}(\Gamma_0) + \overline{F_i^{\text{int}} \cdot q_i}(\Gamma_0)\right] \\ &= \frac{3}{2}P(\Gamma_0)V - \frac{1}{2}\sum_{i=1}^{N}\overline{F_i^{\text{int}} \cdot q_i}(\Gamma_0)\end{aligned} \tag{4.43}$$

which yields

$$PV = Nk_BT + \frac{1}{3}\sum_{i=1}^{N}\overline{F_i^{\text{int}} \cdot q_i} \tag{4.44}$$

where the initial conditions have been eliminated, under the assumption that energy equipartition, hence Eq. (4.43), holds. Recall that if the quantity G defined by Eq. (4.17) is unbounded then a thermodynamically unclear term, $\lim_{\tau\to\infty} G(\Phi^\tau \Gamma_0)/\tau$, would be added to the equation of state (4.44).

The above results concern *time* averaged quantities, not ensemble averaged quantities. Therefore, they apply to a single thermodynamic system and do not need any connection with abstract sets of macroscopically equivalent systems that ensembles seem to require. It is true that energy equipartition was obtained from averaging the kinetic energy with respect to the canonical ensemble, but this step is not really necessary. The canonical ensemble is no more fundamental than the equipartition principle itself: it is one assumption which, in general, is not derived from

[8] The wall produces a very high acceleration, since it inverts the particles velocities in arbitrarily short time intervals. The particle interactions can be either hard, like the wall, in which case a particle colliding with a wall is unlikely to collide at the same time with other particles. Alternatively, when the atomic interactions are not instantaneous, they may still negligibly affect the motion at the wall, since that takes a very short time. Therefore, one may neglect the contributions of the particles interactions to the virial, during collisions with the walls.

the dynamics of particles. Although ensembles are pedagogically useful, they help intuition and, once accepted, they greatly simplify calculations of the most diverse quantities, it is legitimate to adopt the equipartition principle straight away, without passing through an ensemble average. Indeed, in Sect. 6.4, we will justify the notion of ensemble in terms of the dynamics of a single system. As the correction proportional to $\lim_{\tau\to\infty}[G(\Phi^{\tau}\Gamma_0)/2\tau]$ is hard to relate to thermodynamic quantities, the process hardly lies within the realm of equilibrium thermodynamics when such a correction is not negligible.

This is related to pressure, computed as the momentum exchange of particles at walls. The rigid walls adopted here are just an idealization that makes calculations easy. More generally, a wall may be represented by a soft potential that diverges at a finite position in space, so that no particle can go beyond. Confining potentials diverging only at infinite distances are not suitable for this task, because they allow higher energy particles to roam over volumes larger than lower energy particles can explore. The result is vanishing pressure. Clearly, no wall is fully equivalent to an infinite potential barrier; it can be perforated, for instance, by a bullet. However, the energy of the molecules the wall is meant to stop cannot be really infinite. The Maxwell-Boltzmann velocity distribution, which allows arbitrarily large energies, combined with infinite potential barriers produces a very accurate description of particles confined by walls.

4.4 Radial Correlation Function and Virial Expansion

In general, Eq. (4.44) involves time averages that are not manageable in terms of closed-form expressions. On the other hand, if the time average of the last term in Eq. (4.44) equals its canonical ensemble average, explicit expressions can be derived. To proceed in this direction, let us start from a canonical distribution $f_\beta = f_\beta(q_1, p_1; ...; q_N, p_N)$ expressed by

$$f_\beta = \frac{e^{-\beta(K+\Psi)}}{Z} = \frac{e^{-\beta K}}{Z_p}\,\frac{e^{-\beta\Psi}}{Z_q} \tag{4.45}$$

where K is the kinetic energy and Ψ the potential energy, so that momentum variables are separated from position variables, and Z_q and Z_p are defined by

$$Z_p = \int \mathrm{d}p_1 \cdots \mathrm{d}p_N\, e^{-\beta K}\,, \quad Z_q = \int \mathrm{d}q_1 \cdots \mathrm{d}q_N\, e^{-\beta\Psi} \tag{4.46}$$

Because the positions and the momenta distributions are separately normalized, it is convenient to introduce the n-particle reduced distribution as

$$f_\beta^{(n)}(q_1, ..., q_n) = \int \frac{e^{-\beta\Psi}}{Z_q}\,\mathrm{d}q_{n+1}\cdots \mathrm{d}q_N\,, \quad \text{which obeys} \quad \int f_\beta^{(n)}\mathrm{d}q_1\cdots \mathrm{d}q_n = 1 \tag{4.47}$$

where n is any number in $\{1, ..., N\}$. This $f_\beta^{(n)}$ represents the joint probability density of finding particle i near q_i, for $i = 1, ..., n$ independently of the positions of particles $n + 1, ..., N$, and of momenta. According to the standard interpretation of ensembles, this probability density multiplied by the elementary volume $\mathrm{d}q_1 \cdots \mathrm{d}q_n$ represents the fraction of identical systems whose first n particles are located near the coordinates $q_1, ..., q_n$. If such an ethereal notion can be connected with the real mass distribution concerning the first n particles of the given (single) system under investigation, then $f_\beta^{(n)}$ becomes concrete because positions of particles can, to a certain extent, be experimentally observed or inferred.

When particles are indistinguishable, we may associate any n-particle reduced distribution with one that neglects their order:

$$\rho^{(n)}(q_1, ..., q_n) = \frac{N!}{(N-n)!} f^{(n)}(q_1, ..., q_n) \tag{4.48}$$

where the combinatorial factor multiplying $f^{(n)}$ accounts for the possible arrangements of n particles out of N (the first particle can be taken in N different ways, the second in the remaining $(N - 1)$ ways, and the n-th particle in $(N - n + 1)$ ways). This makes the order of particles irrelevant, and one has the probability density of finding any set of n particles about positions $(q_1, ..., q_n)$. The functions $\rho^{(n)}$ reveal the different states of aggregation of matter. For instance, $\rho^{(1)}$ is spatially periodic and sharply peaked at the lattice nodes of a crystal. In a fluid at equilibrium, all points are equivalent, except near the walls (which contribute marginally to the volume), therefore $\rho^{(1)}$ is practically constant. Also, the integral of $\rho^{(1)}$ over the volume divided by the volume yields the number density ρ:

$$\frac{1}{V}\int \rho^{(1)}(q)\mathrm{d}q = \rho^{(1)} = \frac{N}{V} = \rho \tag{4.49}$$

which turns mass density if multiplied by the mass m of one of the identical particles. Continuing, the quantity $\rho^{(2)}(q_1, q_2)\mathrm{d}q_1\mathrm{d}q_2$ is the probability that one particle is found in $\mathrm{d}q_1$ around q_1 and another particle is found in $\mathrm{d}q_2$ around q_2. In a crystal, $\rho^{(2)}$ depends on both q_1 and q_2, even if the two points are widely separated. Given the uniform distribution of particles in the bulk of a fluid at equilibrium, $\rho^{(2)}$ may still depend on the separation r_{12} of the two points q_1 and q_2. For instance, the centers of two hard spheres cannot get closer than the sum of their radii, therefore $\rho^{(2)}$ vanishes at distances smaller than that, while it does not vanish at other distances. An analogous effect on $\rho^{(2)}$ is produced by a wall, which limits the possible arrangements of two particles. In all cases, one can write

$$\int \rho^{(2)}(q_1, q_2)\,\mathrm{d}q_1\,\mathrm{d}q_2 = N(N-1) \tag{4.50}$$

because all $f^{(n)}$ are normalized. Now, call the distribution of particles *completely random* when it is factorized and uniform, i.e., $f^{(n)}(q_1, ..., q_n) = 1/V^n = (\rho/N)^n$ (Hill, 1987). In that case, one obtains

$$\rho^{(n)} = \frac{N!}{(N-n)!}\left(\frac{\rho}{N}\right)^n, \quad \text{hence } \rho^{(1)} = \rho\,, \quad \rho^{(2)} = \rho^2\left(1 - \frac{1}{N}\right) \tag{4.51}$$

Most of the times, however, $f^{(n)}$ and $\rho^{(n)}$ are neither uniform nor factorized, and correlations arise. These can be respectively expressed by a correlation function $C^{(n)}$ defined by Hansen and McDonald (1986)

$$f^{(n)}(q_1, ..., q_n) = f^{(1)}(q_1) \cdots f^{(1)}(q_n)\, C^{(n)}(q_1, ..., q_n), \tag{4.52}$$

i.e.,

$$C^{(n)}(q_1, ..., q_n) = \frac{f^{(n)}(q_1, ..., q_n)}{f^{(1)}(q_1) \cdots f^{(1)}(q_n)} \tag{4.53}$$

and by a correlation function $g^{(n)}$ defined by

$$\rho^{(n)}(q_1, ..., q_n) = \rho^{(1)}(q_1) \cdots \rho^{(1)}(q_n)\, g^{(n)}(q_1, ..., q_n), \tag{4.54}$$

i.e.,

$$g^{(n)}(q_1, ..., q_n) = \frac{1}{\rho^n}\, \rho^{(n)}(q_1, ..., q_n) \tag{4.55}$$

In the case of completely random distribution (e.g., equilibrium homogeneous fluid), one has

$$C^{(n)} = 1 \quad \text{while} \quad g^{(n)} = \frac{N!}{N^n(N-n)!} \tag{4.56}$$

In particular, $g^{(2)}(q_1, q_2)$ is a property that can be experimentally measured, by means of X-ray as well as neutron scattering. In the cases in which $g^{(2)}$ is only a function of the distance between one particle and its companions, we omit the superscript and denote it by $g = g(r)$, which is also called *radial distribution function.*

Let us consider intermolecular forces depending only on the distance between pairs of particles, not affected by the volume V, which derive from a potential of the following form:

$$\Psi(q_1, ..., q_N) = \frac{1}{2}\sum_{\substack{i,j=1 \\ i \neq j}}^{N} u(r_{ij})\,, \quad r_{ij} = |q_{ij}|\,, \quad q_{ij} = q_i - q_j = -q_{ji} \tag{4.57}$$

where $u = u(r)$ is the potential of the forces that a pair of particles at distance r exert on each other and, consequently:

$$F_i^{\text{int}} = \sum_{\substack{j=1 \\ j \neq i}}^{N} F_{ij} \,, \qquad F_{ij} = -\frac{\partial \Psi}{\partial q_{ij}} = -\frac{q_{ij}}{r_{ij}} \frac{\mathrm{d}u}{\mathrm{d}r}(r_{ij}) \,, \qquad F_{ij} = -F_{ji} \tag{4.58}$$

F_{ij} being the force on particle i, located in position q_i that is due to particle j located in position q_j, at distance r_{ij}. With this notation, one may write

$$\sum_{i=1}^{N} F_i^{\text{int}} \cdot q_i = \frac{1}{2} \sum_{\substack{i,j=1 \\ i \neq j}}^{N} F_{ij} \cdot q_{ij} \tag{4.59}$$

where the factor $1/2$ is necessary for each independent pair to be counted only once. Considering that this quantity does not depend on momenta, the corresponding canonical average is expressed by

$$\begin{aligned} \mathbb{E}_\beta \left(\sum_{i=1}^{N} F_i^{\text{int}} \cdot q_i \right) &= \frac{1}{Z} \int \mathrm{d}q_1 \mathrm{d}p_1 \cdots \mathrm{d}q_N \mathrm{d}p_N \, e^{-\beta(K+\Psi)} \sum_{i=1}^{N} F_i^{\text{int}} \cdot q_i \\ &= \frac{1}{Z_q} \int \mathrm{d}q_1 \cdots \mathrm{d}q_N \, e^{-\beta\Psi} \left[\frac{1}{2} \sum_{\substack{i,j=1 \\ i \neq j}}^{N} F_{ij} \cdot q_{ij} \right] = \mathbb{E}_\beta \left(\frac{1}{2} \sum_{\substack{i,j=1 \\ i \neq j}}^{N} F_{ij} \cdot q_{ij} \right) \end{aligned} \tag{4.60}$$

If all particles are indistinguishable, the integrals do not depend on the coordinate labels, and the calculation can be reduced to a sum of $N(N-1)/2$ contributions that are identical after integration, and that correspond to the different pairs of interacting particles. One may thus write

$$\mathbb{E}_\beta \left(\sum_{i=1}^{N} F_i^{\text{int}} \cdot q_i \right) = \frac{1}{Z_q} \int \mathrm{d}q_1 \cdots \mathrm{d}q_N \, e^{-\beta\Psi} \left[\frac{N(N-1)}{2} F_{12} \cdot q_{12} \right] \tag{4.61}$$

$$= \frac{N(N-1)}{2} \int \mathrm{d}q_1 \mathrm{d}q_2 \, F_{12} \cdot q_{12} \int \frac{1}{Z_q} \mathrm{d}q_3 \cdots \mathrm{d}q_N \, e^{-\beta\Psi} \tag{4.62}$$

$$= \frac{1}{2} \int \mathrm{d}q_1 \mathrm{d}q_2 \, F_{12} \cdot q_{12} \, \rho^{(2)}(q_1, q_2) \tag{4.63}$$

$$= \frac{N^2}{2V^2} \int \mathrm{d}q_1 \mathrm{d}q_2 \, F_{12} \cdot q_{12} \, g^{(2)}(q_1, q_2) \tag{4.64}$$

because the inner integral in (4.62) corresponds to the 2-particles reduced distribution function $f^{(2)}$.

As the functions to integrate are here assumed to depend only on the distance between two particles, one may pass to spherical coordinates and, for isotropic systems with short-range interaction, one obtains

$$\mathbb{E}_\beta \left(\sum_{i=1}^{N} F_i^{\text{int}} \cdot q_i \right) = -\frac{N^2}{2\,V} \int_0^\infty 4\pi \, r^3 \, \frac{\mathrm{d}u}{\mathrm{d}r}(r) \, g(r) \, \mathrm{d}r \tag{4.65}$$

where Eq. (4.58) has been used. Then, the pressure, given by Eq. (4.44), is expressed by the following equation of state:

$$\frac{P}{k_B T} = \rho - \frac{2\pi\rho^2}{3k_B T}\int_0^\infty r^3\,\frac{\mathrm{d}u}{\mathrm{d}r}(r)\,g(r)\,\mathrm{d}r \tag{4.66}$$

This expression introduces a correction to the perfect gas law that is proportional to ρ^2. As this must become negligible when the conditions of perfect gas are approached, and g implicitly depends on ρ, one may try to express it as a power series in ρ. Considering that g depends on the temperature T as well, one may thus write

$$g(r;\rho,T) = \sum_{j=0}^{\infty}\rho^j g_j(r;T) \tag{4.67}$$

with $g(r;\rho,T) \approx g_0(r;T)$, at low densities. This leads to the *virial equation of state*:

$$\frac{P}{k_B T} = \rho + \sum_{j=0}^{\infty} B_{j+2}(T)\rho^{j+2} \tag{4.68}$$

where the first term of the power expansion is separated from the other terms, to stress the connection with the perfect gas law. Comparison of Eq. (4.66) with Eq. (4.68) shows that the coefficients known as *virial coefficients* are given by

$$B_{j+2}(T) = -\frac{2\pi}{3k_B T}\int_0^\infty r^3\,\frac{\mathrm{d}u}{\mathrm{d}r}(r)\,g_j(r;T)\,\mathrm{d}r \tag{4.69}$$

At low densities, ρ is fixed, small, and does not depend on r, hence $g_0(r;T) \approx g(r;\rho,T)$. Therefore, the expansion can be truncated to second order, and yields

$$\frac{P}{k_B T} = \rho + B_2(T)\rho^2; \quad \text{with} \quad B_2(T) \approx -\frac{2\pi}{3}\beta\int_0^\infty r^3\,\frac{\mathrm{d}u}{\mathrm{d}r}(r)\,g(r;\rho,T)\,\mathrm{d}r \tag{4.70}$$

In the low density limit, it can also be proven that Tuckerman (2023):

$$g(r) = g(u(r)) \approx e^{-\beta u(r)} \quad \text{so that} \quad \frac{\mathrm{d}}{\mathrm{d}r}g(r) \approx -\beta\frac{\mathrm{d}u}{\mathrm{d}r}(r)g(r) \tag{4.71}$$

where the dependence of ρ and T has been omitted, since at equilibrium these parameters are uniform and do not depend on r. Assume that the potential u is repulsive at short distances, diverging to $+\infty$ below a finite distance $r = \sigma$, and weakly attractive at larger distances, vanishing for r beyond a certain distance L. Then,

$$B_2(T) \approx \frac{2\pi}{3}\int_0^\infty r^3\,\frac{\mathrm{d}g}{\mathrm{d}r}(r)\,\mathrm{d}r = \lim_{L\to\infty}\frac{2\pi}{3}\int_\sigma^L r^3\,\frac{\mathrm{d}g}{\mathrm{d}r}(r)\,\mathrm{d}r \tag{4.72}$$

because g and its derivatives vanish at $r \leq \boldsymbol{\sigma}$. Integration by parts yields

$$\int_{\sigma}^{L} r^3 \frac{\mathrm{d}g}{\mathrm{d}r}(r)\,\mathrm{d}r = r^3\, g(r)\Big|_{\sigma}^{L} - \int_{\sigma}^{L} 3r^2\, g(r)\,\mathrm{d}r = L^3 - \int_{\sigma}^{L} 3r^2\, g(r)\,\mathrm{d}r \quad (4.73)$$

because $g(L) = \exp(-\beta u(L)) = 1$, while $g(\boldsymbol{\sigma}) = \exp(-\beta u(\boldsymbol{\sigma})) = 0$. Assume further that the repulsive part of u is like a hard-core potential, so that

$$u(r) = \begin{cases} u_a(r) < 0 \;\text{ small, with }\; u_a'(r) \geq 0 & \text{for } \; r > \sigma \\ \infty & \text{for } \; r \leq \sigma \end{cases} \quad (4.74)$$

where $\boldsymbol{\sigma}$ is the closest distance the centers of two interacting particles can reach, and u_a is the attractive part of the potential. Then

$$L^3 - \int_{\sigma}^{L} 3r^2\, g(r)\,\mathrm{d}r \approx L^3 - \int_{\sigma}^{L} 3r^2\Big(1 - \beta u_a(r)\Big)\,\mathrm{d}r = \boldsymbol{\sigma}^3 + 3\beta \int_{\sigma}^{L} r^2 u_a(r)\,\mathrm{d}r \quad (4.75)$$

and the second virial coefficient is expressed by

$$B_2(T) \approx \frac{2\pi}{3}\boldsymbol{\sigma}^3 + 2\pi\beta \int_{\sigma}^{\infty} r^2 u_a(r)\,\mathrm{d}r = b - \beta a \quad (4.76)$$

where a and the volume excluded by the hard-core repulsion b are given by

$$a = -2\pi \int_{\sigma}^{\infty} r^2\, u_a(r)\,\mathrm{d}r > 0\,, \qquad b = \frac{2}{3}\pi\sigma^3 \quad (4.77)$$

The volume b can be interpreted this way: a sphere of radius $\boldsymbol{\sigma}$ contains two interacting particles, whose centers are as close they can be. Its volume is $4\pi\boldsymbol{\sigma}^3/3$, therefore a single particle "occupies" half of it. For small densities, this leads to

$$\frac{P}{k_B T} \approx \rho\Big(1 + \rho b\Big) - \frac{a\rho^2}{k_B T} \approx \frac{\rho}{1 - \rho b} - \frac{a\rho^2}{k_B T} \quad (4.78)$$

which becomes exact in the small ρ and weak attraction potential limit, if the short-range repulsion potential is well described by hard-core elastic reflections. Then one obtains the van der Waals equation of state:

$$\frac{P}{k_B T} = \frac{\rho}{1 - \rho b} - \frac{a\rho^2}{k_B T} \quad (4.79)$$

Because of so many approximations, the applicability of the van der Waals equation is limited, although it is pedagogically most effective. Indeed, its derivation highlights the process that leads to a state equation from knowledge of molecular interaction potentials. In most cases of practical interest, interatomic potentials are not explicitly known, or they are too complicated. Therefore, equations of state are postulated and their parameters fitted with experimental data. However, much more robust results are obtained whenever interaction potential can be reasonably well approximated.

4.5 Remarks on the Statistical Hypotheses

To conclude this chapter, it is useful to reflect on the combination of dynamics and statistical hypotheses on which the derivations presented here are based.

In the first place, following the purely mechanical nature of the virial theorem, and the thermodynamic requirement that a single system be treated, various important expressions can be obtained for time averages of mechanical quantities, including expression (4.38) for the pressure. This expression contains merely mechanical quantities, and naturally depends on the initial conditions characterizing the state of the object at the beginning of observations or measurements. The time average is in fact considered to represent an act of measurement: it lasts very long compared to the characteristic microscopic times, and returns the average of the microscopic values taken by a given quantity along the evolution of the system. A connection with thermodynamic properties beyond pressure is not evident. In particular, rather than temperature one finds the virial, whose time average is simply related to the kinetic energy K. One relation between K and T is needed. In equilibrium, this relation is afforded by energy equipartition, a property that can be derived in a suggestive fashion from the validity of the classical ensembles.

Unfortunately, there is no mathematical proof that the microscopic phases of Hamiltonian particle systems are distributed as the classical ensembles prescribe, except in a limited set of special cases. It is even possible that a general proof will never be produced, because of intrinsic limitations of our models of the microscopic world. Therefore, energy equipartition must be taken as a principle, rather than as a consequence of the microscopic dynamics. Nevertheless, this principle and its repercussions are generally found to be in perfect accord with experimentally verifiable properties of systems in thermodynamic equilibrium. Therefore, one can accept equipartition without any reference to the mathematical tool of ensembles. On the other hand, ensembles lead to many more predictions than equipartition. One way or another, to match with thermodynamics, results must apply to a single system, temperature must replace certain mechanical quantities, and the dependence on the initial conditions must be eliminated.

More and more elaborate results turn harder and harder in terms of time averages, hence the use of ensembles appears to be the only viable strategy, apart from direct molecular dynamics simulations. Although not immediately justified from our understanding of the microscopic dynamics, the classical ensembles lead to a countless variety of experimentally confirmed results, becoming in this respect one of the most useful theoretical tools of physics.

Continuing this analysis, one may notice that the number of particles N and the space dimension d apparently played no role in the above calculations, which have been performed for any $N \geq 1$ and any $d \geq 1$. This fact is misleading. Not only the applicability of the classical ensembles withers if N is small (Gallavotti, 1999), but the thermodynamic fields make no sense, in that case, as explained in Sect. 2.5. There, it has also been observed that locality, a fundamental property of thermodynamics, requires a fast decay of correlations both in time and space. In general, correlations decay is harder with $d < 3$, as commonly reported, e.g., in investigations of anomalous transport phenomena in highly confining media.

4.6 Current Research

The present chapter has only scratched the surface of a very active and prolific field of research, concerning matter under many conditions of interest in modern science and technology. In particular, violations of LTE are now established in nonequilibrium and small systems both theoretically and in experiments (Openov and Podlivaev, 2008; Conti et al., 2013b). The very microscopic definition of nonequilibrium temperature poses still unresolved questions (Jou et al., 2010; Jepps and Rondoni, 2010). Moreover, the search for efficient interatomic potentials, that together with the virial theorem are capable of yielding properties of matter in extreme conditions, continues to produce important results (Ree, 1982; Fried and Howard, 1998; Suceska et al., 2019), and recently it benefits from modern machine learning techniques (Wen et al., 2022; Ding et al., 2023; Eyert et al., 2023).

5 Legendre Transforms in Thermodynamics, Statistical Physics, and Large Deviations

And once the storm is over you won't remember how you made it through, how you managed to survive. You won't even be sure, in fact, whether the storm is really over.

Kafka on the Shore, Haruki Murakami

5.1 Legendre Transforms in Thermodynamics

Thermodynamics can be contrived in many different but equivalent guises. For instance, in the *entropy* formulation the equilibrium states are characterized by the maximum of the entropy as a function of *extensive* parameters, subject to the constraints in place, such as that of constant internal energy. The corresponding theoretical framework is equivalent to the *energy* formulation, in which the equilibrium states have minimum energy, and the entropy is constant. In both cases, as well as in any other equivalent formulation of thermodynamics, an equilibrium state is characterized by the extremum of a fundamental quantity, under a given set of constraints.

The equivalence is due to the fact that the fundamental surface of states for a system constituted by r substances, say

$$S = S(U, V, N_1, \dots, N_r) \tag{5.1}$$

is convex and a homogeneous function of extensive parameters, so that

$$S(\lambda U, \lambda V, \lambda N_1, \dots, \lambda N_r) = \lambda S(U, V, N_1, \dots, N_r) \tag{5.2}$$

and the variation of the entropy with respect to the internal energy, when volume and chemical composition are fixed, is strictly positive:

$$\left.\frac{\partial S}{\partial U}\right|_{V,N_1,\dots,N_r} > 0 \tag{5.3}$$

C. Mejía-Monasterio and L. Rondoni, *Nonequilibrium Statistical Mechanics I*, Theoretical and Mathematical Physics,
https://doi.org/10.1007/978-3-031-98249-1_5

This monotonicity of S with respect to U allows the fundamental representation to be inverted, and be expressed in the alternative equivalent form:

$$U = U(S, V, N_1, \dots, N_r) \tag{5.4}$$

The purpose of developing equivalent frameworks is multifaceted. In the first place, choosing the proper formalism may greatly simplify the problems at hand, while an unfortunate formalism may make problems practically insoluble, despite the formal equivalence of the two. This is a common fact in physics, not peculiar to thermodynamics. Furthermore, in cases like the entropy and energy formulations, the extensive parameters play the role of independent variables, whereas intensive parameters are derived concepts. This is in contrast to the experimental practice, which finds often easier to control intensive variables. In particular, there is no tool to measure entropy, while thermometers and thermostats are common laboratory equipment. We may thus like to derive one formulation of thermodynamics expressed in terms of intensive parameters, rather than extensive parameters. Formally, let the fundamental relation to be of the form

$$Y = Y(X_1, X_2, \dots, X_r) \tag{5.5}$$

for the extensive variables Y, $X_i, i = 1, ..., r$, and introduce the intensive quantities

$$P_k = \frac{\partial Y}{\partial X_k} \tag{5.6}$$

where k may run over the whole or just a part of the set of variables on which Y depends.

To introduce the Legendre transform, let us begin with $r = 1$. Then, P is the slope of the curve $Y = Y(X)$, and does not suffice by itself to reconstruct the function Y. However, combining it with the intercept Ψ of the corresponding tangent lines does suffice. For a given differentiable and strictly convex function Y, Ψ is a function of P, and the new function, $\Psi = \Psi(P)$, is totally equivalent to $Y = Y(X)$, cf. Fig. 5.1. The quantity P is now the only independent variable. In other words, $Y = Y(X)$ is a fundamental relation in the Y-formulation, whereas $\Psi = \Psi(P)$ is an equivalent fundamental relation in the Ψ-formulation. How to obtain Ψ if Y is given?

Finding the tangent with slope P means minimizing, at fixed P, the distance $|Y(X) - PX - q|$ between $Y(X)$ and the line of equation $Y = PX + q$. In the case in which the second derivative of Y exists and is positive, $d^2Y/dX^2 > 0$, this leads to a unique solution, solving the simple equation:

$$\frac{\partial Y}{\partial X} - P = 0 \tag{5.7}$$

Given Y, P can be seen as a function of X, and X can be equivalently seen as a function of P. In the second case, the intercept of the tangent is a function of P, $q = \Psi(P)$, and one can write

$$Y(X(P)) = PX(P) + \Psi(P), \quad \text{and} \quad \Psi = Y - PX \tag{5.8}$$

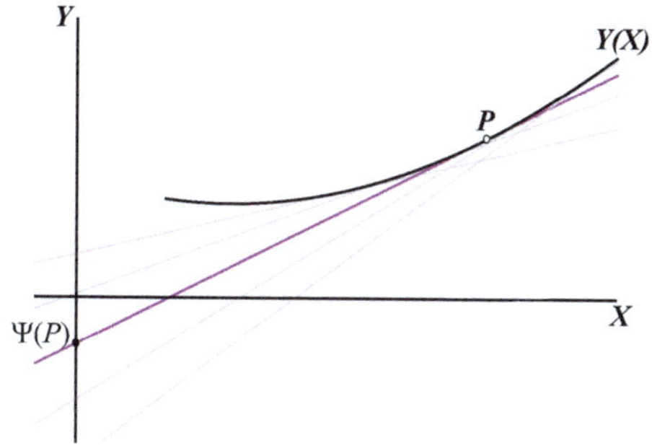

Fig. 5.1 Graphical representation of the Legendre transformation. The fundamental relation $Y(X)$ determines the tangent in each of the points of its graph, which has slope P, hence the intersection $\Psi(P)$ of such lines with the vertical axis. Equivalently, knowledge of the fundamental relation $\Psi(P)$ allows the reconstruction of $Y(X)$

as a function of P. The function $\Psi = \Psi(P)$ is called the Legendre transform of Y, and it is unambiguously identified when Y is strictly convex as a function of X, since the relation $P = P(X)$ can be inverted. In fact, it suffices that the second derivative of Y is either everywhere positive or everywhere negative, in the domain of Y. If it changes sign or vanishes, a different analysis must be performed, but for the moment we only consider strictly convex curves and surfaces.

The above reasoning can be repeated: given $\Psi = Y - PX$, one may move to another fundamental representation, changing variable from P back to X, considering the variations of Ψ with $X = X(P)$ and $Y = Y(X(P))$:

$$d\Psi = dY - PdX - XdP = \frac{dY}{dX}\frac{dX}{dP}dP - P\frac{dX}{dP}dP - XdP = PdX - PdX - XdP = -XdP$$

This yields $-X = d\Psi/dP$, $Y = \Psi + PX$, and shows that the pairs (X, Y) and (P, Ψ) are symmetrical, apart from a sign due to the definition of Legendre transform adopted here. The result is a new representation in a dual space that, thanks to the strict convexity condition, yields the original representation $Y = Y(X)$. One should note that this process does not merely amount to solve the trivial algebraic equation $\Psi = Y - PX$ to obtain $Y = \Psi + PX$. The Legendre transform takes from a description in a given space with given variable X, to a description in a dual space, with different variable P. In a thermodynamic framework, we could say that the fundamental function Y constitutes a formulation in terms of the extensive variable X, while Ψ constitutes a formulation in terms of the intensive variable P, because the proportionality of Y and X to the mass of the system cancels out in the derivative dY/dX.

Extending to a larger set of variables, strict convexity yields the following rule: given the fundamental relation $Y = Y(X_1, ..., X_r)$, consider any two disjoint sets of indices I_P and I_X such that $I_P \cup I_X = \{1, 2, ..., r\}$, and replace the old variables X_k with new variables $P_k = \partial Y/\partial X_k$, $k \in I_P$. Then, the function

$$\Psi(\{P_k\}_{k\in I_P}, \{X_j\}_{j\in I_X}) = Y(\{X_k(P_k)\}_{k\in I_P}, \{X_j\}_{j\in I_X}) - \sum_{k\in I_P} P_k\, X_k(P_k) \quad (5.9)$$

constitutes a new fundamental relation. One may go back to the previous fundamental relation, by changing variables again, i.e., Legendre-transforming the Legendre transform, by replacing P_k with $-X_k = \partial\Psi/\partial P_k$, with $k \in I_P$. This procedure is general, and not restricted to thermodynamics. For instance, take the Lagrangian of a given mechanical system, $\mathscr{L} = \mathscr{L}(v_1, \ldots, v_r, q_1, \ldots, q_r)$, where q_i are the generalized coordinates of a given system, which automatically take into account the constraints acting on it, and v_i the corresponding velocities. Then, changing the velocities with the canonical conjugate momenta $P_k = \partial\mathscr{L}/\partial v_k$, and taking the corresponding partial (because affecting only part of the variables) Legendre transform:

$$- H(P_1, \ldots, P_r, q_1, \ldots, q_r) = \mathscr{L} - \sum_{k=1}^{r} P_k v_k \tag{5.10}$$

one obtains the Hamiltonian of the system.

In Thermodynamics, the most common partial Legendre transforms of the fundamental relation in the energy representation, $U = U(S, V, N_1, \ldots, N_r)$, are the following intensive parameters:

$$T = \frac{\partial U}{\partial S}; \quad -P = \frac{\partial U}{\partial V}; \quad \mu_k = \frac{\partial U}{\partial N_k}, \quad k = 1, ..., r$$

respectively, called temperature, negative pressure, and chemical potentials. In particular, μ_k is the variation of internal energy due to the addition of a unit amount of the k-th chemical, which also corresponds to the work that one should do to perform the operation, when the remaining variables are unchanged. These quantities vanish in the case the particles constituting the object of interest do not interact with each other.

The Legendre transforms of U are called *thermodynamic potentials*. For instance, replacing the old variable S, with the new variable T:

$$F = U - TS$$

one obtains the Helmholtz potential (free energy) $F = F(T, V, N_1, \ldots, N_r)$. Its elementary variation gives

$$dF = -SdT - PdV + \sum_{k=1}^{r} \mu_k dN_k$$

There is no TdS term, here, because S is not an independent variable anymore. One may go back to the variable S by changing variable again, Legendre-transforming F. This yields a new fundamental representation $\tilde{U}$, which, however, equals U:

$$- S = \frac{\partial F}{\partial T}; \qquad \tilde{U} = F + TS = U \tag{5.11}$$

The enthalpy is the partial Legendre transform of U that replaces V by P as independent variable:

$$H = U + PV$$

since the intensive parameter associated with V is $-P = \partial U/\partial V$, and its variations are given by

$$dH = TdS + VdP + \sum_{k=1}^{r} \mu_k dN_k$$

Again, it follows that

$$U = H - PV \ \text{ since } \ V = \frac{\partial H}{\partial P}$$

The Gibbs function (free energy) is the partial Legendre transform of U which replaces S by T and V by P as independent variables. Given:

$$U = U(S, V, N_k)\,; \ \ T = \frac{\partial U}{\partial S}\,, \quad -P = \frac{\partial U}{\partial V} \tag{5.12}$$

one obtains

$$G = U - TS + PV \ \ \text{so that} \ \ G = G(T, P, N_k, k = 1, ..., r) \tag{5.13}$$

and

$$dG = -SdT + VdP + \sum_{k=1}^{r} \mu_k dN_k$$

The Legendre transforms of the fundamental relation $S = S(U, V, N_1, \ldots, N_r)$, in the entropy representation, are called *Massieu functions*. They are particularly useful in irreversible thermodynamics and in the theory of fluctuations. For instance, given:

$$\frac{\partial S}{\partial U} = \frac{1}{T} \ \ \text{(new variable)}$$

one may replace U by $1/T$ to obtain

$$S^*\left[\frac{1}{T}\right] = S - \frac{1}{T}U = -\frac{F}{T}$$

and

$$F + TS = U\,, \quad S = \frac{1}{T}U - \frac{F}{T}$$

Given one expression of S which depends on U and V, e.g., the one derived from Eq. (5.13):

$$S = \frac{1}{T}U + \frac{P}{T}V - \frac{1}{T}G \tag{5.14}$$

one may introduce the new variable

$$\frac{\partial S}{\partial V} = \frac{P}{T} \tag{5.15}$$

so that transforming Eq. (5.14) one gets

$$S^*\left[\frac{P}{T}\right] = S - \frac{P}{T}V = \frac{1}{T}U - \frac{1}{T}G \tag{5.16}$$

Replacing U by $1/T$ and V by P/T, one gets

$$S^*\left[\frac{1}{T}, \frac{P}{T}\right] = S - \frac{1}{T}U - \frac{P}{T}V = -\frac{1}{T}G \tag{5.17}$$

One goes back to the S-representation by transforming again, changing variable from P/T to $-V = \partial S^*/\partial(P/T)$, and from $1/T$ to $-U = \partial S^*/\partial(1/T)$. The above are among the most common Massieu functions.

5.2 Legendre Transforms in Statistical Physics

Let us now illustrate the use of Legendre transforms in statistical mechanics. A general theory would require a whole book, but a few meaningful examples suffice to understand the subject. Suppose X is a fluctuating real quantity, which is only known statistically. For instance, it could be an extensive quantity, measured within a small volume, i.e., a local variable, rather than a global one. Let its range of values be characterized by a probability distribution, which can be very generally expressed as

$$f_X(x)\, \mathrm{d}M(x) = \frac{1}{e^{\Psi(h)}} e^{-hx} \mathrm{d}M(x)\,, \quad e^{\Psi(h)} = \int_{-\infty}^{\infty} e^{-hx}\, \mathrm{d}M(x) \tag{5.18}$$

where h is a positive parameter, and $\mathrm{d}M(x)$ can be chosen so that the probability distribution of the values of X is correctly expressed. For instance, it could be the number or the volume of states in an elementary interval $\mathrm{d}x$. Obviously, whenever the probability can be expressed by a probability density, this construction is allowed, and the following expression

$$P(X < x_0) = \int_{-\infty}^{x_0} e^{-\Psi(h)-hx}\, \mathrm{d}M(x) \tag{5.19}$$

gives the probability that the variable X takes values not larger than x_0. The corresponding moments and cumulant generating function, cf./ Appendix C are, respectively, defined by

$$\mathbb{E}\left(e^{\alpha X}\right) = e^{-\Psi(h)} \int_{-\infty}^{\infty} e^{-(h-\alpha)x}\, \mathrm{d}M(x) = e^{-\Psi(h)} e^{\Psi(h-\alpha)} \tag{5.20}$$

where α is a parameter in $\mathbb{R}$, and by

$$\ln \mathbb{E}\left(e^{\alpha X}\right) = -\Psi(h) + \Psi(h-\alpha) = \sum_{n=1}^{\infty} \frac{(-\alpha)^n}{n!}\Psi^{(n)}(h) = \sum_{n=1}^{\infty} \frac{\alpha^n}{n!}\mathbb{E}\left[X^n\right]_c \tag{5.21}$$

where $\Psi^{(n)}$ is the n-th derivative of Ψ, the expansion is performed about $\alpha = 0$, and its coefficient $\mathbb{E}\left[X^n\right]_c$ is the n-th cumulant of X: $\mathbb{E}\left[X^n\right]_c = (-1)^n \Psi^{(n)}(h)$.

Whatever the variable X represents, let us introduce the following quantity:

$$S = -k_B \int f_X(x) \ln f_X(x)\, \mathrm{d}M(x) = -k_B \mathbb{E}\,(\ln f_X) = k_B \mathbb{E}\,(\Psi(h) + hx) = k_B \Psi(h) + k_B h\, \mathbb{E}(X) \tag{5.22}$$

and let us call it *"entropy."* Then, $k_B \Psi$ is the "Massieu function" of S in which the variable $\mathbb{E}(X)$ is replaced by the variable $k_B h$, since

$$k_B \Psi(h) = S - k_B h\, \mathbb{E}(X) \quad \text{and} \quad k_B h = \frac{\partial S}{\partial \mathbb{E}(X)} \tag{5.23}$$

can be interpreted as the "intensive" variable conjugated to the "extensive" variable $\mathbb{E}(X)$ which is given by

$$\mathbb{E}(X) = e^{-\Psi(h)} \int_{-\infty}^{\infty} x\, e^{-hx} \mathrm{d}M(x) = e^{-\Psi(h)} \left(-\frac{\partial}{\partial h} \int_{-\infty}^{\infty} e^{-hx} \mathrm{d}M(x) \right) \tag{5.24}$$

$$= -e^{-\Psi(h)} e^{\Psi(h)} \frac{\partial}{\partial h}\Psi(h) = -\frac{\partial(k_B \Psi(h))}{\partial(k_B h)} \tag{5.25}$$

Now, let us call *"internal energy"* U the average $\mathbb{E}(X)$, and *"inverse temperature"* $1/T$ the quantity $k_B h$. Then, one formally obtains thermodynamic-like relations, such as those derived in the previous section:

$$k_B \Psi(h) = S - \frac{1}{T} U \quad i.e. \quad -\frac{F}{T} = S - \frac{1}{T} U \text{ by simply taking } k_B \Psi(h) = -\frac{1}{T} F \tag{5.26}$$

which defines the *"free energy"* $F = F(k_B h) = F(1/T)$. Moreover, the cumulants represent other "thermodynamic" quantities, and produce a completely consistent rational *"thermodynamics"*. For instance:

$$\mathbb{E}\,[X]_c = -\frac{d}{d(1/T)}\left(-\frac{1}{T}F\right) = F + \frac{1}{T}\left(\frac{\partial F}{\partial T}\right)\frac{dT}{d(1/T)} = U - TS + \frac{1}{T}(-S)\left(-\frac{1}{T^2}\right)^{-1} = U \tag{5.27}$$

where $F = U - TS$, hence $-S = \partial F / \partial T$, as in Eq. (5.11). Analogously, the second cumulant yields

$$\mathbb{E}\left[X^2\right]_c = k_B \frac{d^2}{d(1/T)^2}\left(-\frac{1}{T}F\right) = -k_B \frac{dU}{d(1/T)} = -k_B \frac{dU}{dT}\frac{d(1/T)^{-1}}{d(1/T)} = k_B T^2 C_v \tag{5.28}$$

where the derivative of the *"internal energy"* with respect to the *"temperature"* has been denoted by the symbol of the *"heat capacity."*

The terms "entropy," "Massieu function," etc. are in quotes, because they do not need to be thermodynamic quantities. In general, X may represent any random variable, whose probability distribution is expressed by Eq. (5.18). It could be the mass of potatoes falling from a truck, the length of text messages exchanged in a network of mobile phones, or the hypothetical number of fingers of aliens from another galaxy. Although mathematically consistent, the thermodynamic identifications would be merely formal. This does not exclude they be useful in some contexts. Attempting to use mathematical symbols beyond the realm in which they are known to work, to understand new phenomena, is natural. Strict obedience to the facts is still necessary, and not automatically granted. As a matter of fact, we have observed that thermodynamics emerges from molecular dynamics only under rather stringent conditions, and yet the identification of purely mechanical quantities with thermodynamic quantities has a celebrated history. Consider, for instance, the Helmholtz heat theorem, concerning the periodic motion of the Moon around the Earth. It proves that the classical mechanics and the thermodynamic formalisms are not incompatible, although they concern substantially different, irreducible phenomena (Chibbaro et al. 2014). Our illustration of Legendre transforms also supports this point of view.

Of course, if the analogy is purely formal, not all consequences of thermodynamic relations may apply. Typically, this is immediately realized when the system of interest is driven away from its formal thermodynamic equilibrium state. Even physically relevant particle systems, not just potatoes falling from a truck, may reveal themselves as non-thermodynamic. For instance, a gas of non-interacting particles, such as photons, may acquire thermal statistics from the walls of a black-body cavity, but if two such gases from two cavities at different temperatures meet, they will not equilibrate: their energy statistic will be a bimodal distribution, rather than a Planckian at an intermediate temperature.

In the cases in which the conditions of local thermodynamic equilibrium are satisfied, the identification of the mechanical quantities with the thermodynamic ones is real, and the above results allow very interesting conclusions. For instance, let X be the energy of a system of a large number N of molecules which properly interact, and let the observation space and time scales correspond to thermodynamic measurements. Then, $\mathbb{E}[X]_c = O(N)$, like $\mathbb{E}[X^2]_c = O(N)$, because C_v is an extensive quantity, while T is intensive and k_B is a constant. This implies that the ration of the average size of the energy fluctuations with respect to the average energy of the system itself goes like:

$$\frac{\sqrt{\mathbb{E}[X^2]_c}}{\mathbb{E}[X]_c} = O\left(\frac{1}{\sqrt{N}}\right) \tag{5.29}$$

hence it vanishes in the large N limit, although fluctuations themselves grow with N. This corresponds to the observation that in macroscopic objects the thermodynamic quantities do not fluctuate. The point is that thermodynamic observations take place on a scale at which the fluctuations due to the molecular motions are negligible compared to the observed signal. In fact, not only fluctuations exist, but they are even observable with proper measurement tools, like those used to detect gravitational waves. Furthermore, and most importantly, Eqs. (5.28), (5.29) allow us to quantify the size of *microscopic* fluctuations through macroscopic measurements.

The above calculations can be repeated for all cumulants of the energy, leading to an accurate knowledge of the full PDF, which is yet another impressive success of the atomistic hypothesis in describing macroscopic objects, within the framework of statistical mechanics. In probabilistic terms, one may say that the atomistic approach does not exclude that fluctuations be observed on the macroscopic scale; it simply implies that they are too rare to be observed, because they occur many standard deviations away from the observed (mean) signal: they constitute large deviations.

5.3 From Large Deviations to Legendre Transforms

The theory of large deviations has motivated a vast literature in all kinds of applications of probability theory, ranging from economy and finance to social sciences, to physics and biology. Indeed, this theory generalizes that of small fluctuations for which the Central Limit Theorem applies, cf. Appendix C. In this section, we provide a concise introduction to the subject, by explicitly solving a simple example which, however, suffices to grasp the fundamental concepts of this most interesting section of probability theory. We will conclude with an application to a random walk and the associated diffusion process in a one-dimensional space.

Take a biased coin,[1] which yields head with probability $\alpha \in [0, 1]$, and tail with probability $1 - \alpha$. Let the random variable X describing a single tossing take the value 1 for head, and value -1 for tail, so that its first and second moments are given by

$$\mathbb{E}(X) = \alpha(+1) + (1 - \alpha)(-1) = 2\alpha - 1$$

and

$$\mathbb{E}(X^2) = \alpha(+1)^2 + (1 - \alpha)(-1)^2 = 1$$

Consequently, the variance of X is expressed by $\boldsymbol{\sigma}^2 = \mathbb{E}(X^2) - \mathbb{E}(X)^2 = 4\alpha(1 - \alpha)$. Toss the coin N times, and assume that the random variable expressing the outcome at the n-th tossing, X_n, has the same statistical properties of X: in symbols $X_n \sim X$, for all $n = 1, ..., N$. Introduce the empirical mean:

$$Y_N = \frac{1}{N} \sum_{n=1}^{N} X_n \tag{5.30}$$

[1] A coin toss with independent subsequent trials is an example of a Bernoulli stochastic process.

This is a random variable taking $N+1$ values: $\{-1, -1+2/N, ..., 1-2/N, 1\}$, whose average is

$$\mu = \mathbb{E}(Y_N) = \frac{1}{N}\sum_{n=1}^{N}\mathbb{E}(X_n) = \mathbb{E}(X) = 2\alpha - 1 \tag{5.31}$$

If the consecutive tossings are also independent, the variance of Y_N is σ^2/N, and the Central Limit Theorem ensures that the distribution of Y_N for large N and within a distance of order $O(\sigma/\sqrt{N})$ from the mean μ is well approximated by the density:

$$f_N(y) = \sqrt{\frac{N}{2\pi\sigma^2}}e^{-N(y-\mu)^2/2\sigma^2}, \quad \text{for } |y-\mu| \text{ within order } O(\sigma/\sqrt{N}) \tag{5.32}$$

What is the distribution at a larger distance from μ? In our case, the exact answer is known. Let K be the number of heads obtained in N tossings, and express Y_N in terms of K at fixed N:

$$Y_N = \frac{1}{N}[K(+1) + (N-K)(-1)] = \frac{2K}{N} - 1 \tag{5.33}$$

Then, counting how many possible cases occur for a given K, and considering that the probability of each one of them is given by $\alpha^K(1-\alpha)^{N-K}$, one obtains

$$P\left(Y_N = \frac{2K}{N} - 1\right) = \frac{N!}{K!(N-K)!}\alpha^K(1-\alpha)^{N-K} \tag{5.34}$$

In the case in which N, K, and $(N-K)$ are relatively large, one may apply Stirling approximation:

$$\lim_{n\to\infty}\frac{n!e^n}{n^n\sqrt{2\pi n}} = 1\,, \quad i.e. \quad \ln n! = n\ln n - n + O(\ln n) \tag{5.35}$$

to all factorials in Eq. (5.34). Moreover, in view of a continuum limit, one may write $K = pN$, with $p \in [0, 1]$, so that Eq. (5.34) can be rewritten as

$$P(Y_N = 2p-1) = \frac{N!\alpha^{pN}(1-\alpha)^{(1-p)N}}{(pN)![(1-p)N]!} = e^{-NI_N(p;\alpha)} \tag{5.36}$$

where a suitable functional I_N has been introduced. Note that because these are probabilities, the sum of $\exp(-NI_N(p;\alpha))$ over all possible values of p is 1.

Taking the logarithm of the above expression, dividing by N, and applying the Stirling approximation, I_N tends to an asymptotic functional I:

$$I_N(p;\alpha) = -\frac{1}{N}\ln P(Y_N = 2p-1) \overset{N\to\infty}{\longrightarrow} I(p;\alpha) = p\ln\frac{p}{\alpha} + (1-p)\ln\frac{1-p}{1-\alpha} \tag{5.37}$$

This expression holds for all $p \in (0, 1)$, because for every p in this open interval and for any chosen tolerance, there is sufficiently large N that makes the Stirling approximation more accurate than required for that p. The functional I is known as a Kullback-Leibler divergence or relative entropy of the probability measures $(p, 1-p)$ and $(\alpha, 1-\alpha)$. It obeys $I(p; \alpha) \geq 0$ with equality if and only if $p = \alpha$.

The density of the $N+1$ values of Y_N in $[-1, 1]$ is uniform and equal to $(N+1)/2$. Then, for large N, the number of values in any set $(y, y+\delta) \subset [-1, 1]$ is approximately equal to $(N+1)\delta/2$, and the probability that Y_N falls within $(y, y+\delta)$ is given by the sum of the probabilities, expressed by Eq. (5.36), of the values in the interval. Searching for a way to obtain better and better accuracy, we can keep in mind that the role of probabilities is yielding averages of observables of interest, and that a continuous approximation of the discrete probability distribution produces more accurate results when applied to observables with smaller total variation. For such functions, taking δ sufficiently small, sufficiently large N eventually makes indistinguishable the averages produced by the discrete and the continuous distributions, on the scale of the accepted accuracy. In this sense of considering their action on observables, we can speak of probabilities turning as close to each other as one likes. Then, adopting a reasoning analogous to one of Sect. 3.3, we can write

$$P\left(Y_N \in (y, y+\delta)\right) \approx \frac{(N+1)}{2} e^{-NI(p;\alpha)} \delta\,, \quad \text{with} \quad p = \frac{y+1}{2} \tag{5.38}$$

where $(N+1)e^{-NI((y+1)/2;\alpha)}/2$ plays the role of the probability density of the continuous variable y, which is constructed in the large N limit.

Taking a small ϵ and $p = \alpha + \epsilon$, which means $y = 2p - 1 = 2\alpha - 1 + 2\epsilon = \mu + 2\epsilon$, I can be expanded to second order in ϵ obtaining:

$$\begin{aligned} I(p; \alpha) = {} & (\alpha+\epsilon)\left[\frac{\epsilon}{\alpha} - \frac{\epsilon^2}{2\alpha^2} + O\left(\epsilon^3\right)\right] + (1-\alpha-\epsilon)\left[-\frac{\epsilon}{1-\alpha} - \frac{\epsilon^2}{2(1-\alpha)^2} + O\left(\epsilon^3\right)\right] \\ = {} & \frac{\epsilon^2}{2\alpha(1-\alpha)} + O(\epsilon^3) = \frac{[(y-\mu)/2]^2}{2\alpha(1-\alpha)} + O(\epsilon^3) = \frac{(y-\mu)^2}{2\sigma^2} + O(\epsilon^3) \end{aligned} \tag{5.39}$$

so that

$$e^{-NI(p;\alpha)} \approx \frac{N+1}{2} e^{-N(y-\mu)^2/2\sigma^2} \tag{5.40}$$

As the prefactor $(N+1)/2$ is absorbed by the normalization, this reproduces the Gaussian derived from the Central Limit Theorem, which reveals that Eq. (5.38) generalizes such a theorem to *large deviations*. Indeed, given any p in the open interval $(0, 1)$, the Stirling approximation used above is accurate for all sufficiently large N, because both $K = pN$ and $N - K = (1-p)N$ are large numbers. The Gaussian expression, instead, only holds for p close to α.

The above is often directly formulated in terms of the quantity $y = 2p - 1$, so that Eq. (5.36) writes

$$P(Y_N = y) \approx \exp\left[-NS(y; \alpha)\right] \tag{5.41}$$

where

$$S(y;\alpha) = \frac{1+y}{2}\ln\frac{1+y}{2\alpha} + \frac{1-y}{2}\ln\frac{1-y}{2(1-\alpha)} \tag{5.42}$$

is called *Cramér's rate function* (or simply *rate function*).

Consider now the non-normalized variable $NY_N = \sum_{n=1}^{N} X_n$, in which the X_n's are independent and distributed like the random variable X. In symbols, $X_n \sim X$, i.e., $P(X_n \le x) = P(X \le x)$, for all $n \in \{1, ..., N\}$ and all values x in the range of X. Introduce the cumulant generating function of X:

$$L(q) = \ln \mathbb{E}\left(e^{qX}\right) \tag{5.43}$$

By independence, we have

$$\mathbb{E}\left(e^{qNY_N}\right) = \mathbb{E}\left(e^{q(X_1+\cdots+X_N)}\right) = \Pi_{n=1}^{N}\mathbb{E}\left(e^{qX_n}\right) = \mathbb{E}\left(e^{qX}\right)^N = e^{NL(q)} \tag{5.44}$$

and its logarithm divided by N gives

$$\frac{1}{N}\ln \mathbb{E}\left(e^{qNY_N}\right) = L(q) \tag{5.45}$$

which is the quantity of interest in the large N limit, because the probability becomes singular, hence problematic to handle. Moreover, using (5.38), one can write

$$\mathbb{E}\left(e^{qNY_N}\right) = \sum_{k=0}^{N} e^{qNy_k}P(y_k) \approx \frac{(N+1)}{2}\delta \sum_{j=-1/\delta}^{(1/\delta)-1} e^{qN\hat{y}_j}e^{-NS(\hat{y}_j;\alpha)} = \frac{(N+1)}{2}\delta \sum_{j=-1/\delta}^{1/\delta-1} e^{N[q\hat{y}_j - S(\hat{y}_j)]} \tag{5.46}$$

where the $N+1$ values y_k in the first sum have been ordered so that

$$y_0 = -1,\ y_1 = -1 + 2/N, \ldots,\ y_N = 1$$

and in the second sum they have been collected in a number $(2/\delta) \in \mathbb{N}$ of small intervals $[j\delta, (j+1)\delta)$, with

$$j \in \left\{-\frac{1}{\delta}, -\frac{1}{\delta}+1, ..., \frac{1}{\delta}-2, \frac{1}{\delta}-1\right\}$$

within which y varies so little that S can be considered practically constant. Then, $S(\hat{y}_j)$ closely approximates the values of S in the j-th interval for any $\hat{y}_j$ in $[j\delta, (j+1)\delta)$. This justifies the approximate equality of the first and second sums in Eq. (5.46), as well as the following:

$$\frac{1}{N}\ln \mathbb{E}\left(e^{qNY_N}\right) \approx \frac{1}{-}N \ln\left(\frac{(N+1)}{2}\delta \sum_{j=-1/\delta}^{1/\delta-1} e^{N[q\hat{y}_j - S(\hat{y}_j)]}\right) \tag{5.47}$$

For sake of concreteness, let us take the mid-point of the interval, $\hat{y}_j = (j + 1/2)\delta$.

Equation (5.46) is thus accurate for large N, and $N\hat{y}_j$ bounded away from $-N$ and N, i.e., $\hat{y}_j$ away from -1 and 1. Given the required accuracy, smaller δ and $\hat{y}_j$ closer to -1 and 1 can be included, simply taking larger N. Alternatively, given $y \in (-1, 1)$, there is sufficiently small δ that the interval containing y is contained in $(-1, 1)$, and then the Stirling approximation at the corresponding $\hat{y}_j$ improves with growing N, because the values k and $N - k$ needed for y_k and y_{N-k} to lie within the j-th interval grow, making Eq. (5.38) more accurate. To estimate the required values of δ and N, observe that

$$F(y; q) = qy - S(y) \quad \text{with} \quad q \in \mathbb{R} \tag{5.48}$$

is a strictly convex function of y in $(-1, 1)$:

$$\frac{\mathrm{d}^2 F}{\mathrm{d}y^2} = -\frac{1}{1 - y^2} < 0 \tag{5.49}$$

hence, for each q, it has a single maximum in $(-1, 1)$. For simplicity, take $\alpha = 1/2$. Then, this maximum is located at

$$y_{\max}(q) = \frac{e^{2q} - 1}{e^{2q} + 1} \tag{5.50}$$

and

$$F(y_{\max}(q); q) = \ln\left(e^{2q} + 1\right) - q - \ln 2\,, \quad \text{so that} \quad F(y_{\max}(q); q) > q - \ln 2 \tag{5.51}$$

For any $\delta > 0$, $y_{\max}(q)$ belongs to one of the sub-intervals partitioning $[-1, 1]$, say the k-th sub-interval $u_k = [k\delta, (k + 1)\delta)$ and, by definition, one has

$$F(y_{\max}(q); q) \geq \max_j F\left(\hat{y}_j; q\right) \tag{5.52}$$

Furthermore, $|\hat{y}_j - y_{\max}(q)| \geq \delta/2$ for $j \neq k$, and the fact that F is also analytic, allows us to write

$$F(y_{\max}(q) \pm \delta/2; q) = F(y_{\max}(q); q) - \frac{\delta^2/4}{2(1 - y_{\max}(q)^2)} + O(\delta^3) \geq F(\hat{y}_j; q)\,, \quad \forall\, j \neq k \tag{5.53}$$

where Eq. (5.49) has been used. Therefore, considering that there are $N(1 - O(\delta))$ addends in the first sum of Eq. (5.46) outside the k-th sub-interval, one obtains

$$\frac{N(1 - O(\delta))e^{NF(y_{\max}(q)\pm\delta/2;q)}}{e^{NF(y_{\max}(q);q)}} = N(1 - O(\delta)) \exp\left\{-N\left[\frac{\delta^2}{8(1 - y_{\max}(q)^2)} + O(\delta^3)\right]\right\} \tag{5.54}$$

which means that the maximum value of $\exp(NF)$ dominates, for large N, the sum of the values taken by that function at a distance $\delta/2$ or larger from $y_{\max}(q)$.

Indeed, their values are smaller than any value $\exp[NF(y;q)]$ for $y \in u_k$. Then, for large N, the first equality of Eq. (5.46) may be rewritten including only the $O(\delta)N$ terms within u_k. Among them, at fixed δ and sufficiently large N, some lie at distances $O(1/N)$ from $y_{\max}(q)$, and others at larger distances. The first, according to Eq. (5.53), and given the accepted accuracy, yield values indistinguishable from $\exp[NF(y_{\max}(q);q]$ at large N. Let their number be b_N. The others can be neglected, like those belonging to sub-intervals other than the k-th. Therefore, we eventually get

$$\frac{1}{N}\ln \mathbb{E}\left(e^{qNY_N}\right) = \frac{1}{N}\ln \sum_{k=0}^{N} e^{qNy_k} P(y_k) \approx \frac{1}{N}\ln \sum_{y_k\in u_k} e^{NF(y_k;q)} \approx \frac{1}{N}\ln b_N + F(y_{\max}(q);q) \tag{5.55}$$

where $b_N \leq \delta N$. Now, we may take $\sqrt{N}\delta = \kappa > 0$, and let N grow without bounds, so that the number of points in an interval of width δ also grows without bounds while δ tends to zero. This makes the approximations introduced above turn exact and Eq. (5.47) takes the following form:

$$\lim_{N\to\infty\,;\,\delta\sqrt{N}=\kappa} \frac{1}{N}\ln \mathbb{E}\left(e^{qNY_N}\right) = F(y_{\max}(q);q) + \lim_{N\to\infty\,;\,\delta\sqrt{N}=\kappa} \frac{1}{N}\ln b_N = F(y_{\max}(q);q) \tag{5.56}$$

More in general, one cannot guarantee that the maximum of F is reached, and one obtains

$$L(q) = \sup_y\,[qy - S(y)] \tag{5.57}$$

Because we are not treating thermodynamic variables, here, we can adopt the usual mathematical definition of Legendre transforms, given in Appendix A. That differs from the thermodynamic expression (5.9) by a sign, and makes L the Legendre transform of S. The Legendre transform S of L is analogously expressed by

$$S(y) = \sup_q\,[qy - L(q)] \tag{5.58}$$

which is exactly S if S is strictly convex, otherwise it yields the convex envelope of S.

In the case that the random variables X_n are not independent, one may define

$$L(q) = \lim_{N\to\infty} \frac{1}{N}\ln \mathbb{E}\left(e^{q\sum_{n=1}^{N} X_n}\right) \tag{5.59}$$

and see what happens case by case.

For systems of N interacting particles, the identification of macroscopic (non-fluctuating) thermodynamic quantities with microscopic (fluctuating) mechanical quantities follows from the formalism that has just been illustrated. Because both the internal energy U and the entropy S are extensive quantities, and because interactions make the microscopic motions quite disordered, one may assume that the contribution Z_n of a single particle to the total, say

$$X = \sum_{n=1}^{N} Z_n = NY_N \tag{5.60}$$

does not depend on n, and is equally distributed as the other particles contributions: $Z_n \sim Z$, with $n = 1, ..., N$, where Z is the random variable of interest. Therefore, one may write

$$\mathbb{E}\left(e^{qX}\right) = \int e^{qx} f_N(x)\mathrm{d}x = \mathbb{E}\left(e^{qNY}\right) = \mathbb{E}\left(e^{qZ}\right)^N = e^{NL(q)} \tag{5.61}$$

Then, following Einstein's inversion of the celebrated Boltzmann bridge law

$$S_B = k_B \ln W \tag{5.62}$$

and recalling that equilibrium corresponds to the maximum entropy, $S_{B,\max}$, we write $S_B = S_{B,\max} - \Delta S_B = S_{B,\max} + Ns$, where s is the (negative) entropy variation per particle, which yields $W = \text{const}\cdot \exp(Ns/K_B)$, i.e.:

$$W(S_B) \sim e^{Ns/k_B} \tag{5.63}$$

This quantity represents the (finite) number of microstates corresponding to the equilibrium macrostate having entropy S_B. It yields a probability if it is normalized. Repeating the calculations performed above, one obtains the internal energy as the Legendre transform of the entropy variation defined by

$$U = L(q) = \sup_y \left[qy - \frac{1}{k_B} S_B \right] \tag{5.64}$$

To prove the validity of our i.i.d. assumption on the grounds of the microscopic dynamics is far from trivial and still object of research. The only cases that have been thoroughly mathematically cleared are certain variations of rarefied gases, whose molecules have simple short-range repulsion potentials. There could indeed be fundamental reasons for which Hamiltonian dynamics cannot yield the desired results in general, not just technical difficulties. At present, we do not know. No doubt, a rigorous derivation of, say, the Navier-Stokes equations from some Hamiltonian dynamics would be extremely interesting and quite a feat. It would also support our picture of the microscopic realm. On the other hand, the applicability of physical theories is limited to certain ranges of space and time scales, hence it is not necessary to have one such derivation, if our goal is to describe fluids in normal conditions. Navier-Stokes equations have been working perfectly well without any recourse to the atomic picture for two centuries. Should one day be proven that they cannot be derived from a Hamiltonian microscopic theory, we will not dispose of them. On the contrary, our microscopic theory will be challenged and required to be improved.

5.3.1 Large Deviations and Diffusion

Consider the diffusion equation in the real line, with localized initial condition:

$$\frac{\partial p}{\partial t} = D\frac{\partial^2 p}{\partial x^2}, \quad p_0(x) = \delta(x) \tag{5.65}$$

where the diffusion coefficient is $D = 1/2$, and δ is the Dirac delta function. Its solution is expressed by

$$p_t(x) = \frac{1}{\sqrt{2\pi t}}\, e^{-x^2/2t} \tag{5.66}$$

which formally means that, whatever diffuses according to this law must be made of something that travels arbitrarily fast. In fact, at any positive, even very small time t, the density $p_t(x)$ is greater than 0 for all $x \in \mathbb{R}$.

This law can be obtained in various ways, for instance, assuming the diffusing substance is a continuum driven by gradients of concentration, and obeys linear response. Another interesting derivation is afforded by certain random walks on the line, taken at times as a kind of microscopic description. Following this approach, we consider particles having two velocities only, $v_\pm = \pm 1$, which are fished at random with probability 1/2 each, at every time step. Let the time unit be $\tau = 1$. Starting from $x = 0$ at $t = 0$, a velocity is randomly chosen, and the particle moves with constant speed, but random direction, for a time τ. At time $\tau = 1$, the process is repeated, independently of the previous step, and the process continues forever like that.

A direct application of the Central Limit Theorem for large t shows that the sum of the steps in the real line is a random variable X_t with Gaussian probability density of vanishing mean and variance equal to t, which is the solution (5.66) of the diffusion equation (5.65). This may appear paradoxical, because at time t the particles positions are confined within $[-t, t]$. However, the result looks immediately less surprising, recalling that the Central Limit Theorem yields the normal distribution as a good asymptotic approximation of the real distribution only within a distance $x = O(\sqrt{t}\,)$. Moreover, in case one really wants to keep a nonequilibrium thermodynamic interpretation of the result, and Eq. (5.65) is taken as Fick's second law of mass diffusion, one should also recall that Eq. (5.65) becomes thermodynamically nonsensical if the density is too small. Indeed, a few particles in a given volume do not allow LTE, cf. Section 2.5.

Both mathematically and physically, there is thus no reason to invoke any kind of paradox. Nevertheless, it is interesting to ask whether there is a differential equation that constitutes a better limiting description of our random walk than the diffusion equation. In this respect, Ref. Keller (2004) suggests an informative approach, which we summarize. Let j denote the number of steps to the right, taken by a particle in a time t. Then, the step taken to the left is $t - j$, and the position reached is $x_{tj} = j - (t - j) = 2j - t$. Because the steps are independent, the probability of this walk is the product of the probabilities of the single steps, $1/2^t$, multiplied by the number of walks corresponding to x_{tj}. The result is

$$P(x_{tj}, t) = \begin{cases} \dfrac{t!\,2^{-t}}{j!(t-j)!} & \text{for } j = 0, 1, ..., t \\ 0 & \text{for } j < 0 \text{ or } j > t \end{cases} \tag{5.67}$$

Applying Stirling's formula, one obtains a result that extends much beyond $O(\sqrt{t})$, as long as both j and $t - j$ are not small, analogously to the calculations of the previous subsection. The result is the following:

$$p(x,t) \sim \begin{cases} \dfrac{1}{\sqrt{2\pi t}}\left(1+\dfrac{x}{t}\right)^{-(t+x+1)/2}\left(1-\dfrac{x}{t}\right)^{-(t-x+1)/2} & \text{if } \left|\dfrac{x}{t}\right| < 1\,,\ t \gg 1 \\ 0 & \text{if } |x/t| > 1 \end{cases} \tag{5.68}$$

However, one can do even better, considering that small j implies large $t - j$ and vice versa. Applying Stirling's approximation to that factorial in the denominator of Eq. (5.67) which is not small, and keeping the exact value for the other, the probability density[2] p of finding a particle at x is found to obey (Kuske and Keller 1997):

$$p(x,t) \sim \begin{cases} \dfrac{1}{\sqrt{2}}\left(1+\dfrac{x}{t}\right)^{-(t+x+1)/2} \dfrac{(t/2)^{(t-x)/2}\,e^{-(t-x)/2}}{\left(\dfrac{t}{2}-\dfrac{x}{2}\right)!}\,, & \text{if } 1 \geq \dfrac{x}{t} > -1\,,\ t \gg 1 \\ 0 & \text{if } |x/t| > 1 \end{cases} \tag{5.69}$$

One may analogously treat the case with $-1 \leq x/t < 1$. The fact that the standard large deviation result applies to distances of order $O(t)$ but not up to t is evidenced by one characteristic feature of this approximation: the solution develops in time a finite speed traveling front that gets narrower and narrower, while it diverges in height. This is not a feature of the exact density, which decreases monotonically up to $|x| = t$, where it discontinuously vanishes, rather than diverging. Nevertheless, the approximation we have obtained works very well up to quite close to the front.

A continuum model may also be derived from the discrete stochastic process as follows. Let $x_j = j$ with $j \in \mathbb{Z}$ and $t \in \mathbb{N} \cup \{0\}$. Then, the probability density in x_j at time $t + 1$ is exactly given by

$$p\left(x_j, t+1\right) = \frac{1}{2} p\left(x_{j-1}, t\right) + \frac{1}{2} p\left(x_{j+1}, t\right) \tag{5.70}$$

which implies

$$p\left(x_j, t+1\right) - p\left(x_j, t\right) = \frac{1}{2}\left[p\left(x_{j-1}, t\right) - 2p\left(x_j, t\right) + p\left(x_{j+1}, t\right)\right] \tag{5.71}$$

[2] That is obtained dividing the probability by 2, for the distance between consecutive x_{tj}'s is 2.

Now, one may wish that, in a suitable limit, p obeys a differential law analogous to the diffusion equation. Therefore, it must be smooth and can be expanded in a power series, in both variables x and t. Assuming that the units of space and time are in fact very small (microscopic) from the point of view of the continuous description, which corresponds to our (macroscopic) observations, the expansion can be truncated at small orders. This wish is not necessarily realized, but one can try obtaining:

$$p\left(x_j, t+1\right) = p\left(x_j, t\right) + \left(\left.\frac{\partial p}{\partial t}\right|_{(x_j,t)}\right) \cdot 1 + \frac{1}{2}\left(\left.\frac{\partial^2 p}{\partial t^2}\right|_{(x_j,t)}\right) \cdot 1^2 \quad (5.72)$$

$$p\left(x_j + 1, t\right) = p\left(x_j, t\right) + \left(\left.\frac{\partial p}{\partial x}\right|_{(x_j,t)}\right) \cdot 1 + \frac{1}{2}\left(\left.\frac{\partial^2 p}{\partial x^2}\right|_{(x_j,t)}\right) \cdot 1^2 \quad (5.73)$$

Substituting these quantities in Eq. (5.71), and removing the subscript j from x_j, leads to

$$\frac{\partial p}{\partial t} + \frac{1}{2}\frac{\partial^2 p}{\partial t^2} = \frac{1}{2}\frac{\partial^2 p}{\partial x^2} \quad (5.74)$$

which is known as the telegrapher's equation.

These issues remind us that use of a model must be limited within its proper bounds, although it is not forbidden to try it beyond. This is a conceptual message recalled throughout this book, and in the following we illustrate it with one example related to the random walk discussed above. The approach that led to Eq. (5.74), via Eqs. (5.72), (5.73) looks curious, because 1 is not an infinitesimal quantity, hence it is not clear why higher order terms can be neglected in the Taylor expansion. Had we written Δt and $(\Delta t)^2$ instead of 1 and 1^2 in Eq. (5.72), we would have found that the $\Delta t \to 0$ limit either makes the second-order term vanish, recovering the standard diffusion equation or that the first-order term explodes, making senseless the resulting equation. Indeed, we would write

$$\frac{\partial p}{\partial t}\Delta t + \frac{1}{2}\frac{\partial^2 p}{\partial t^2}(\Delta t)^2 = \frac{1}{2}\frac{\partial^2 p}{\partial x^2}(\Delta x)^2 \quad (5.75)$$

and, dividing by Δt we would have

$$\frac{\partial p}{\partial t} + \frac{1}{2}\frac{\partial^2 p}{\partial t^2}\Delta t = \frac{1}{2}\frac{\partial^2 p}{\partial x^2}\frac{(\Delta x)^2}{\Delta t} \quad (5.76)$$

which makes sense only if $\Delta t \to 0$ and $[(\Delta x)^2/\Delta t] \to D'$, a constant. This scaling, for which $|\Delta x|$ cannot grow faster than $\sqrt{\Delta t}$, produces the diffusion equation with diffusion constant half of D'. On the other hand, dividing by $(\Delta t)^2$, we would obtain

$$\frac{1}{\Delta t}\frac{\partial p}{\partial t} + \frac{1}{2}\frac{\partial^2 p}{\partial t^2} = \frac{1}{2}\frac{\partial^2 p}{\partial x^2}\left(\frac{\Delta x}{\Delta t}\right)^2 \quad (5.77)$$

In this case, the $\Delta t \to 0$ limit makes no sense, although required by the theory. In a different framework, one could make sense of Eq. (5.77) by taking $\Delta t \to \infty$ while $(\Delta x/\Delta t) \to c'$, another constant, which leads to a wave equation of speed c':

$$\frac{\partial^2 p}{\partial t^2} = c'^2 \frac{\partial^2 p}{\partial x^2} \tag{5.78}$$

So, apparently, the random walk considered above can be macroscopically described only by a diffusion equation.

A more general setting still yields Eq. (5.75). For instance, let the fixed space steps be replaced by a real random variable S, whose values are sampled with distribution ρ, such that $\mathbb{E}_\rho(S) = 0$ and $\mathbb{E}_\rho(S^2) = 1$. In this case, one may write

$$p(x, t + \tau) = \int p(x - s, t)\rho(s; \tau)\, \mathrm{d}s \tag{5.79}$$

since in a time τ any step of size s can be taken. Then, expanding p as a Taylor series in x with t fixed, and then in t with x fixed, one obtains

$$p(x, t + \tau) = p(x, t) + \tau \left(\left.\frac{\partial p}{\partial t}\right|_{(x,t)} \right) + \frac{\tau^2}{2} \left(\left.\frac{\partial^2 p}{\partial t^2}\right|_{(x,t)} \right) + O(\tau^3) \tag{5.80}$$

$$p(x - s, t) = p(x, t) + s \left(\left.\frac{\partial p}{\partial x}\right|_{(x,t)} \right) + \frac{s^2}{2} \left(\left.\frac{\partial^2 p}{\partial x^2}\right|_{(x,t)} \right) + O(s^3) \tag{5.81}$$

and the integral (5.79) yields

$$\begin{aligned} p(x, t + \tau) = &\ p(x, t) - \mathbb{E}_\rho(S) \left(\left.\frac{\partial p}{\partial x}\right|_{(x,t)} \right) + \frac{\mathbb{E}_\rho\left(S^2\right)}{2} \left(\left.\frac{\partial^2 p}{\partial x^2}\right|_{(x,t)} \right) + o\left(\mathbb{E}_\rho\left(s^3\right)\right) \\ = &\ p(x, t) + \frac{1}{2} \left(\left.\frac{\partial^2 p}{\partial x^2}\right|_{(x,t)} \right) + o\left(\mathbb{E}_\rho\left(s^3\right)\right) \end{aligned} \tag{5.82}$$

Equating this to the expression (5.80) and neglecting terms of third order, one obtains

$$\tau \frac{\partial p}{\partial t} + \frac{\tau^2}{2} \frac{\partial^2 p}{\partial t^2} = \frac{1}{2} \frac{\partial^2 p}{\partial x^2} \tag{5.83}$$

So, no progress has really been done. Nevertheless, numerically the validity of the telegrapher's equation is confirmed, showing that its solution differs only negligibly, up to relatively long times and large $|x|$, from the standard large deviation result (5.68) and, in turn, this differs negligibly over the same space and time scales from the indistinguishable improved large deviation result and exact result. Therefore, there must be alternative ways of deriving Eq. (5.74), which are consistent with our random walk.

Analyzing Eq. (5.83) we note that the symbol τ instead of Δt more clearly suggests that the time increment and the time derivatives do not need to refer to the same variable t. This indeed is desired if one wants to separate two time scales, a microscopic scale and a macroscopic scale, the first of which looks infinitesimal from the point of view of the second, while the second looks infinite from the point of view of the first. Letting τ be the microscopic characteristic time and t that of our observations, a dimensionless time $r = t/\tau$ can be introduced, which is large when τ is small, for fixed t. Then we have $\tau \partial p/\partial t = \partial p/\partial r$ and $\tau^2 \partial^2 p/\partial t^2 = \partial^2 p/\partial r^2$, and the telegrapher's equation:

$$\frac{\partial p}{\partial r} + \frac{1}{2}\frac{\partial^2 p}{\partial r^2} = \frac{1}{2}\frac{\partial^2 p}{\partial x^2} \tag{5.84}$$

arises without further assumptions, in terms of the dimensionless time r. This provides a justification of the accuracy with which the telegrapher's equation describes our random walk, when sufficiently long times are considered. It corresponds to a change of perspective: τ may be fixed and finite, but the time an observer takes to measure what happens may be extremely larger. Then, τ looks an infinitesimal instant to the observer. Various things can be learned through this simple exercise.

1. What initially seemed to be a good model for macroscopic finite speed transport, i.e., the unbiased random walk with constant or random space steps and with discrete or continuous time, does not achieve the goal by itself. A change of perspective, which is not part of the model, but is—in a sense arbitrarily added to the model—necessary to obtain the proper macroscopic description.
2. Arbitrariness in this approach is revealed by the fact that different scalings are possible, and lead to different macroscopic equations. For instance, one scaling yields the diffusion equation, another scaling yields the wave equation, still another one is required to obtain the telegrapher's equation. Furthermore, one may ask why should one stop the expansion of the probability distribution to the second order in t and x, rather than continuing to higher orders.
3. Arbitrariness is evident also in the fact that totally different microscopic models could have been devised, so that no extra ingredients, or different scalings be required to recover the desired macroscopic relations. The choice of one microscopic model or another, and of certain scalings rather than others often requires information from the macroscopic realm. Experimental evidence limits arbitrariness.
4. Nevertheless, even such a simple example as the random walk on a line shows that the details of the microscopic dynamics are by and large irrelevant for the macroscopic, or collective behavior determined by a large number of elements and of time units. The equivalence of equilibrium ensembles, recently extended to non-equilibrium phenomena, Gallavotti et al. (2004), is a consequence of that. Also the theory of equilibrium universality classes rests on these facts (Vollmer et al. 2021a).
5. The possibility of changing perspective does not imply subjectivity in physics. The physical properties of a given object do not depend on our point of view. A

cathedral observed from a distance of 20 cm does not look a cathedral, but a stone next to other stones. Still, this does not stop it from objectively being a cathedral. And looking at it from a distance of 100 m does not stop it from being objectively made of stones. The microscopic and the macroscopic descriptions can be seen under the same light.

This helps understand, for instance, the success of thermodynamics. It remains far from trivial to delimit the range of applicability of a given theory, beyond the realm in which it has been derived and tested. Often, it suffices to change interpretation of symbols to obtain a theory of a different phenomenon. For instance, the same equation describes such diverse situations as pollen in water, diffusion of heat, transport of electrons in electrical conductors, etc.; it suffices to call with different names the mathematical symbols in a given equation.

5.4 Current Research

Large deviations and the thermodynamic formalism constitute a very active research field. Some applications will be described in Chap. 9. Here, we mention recent applications of machine learning and artificial intelligence, where large deviations and the thermodynamic formalism provide efficient statistical tools (Rotskoff and Vanden-Eijnden 2018; Carbone et al. 2024). Investigations are also underway to improve the simulation of rare events using large deviations in general (Giardina et al. 2011). Rare events are indeed important for phenomena such as the onset of chemical reactions, the stability of planetary systems, the development of rouge waves and turbulence, phase transitions, and countless many more, including subjects quite remote from hard sciences, such as opinion dynamics (Vulpiani et al. 2014; Harris and Touchette 2017; Balducci et al. 2023; Oliveira et al. 2024). This is due to the fact that formal (not necessarily thermodynamic) entropy productions are generally expressed in terms of large deviation functionals and often constitute good estimators for rare events. Among physical phenomena dominated by rare events, one of the most interesting are those concerning transport of mass and energy in highly confining media, such as carbon nanotubes. These can be modeled to a good extent as one-dimensional objects in which transport can be split into two main contributions: normal diffusion and ballistic transport (Dematteis et al. 2020). When the second dominates in such a way that a situation named *single big jump* is realized (Vezzani et al. 2019), transport is anomalous and belongs to a certain universality class (Vollmer et al. 2021b). This situation is analogous to that of Eq. (5.54), in which a small number of events, in fact even a single one, dominates over all the others.

Probabilities in Phase Space 6

I close my eyes and see a flock of birds. The vision lasts a second or perhaps less; I don't know how many birds I saw. Were they a definite or an indefinite number?

Dreamtigers, Jorge Luis Borges

6.1 Dynamical Systems

The dynamics of classical particles can be represented as a dynamical system acting on their phase space. There are numerous properties of generic dynamical systems that apply directly to systems of particles. In particular, those that concern probability distributions are of interest in response theory and its modern applications. At the same time, properties such as *ergodicity*, defined below, must be taken with a grain of salt in Physics. In this chapter, we summarize the statistical theory of dynamical systems, with an eye to their physical relevance.

Let $\mathcal{M} \subset \mathbb{R}^n$ be a phase space (usually a manifold) in which a dynamical evolution takes place obeying:

$$\dot{x} = G(x) , \quad \text{with } x \in \mathcal{M} \tag{6.1}$$

In the case of particles, the points $x \in \mathcal{M}$, called phases, represent all the microscopic states of the system, and the differential equations (6.1) are the corresponding equations of motion. Let

$$\Phi^t : \mathcal{M} \longrightarrow \mathcal{M} ; \quad \Phi^t : x \longmapsto \Phi^t x , \quad \text{with} \quad x, \Phi^t x \in \mathcal{M} , \tag{6.2}$$

denote the time evolution operator up to time t, meaning that $\Phi^t x_0$ represents the solution, $x(t; x_0)$, of the dynamical equations with initial condition $x_0 \in \mathcal{M}$. By definition, one has $\Phi^0 = \mathbb{1}$ and $\Phi^t \Phi^s = \Phi^{t+s}$. Furthermore, assuming the solution

C. Mejía-Monasterio and L. Rondoni, *Nonequilibrium Statistical Mechanics I*, Theoretical and Mathematical Physics,
https://doi.org/10.1007/978-3-031-98249-1_6

exists for all $t \in \mathbb{R}$ and is unique, the collection of evolution operators forms a one-parameter group, $\Phi = \{\Phi^t\}_{t \in \mathbb{R}}$, of automorphisms of $\mathscr{M}$. In fact, for every $t \in \mathbb{R}$, the function $\Phi^t x$ is also differentiable with respect to x, at every $x \in \mathscr{M}$, if it is solution of Eq. (6.1) with a smooth G.

In the case of Hamiltonian, as well as of Gaussian, Nosé-Hoover and many other dynamics, time reversal invariance (TRI) holds. This means that there exists an involution $\mathscr{I} : \mathscr{M} \longrightarrow \mathscr{M}$, i.e., a linear unitary operator on $\mathscr{M}$ obeying $\mathscr{I}^2 = \mathbb{1}$ (the identity operator), that anti-commutes with the time evolution, cf. Fig. 6.1:

$$\mathscr{I} \, \Phi^{-t} \, x = \Phi^t \, \mathscr{I} \, x \,, \quad \text{for all } t \in \mathbb{R} \text{ and all } x \in \mathscr{M} \tag{6.3}$$

This operator is associated with time reversal because it yields an earlier point in a given trajectory without having to reverse the direction of time:

$$\Phi^{-t} \, x = \mathscr{I} \, \Phi^t \, \mathscr{I} \, x \tag{6.4}$$

while going backward in time is something we cannot realize, in the macroscopic world at least. It should be noted that TRI is a property stronger than invertibility, which is the existence of the inverse, Φ^{-t}. The inverse exists in all cases in which the solution of the equations of motion is unique, but TRI also requires the existence

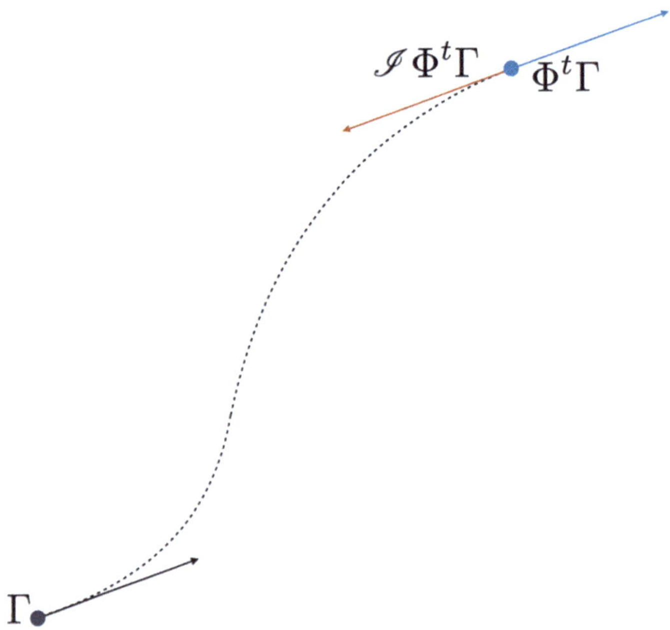

Fig. 6.1 Schematic representation of TRI dynamics in phase space. The points of the dashed curve represent the positions $\mathbf{q}$, the arrows represent the momenta $\mathbf{p}$; $\Gamma = (\mathbf{q}, \mathbf{p})$. Time reversibility holds if the forward time trajectory with initial condition $\Gamma' = \mathscr{I}\Phi^t\Gamma$ traces backward with opposite momenta the set of $\mathbf{q}$'s traced in the same time by the trajectory starting at Γ. The trajectories appear to overlap, but in phase space they do not, because they have different momenta

of at least one involution $\mathcal{I}$ for which Eq. (6.3) holds. Furthermore, the *backward time* operation Φ^{-s}, with $s \in [0, t]$, starting from the point $\Phi^t x$, traces backward the trajectory that led from x to $\Phi^t x$: all the coordinates of the generic point of the trajectory take the same values they took in the forward time evolution, following the opposite chronological order, like rewinding a video tape. This is not what happens with Eq. (6.3), where every initial condition x that leads forward in time to a point $y = \Phi^t x$ is associated with another initial condition, $z = \mathcal{I} y = \mathcal{I} \Phi^t x$, such that *forward* time evolution for a time t starting at z leads to $\mathcal{I} x$. This is not the origin x of the first trajectory. Evolving forward the point $\mathcal{I} y$, a totally different trajectory is spanned in phase space; nevertheless, a trajectory with interesting properties, compared to the original one. Consider for simplicity the best known reversal operator, which takes a point with conjugate positions and momenta, say, $x = (q, p)$, and reverses the momenta: $\mathcal{I} x = (q, -p)$. When the forward time operator is applied, only configurations are traced in opposite chronological order, compared to the first trajectory; the momenta take opposite signs. However, in case the first trajectory produced a positive current, the reverse trajectory produces exactly the opposite current, which is interesting in many applications.

6.1.1 Observables and Probabilities

In the framework of the statistical mechanics of many particle systems, let $n = 2dN$ be the total number of variables, with d the number of degrees of freedom of each particle, and N the number of particles. Then, functions of phase

$$\mathcal{O} : \mathcal{M} \longrightarrow \mathbb{R}$$

are considered as microscopic properties whose time averages

$$\overline{\mathcal{O}}(x) = \lim_{t \to \infty} \frac{1}{t} \int_0^t \mathcal{O}(\Phi^s x) ds \tag{6.5}$$

are postulated to represent measurements of macroscopically observable quantities. The idea is that the time average takes into account the exchange between a measurement tool and the macroscopic object of interest. The theory implies that such an exchange is described at the microscopic level, because the phase space dynamics corresponds to that level; hence, it involves a very large number of events, provided the microscopic and the macroscopic scales are widely separated. As discussed in relation to the notion of local thermodynamic equilibrium, this requires N and the observation times to be very large. Then, assuming that one is treating objects amenable to thermodynamic analysis, the $t \to \infty$ limit results in a very good mathematical idealization of the measurement time, since it may last $10^9 - 10^{12}$ or more microscopic units.

Typically, a handful of observables suffice to describe the thermodynamic state of a macroscopic object; consequently, the microscopic details bear little relevance for macroscopic observations. In particular, the dependence on the initial condition x of a trajectory, $\gamma(x) = \{\Phi^t x\}_{t\in\mathbb{R}}$, representing the microscopic history of a given (single) object of interest, is irrelevant.[1] This suggests that the time average (6.5) may be obtained replacing the to date impossible to compute[2] phase space trajectory with a phase space probability distribution μ that weighs the different regions of $\mathcal{M}$ proportionally to the frequency with which $\gamma(x)$ visits them:

$$\overline{\mathcal{O}}(x) = \int_{\mathcal{M}} \mathcal{O} d\mu = \mathbb{E}_\mu[\mathcal{O}] \tag{6.6}$$

where the right-hand side is called phase space average. Because the left-hand side of Eq. (6.6) depends on x, while the right-hand side does not, one must specify that the equality is not expected to hold for all $x \in \mathcal{M}$. Indeed, it suffices that it holds for *μ-almost every $x \in \mathcal{M}$*, i.e., as long as its possible violations are confined within a set $E \in \mathcal{M}$ of vanishing probability:

$$\mu(E) = \int_E d\mu = 0$$

which by definition does not affect the integral in Eq. (6.6). The question is: Given the dynamics Φ^t, what probability μ will do? The answer to this question requires some elementary notions of probabilities in phase space.

Given $\mathcal{M}$, a probability on it can be defined as an operator $\mu : \sigma(\mathcal{M}) \longrightarrow [0, 1]$ that assigns non-negative numbers smaller than 1 to the elements E of a collection $\sigma(\mathcal{M})$ of its subsets:

$$0 \leq \mu(E) = \int_E d\mu \leq 1, \quad \forall E \in \sigma(\mathcal{M}) \tag{6.7}$$

so that

$$\mu(\mathcal{M}) = \int_{\mathcal{M}} d\mu = 1; \qquad \mu(\emptyset) = 0 \tag{6.8}$$

and

$$\mu(E_1 \cup E_2) = \mu(E_1) + \mu(E_2) - \mu(E_1 \cap E_2) \tag{6.9}$$

One also understands that this structure implicitly assumes that $\emptyset$, $\mathcal{M}$, unions and intersections of elements of $\sigma(\mathcal{M})$ must belong to $\sigma(\mathcal{M})$. Thus, a probability in the phase space looks like the "*mass*" of subsets of the phase space. However, paradoxes arise unless some restriction is imposed on the collection $\sigma(\mathcal{M})$. The most common choice is that $\sigma(\mathcal{M})$ be a σ-algebra, i.e., closed under countable

---

[1] This is not necessarily true in small systems, in which local thermodynamic equilibrium is not verified, hence thermodynamic relations may be inapplicable.

[2] Unless N is relatively small.

unions and complement. In other words, given a set $\{E_\ell\}_{\ell=1}^{\infty}$ of sets $E_\ell \in \sigma(\mathscr{M})$, for all $\ell \in \mathbb{N}$, one has

$$\bigcup_{\ell=1}^{\infty} E_\ell \in \sigma(\mathscr{M}); \quad E_\ell^C \in \sigma(\mathscr{M}) \tag{6.10}$$

The sets belonging to $\sigma(\mathscr{M})$ are called *measurable* sets, to distinguish them from those that are not taken into consideration. Given $\mathscr{M}$, there are many different σ-algebras of measurable sets that are chosen depending on the needs. Although this method solves numerous mathematical problems, it introduces a sophisticated structure whose impact on the definition of physical measurements will have to be assessed in applications. In particular, this structure defines a so-called probability measure, which corresponds to the extremely successful Kolmogorov's axiomatic approach to probability.

The set $E \in \sigma(\mathscr{M})$ may be called an *event*, in the sense that finding the object of interest with phase $x \in E$ is given probability $\mu(E)$. In a frequentist approach, this also represents the frequency with which the system can be found in E.

An equivalent and physically more revealing way to define a probability measure μ on $\mathscr{M}$ refers to the set of observables. One selects a class $\mathscr{F}$ of functions on $\mathscr{M}$, and associates their time averages with the action of the measure imposing the validity of Eq. (6.6) as the definition of μ. As for measurable sets, also in this case mathematical consistency requires an informed choice: that of the functions of interest. In particular, one may start from the consideration that *indicator functions* of subsets of $\mathscr{M}$, (cf. Appendix C), defined by

$$\chi_E(x) = \begin{cases} 1 & \text{if } \; x \in E \\ 0 & \text{else} \end{cases} \tag{6.11}$$

and their linear combinations known as simple functions are particularly interesting, because they are *"dense"* in the space of Lebesgue integrable functions. Therefore, they can approximate as closely as one likes, in the L^1-norm, any other such function. Moreover, they yield

$$\mu(E) = \int_E \mathrm{d}\mu(x) = \int_{\mathscr{M}} \chi_E(x)\,\mathrm{d}\mu(x) \tag{6.12}$$

which explains why the definition of a probability measure on $\mathscr{M}$ in terms of probabilities attributed to sets of a σ-algebra, or in terms of averages attributed to the elements of a given function space are mathematically equivalent. However, the phase space and its subsets, which at times are very strange indeed, are quite abstract concepts, while the result of a measurement is of immediate physical relevance. Therefore, it is useful to think of probabilities as defined by the averages of the functions that belong to a relevant function space $\mathscr{F}$ on $\mathscr{M}$. One way or another, a probability measure is not defined by itself: it is defined by the set of values $\{\mathbb{E}(\mathscr{O})\}_{\mathscr{O}\in\mathscr{F}}$ it attributes to certain functions, which are called averages of those functions. Other notions of probability on a phase space sooner or later lead to contradictions. Physically, this is just right; since observers have no access to phase space events, they may only rely on measurements.

As the dynamics move the points of $\mathcal{M}$, in a sense, it moves the sets $E \subset \mathcal{M}$ as well. Therefore one may say that the measure of E is the measure of the set of points that a time t earlier were somewhere else. This set is called preimage of E under the transformation Φ^t, and is denoted by

$$\Phi^{-t} E = \left\{ X \in \mathcal{M} : \Phi^t X \in E \right\} \tag{6.13}$$

Given μ, one may construct another measure, ν, using the dynamics as follows:

$$\nu(E) = \int_E \mathrm{d}\nu = \int_{\Phi^{-t} E} \mathrm{d}\mu = \mu(\Phi^{-t} E) \tag{6.14}$$

There is no reason to expect $\mu(E) = \nu(E)$ or $\mu(\Phi^t E) = \mu(E)$, therefore, one generically obtains $\nu \neq \mu$. While suggestive, this reasoning should be taken with care, because the points of phase space actually do not move! For example, consider the solution

$$\Phi^t x = e^{-t} x \,, \quad x \in \mathbb{R} \tag{6.15}$$

of the dynamical system on $\mathbb{R}$ defined by

$$\dot{x} = -x$$

Let $x = 1$ and $t = -\ln 10$. Then $\Phi^t x = 10$, but neither 1 nor 10, nor any other point in the real line has gone anywhere leaving a hole behind: they all remain where have always been: 1 at unitary distance from the origin to the left; 10 at a distance of 10 units from the origin, also to the left. What is actually meant by Eq. (6.15) is that there is some property, e.g., a sort of mass, that at time 0 is in 1, and that moves in the real line as prescribed by Eq. (6.15).

Leaving aside issues concerning the interpretations of probabilities, one may imagine that they belong to subsets of $\mathcal{M}$, and that they are conserved and transported in phase space, like the mass of fluids within given volumes is conserved in time and is transported in real space. Keeping in mind that the abstract concept of probability in the abstract $2dN$ phase space is obviously immensely different from mass, a concrete property of real objects that moves in the familiar three-dimensional space in which humans also move, one may advance the following definition of probability evolution in phase space:

$$\int_E d\mu_t = \mu_t(E) = \mu_0 \left(\Phi^{-t} E\right) = \int_{\Phi^{-t} E} d\mu_0 \,, \quad \text{for all} \quad E \in \sigma(\mathcal{M}) \tag{6.16}$$

which is equivalent to

$$\mathbb{E}_t \left(\chi_E\right) = \int_{\mathcal{M}} \chi_E \, \mathrm{d}\mu_t = \int_{\mathcal{M}} \chi_{\Phi^{-t} E} \, \mathrm{d}\mu_0 = \mathbb{E}_0(\chi_{\Phi^{-t} E}) \tag{6.17}$$

where the index on both the symbol of phase space average and of probability measure denotes the time at which they are considered. For the phase space averages of the functions of $\mathscr{F}$, in general, one has

$$\mathbb{E}_t(\mathscr{O}) = \int_{\mathscr{M}} \mathscr{O}\, \mathrm{d}\mu_t \neq \int_{\mathscr{M}} \mathscr{O}\, \mathrm{d}\mu_0 = \mathbb{E}_0(\mathscr{O}) \tag{6.18}$$

which can be viewed as a definition of the evolution up to time t of the observable associated with $\mathscr{O}$, hence of the initial probability measure μ_0 on $\mathscr{M}$. A stationary state can then be defined as that in which the following holds:

$$\mathbb{E}_t(\mathscr{O}) = \mathbb{E}_0(\mathscr{O}), \quad \text{for every } \mathscr{O} \in \mathscr{F}, \quad \text{and every} \quad t \in \mathbb{R} \tag{6.19}$$

Using Eqs. (6.18) and (6.19), and taking $\mathscr{O} = \chi_E$, for any $E \in \sigma(\mathscr{M})$, this implies

$$\mathbb{E}_0(\chi_E) = \mathbb{E}_t(\chi_E) = \mathbb{E}_0(\chi_{\Phi^{-t}E}) \tag{6.20}$$

which is to say

$$\mu_0(E) = \mu_0(\Phi^{-t}E), \quad \forall E \in \sigma(\mathscr{M}) \tag{6.21}$$

When this is the case, μ_0 is called invariant for the dynamics Φ^t. Clearly, measures defined by Eq. (6.6) are invariant, because the dependence on t has been eliminated by the $t \to \infty$ limit.

Although it is a very special case, it may happen that a probability measure μ_t has a *"probability density"* f_t, which means that it is expressed by an integrable non-negative function $f_t : \mathscr{M} \to [0, \infty)$ such that

$$\mathrm{d}\mu_t(x) = f_t(x)\,\mathrm{d}x, \quad \int_{\mathscr{M}} f_t(x)\,\mathrm{d}x = 1 \tag{6.22}$$

This is the case, for instance, of the celebrated microcanonical, canonical, and grand canonical ensembles, on which the success of equilibrium statistical mechanics is based. In this case, μ_t is called *regular*, or absolutely continuous with respect to the Lebesgue measure. If such a density does not exist, μ_t is called singular (with respect to the Lebesgue measure). If μ_t evolves in time, f_t does. In particular, Eq. (6.16) is now written as

$$\int_E f_t(x)\,\mathrm{d}x = \mu_t(E) = \mu_0\left(\Phi^{-t}E\right) = \int_{\Phi^{-t}E} f_0(x)\,\mathrm{d}x \tag{6.23}$$

The Jacobian of the coordinates change $x = \Phi^{-t}y$ is here denoted by

$$J^{-t}(y) = \left|\frac{\partial \Phi^{-t}}{\partial y}\right|(y), \quad \text{so that} \quad \mathrm{d}x = J^{-t}(y)\,\mathrm{d}y \tag{6.24}$$

and

$$\int_E f_t(x)dx = \int_{\Phi^{-t}E} f_0(x)\,\mathrm{d}x = \int_E f_0(\Phi^{-t}y)J^{-t}(y)\,\mathrm{d}y \tag{6.25}$$

Provided this holds for all sets of the chosen σ-algebra, one can write

$$f_t(x) = f_0(\Phi^{-t}x)J^{-t}(x) \tag{6.26}$$

Then, in a stationary state one may write $f_t = f_0 = f$, and

$$f(x) = f(\Phi^{-t}x)J^{-t}(x) \tag{6.27}$$

meaning that the density varies along the points of a trajectory in order to balance the compression or expansion of phase space volumes. If the dynamics is Hamiltonian, Liouville's theorem states that

$$J^{-t}(x) = 1\,, \quad \forall x \in \mathcal{M} \tag{6.28}$$

so that

$$f_t(x) = f_0(\Phi^{-t}x) \tag{6.29}$$

showing that the density evolving in phase space is nevertheless constant along a trajectory, analogously to the evolution of an incompressible fluid in real space. Phase space densities may change over time, but not along the trajectory of any such "fluid" phase element. If the dynamics is Hamiltonian and f is stationary for it, one gets

$$f(x) = f(\Phi^{-t}x) \tag{6.30}$$

In other words, the trajectories are level curves for f.

The evolution of the probability density affects the average values of phase functions computed with it. In practice, one gets

$$\mathbb{E}_t(\mathcal{O}) = \int_{\mathcal{M}} \mathcal{O}(x)f_t(x)\,\mathrm{d}x = \int_{\mathcal{M}} \mathcal{O}(x)f_0(\Phi^{-t}x)J^{-t}(x)\,\mathrm{d}x \tag{6.31}$$

which, introducing y as $x = \Phi^t y$, yields

$$\mathbb{E}_t(\mathcal{O}) = \int_{\mathcal{M}} \mathcal{O}(\Phi^t y)f_0(y)J^{-t}(\Phi^t y)J^t(y)\,\mathrm{d}y \tag{6.32}$$

Clearly, when volumes expand/contract by a factor $J^t(y)$ along a trajectory starting at y and ending in $\Phi^t y$, they contract/expand by the same factor in the reverse motion proceeding from $\Phi^t y$ for the time t, which leads to y. One then obtains

$$J^{-t}(\Phi^t y) = \frac{1}{J^t(y)} \tag{6.33}$$

in perfect analogy with the evolution of compressible fluids, mirrored by our treatment of probabilities in phase space. As a consequence:

$$\mathbb{E}_t(\mathscr{O}) = \int_{\mathscr{M}} \mathscr{O}(\Phi^t x) f_0(x)\, \mathrm{d}x = \mathbb{E}_0(\mathscr{O} \circ \Phi^t) \tag{6.34}$$

which shows that evolution of observables con be equivalently obtained either evolving the microscopic observable or the probability density. This is reminiscent of the equivalence of Schrödinger to Heisenberg picture of quantum mechanics. The next section is devoted to the derivation of an explicit expression for J^t, thus also vindicating Eq. (6.28).

Before proceeding with the calculations, it is useful to ponder. While mathematically consistent, the above requires more than physics commonly needs, because physics is typically satisfied with a handful of microscopic properties and associated macroscopic observables. Moreover, the frequentist notion that has justified our use of probabilities so far is now lost: as mentioned above, that notion of probabilities is limited to stationary states, both mathematically, because of the $t \to \infty$ limit, and physically, because a measurement performed while the state of the systems shifts yields an average of the properties of the different transient states, rather than one characterizing a given state. To make sense of Eq. (6.34) in a time-dependent setting, one needs to invoke a vast separation of time scales, so that what appears to be an instant at the level of observations still corresponds to very many microscopic time units. This requires the object of interest to consist of very many elementary constituents that properly interact with each other much more rapidly than the response times of the measurement tools. Although fine with thermodynamics, whose scales are much larger than the microscopic scales, this is not so in general.

6.2 Solution of Liouville's Equation

Proceeding with the formal analogy between probability in the phase space of a classical particle system and the mass of a fluid in real space, one can derive the phase space continuity equation for such a peculiar $2dN$-dimensional compressible fluid, whose density f is the probability density on $\mathscr{M}$. Let

$$\dot{\Gamma} = G(\Gamma)\,; \quad \Gamma = (Q, P) \in \mathscr{M} \tag{6.35}$$

represent the equations of motion, with Q the set of configuration coordinates and P the corresponding momentum coordinates. Under the same regularity assumptions that led to the continuity equation for a fluid in three dimensions, the phase space continuity equation in $\mathscr{M}$ may be written as

$$\frac{\partial f}{\partial t} = -\mathrm{div}_\Gamma(f\dot{\Gamma}) = -\mathrm{div}_\Gamma(fG) = -G \cdot \nabla_\Gamma f - f\, \mathrm{div}_\Gamma G \tag{6.36}$$

which is called *Liouville's equation*. Here, div_Γ denotes the divergence in $\mathscr{M}$, and ∇_Γ the gradient in $\mathscr{M}$. Using the total derivative, one can write

$$\frac{df}{dt} = \frac{\partial f}{\partial t} + G \cdot \nabla_\Gamma f = -f \, \mathrm{div}_\Gamma G = -f \Lambda \tag{6.37}$$

where $\Lambda = \mathrm{div}_\Gamma G$ is the phase space variation rate, that amounts to volumes expansion, when positive, and to volumes contraction when negative. In certain studies, in which the average variation is expected to be negative, its opposite, $-\Lambda$, is called phase space contraction rate. Equation (6.37) may also be written as

$$\frac{d}{dt} \ln f = -\Lambda \tag{6.38}$$

For Hamiltonian systems, one has

$$\Lambda(\Gamma) = \sum_{i=1}^{dN} \left[\frac{\partial}{\partial q_i} \dot{q}_i + \frac{\partial}{\partial p_i} \dot{p}_i \right] = \sum_{i=1}^{dN} \left[\frac{\partial}{\partial q_i} \frac{\partial H}{\partial p_i} - \frac{\partial}{\partial p_i} \frac{\partial H}{\partial q_i} \right] = 0 \tag{6.39}$$

Hamiltonian dynamics is thus a sufficient (not necessary) condition for the validity of the *Liouville's theorem*, i.e.:

$$\frac{\mathrm{d}f}{\mathrm{d}t} = 0 \tag{6.40}$$

Note that Hamiltonian dynamics means forces are conservative, hence a potential function exists, the Schwartz theorem on cross derivatives holds allowing derivatives to be swapped, and the expressions in square brackets in Eq. (6.39) yield 0.

A suggestive formalism, for the solution of Liouville's equation is based on the Liouvillian operator:

$$\mathscr{L} = -i \left(\mathrm{div}_\Gamma \, G + G \cdot \nabla_\Gamma \right) \tag{6.41}$$

and write

$$\frac{\partial f}{\partial t} = -i \mathscr{L} f \tag{6.42}$$

whose solution for time-independent $\mathscr{L}$ is formally given by

$$f_t(\Gamma) = e^{-i\mathscr{L}t} f_0(\Gamma) \,, \qquad \Gamma \in \mathscr{M} \tag{6.43}$$

More explicitly, the solution for a given initial condition $\Gamma \in \mathscr{M}$, hence the evolution operator, can be constructed starting from a discretization of Eq. (6.42) as

$$\frac{f_{t+\Delta t}(\Gamma) - f_t(\Gamma)}{\Delta t} = -i \mathscr{L} f_t(\Gamma) + O(\Delta t) \tag{6.44}$$

using the definition and the existence of derivatives, and the consequent smoothness of the solution, here assumed to exist globally. This yields

$$f_{\Delta t}(\Gamma) = (1 - i \mathscr{L} \Delta t) f_0(\Gamma) + O(\Delta t^2) \tag{6.45}$$

and

$$f_{2\Delta t} = (1 - i\mathscr{L}\Delta t) f_{\Delta t}(\Gamma) + O(\Delta t^2) = (1 - i\mathscr{L}\Delta t)\Big[(1 - i\mathscr{L}\Delta t) f_0(\Gamma) + O(\Delta t^2)\Big] + O(\Delta t^2)$$
$$= (1 - i\mathscr{L}\Delta t)^2 f_0(\Gamma) + 2\, O(\Delta t^2) \tag{6.46}$$

Then, iterating n times, one obtains

$$f_{n\Delta t}(\Gamma) = (1 - i\mathscr{L}\Delta t)^n f_0(\Gamma) + n\, O(\Delta t^2) \tag{6.47}$$

Now, fix t, take $\Delta t = t/n$ and let $n \to \infty$, so that $O(n\Delta t^2) = O(t^2/n) \to 0$. The result is

$$f_t(\Gamma) = \lim_{n\to\infty} \left(1 - \frac{i\mathscr{L}t}{n}\right)^n f_0(\Gamma) \equiv e^{-i\mathscr{L}t} f_0(\Gamma) \tag{6.48}$$

which defines the propagator $e^{-i\mathscr{L}t}$, in its Eulerian form, solving Eq. (6.36). Liouville's equation in Lagrangian form (6.38) can instead be solved integrating along a trajectory starting at Γ and ending at $\Phi^t\Gamma$, obtaining

$$\int_{f_0(\Gamma)}^{f_t(\Phi^t\Gamma)} \mathrm{d}\ln f = \ln f_t(\Phi^t\Gamma) - \ln f_0(\Gamma) = -\int_0^t \Lambda(\Phi^s\Gamma)\,\mathrm{d}s \tag{6.49}$$

and can be rewritten as

$$f_t(\Phi^t\Gamma) = f_0(\Gamma)e^{-\int_0^t \Lambda(\Phi^s\Gamma)ds}\,, \quad \text{or} \quad f_t(\Gamma) = f_0(\Phi^{-t}\Gamma)e^{-\int_0^t \Lambda(\Phi^{s-t}\Gamma)ds} \tag{6.50}$$

Comparing with Eq. (6.26) one identifies

$$J^{-t}(\Gamma) = e^{-\int_0^t \Lambda(\Phi^{s-t}\Gamma)ds} \tag{6.51}$$

and replacing t with $-t$ yields

$$J^t(\Gamma) = e^{-\int_0^{-t} \Lambda(\Phi^{s+t}\Gamma)ds} = e^{\int_{-t}^0 \Lambda(\Phi^s(\Phi^t\Gamma))ds} = \frac{1}{J^{-t}(\Phi^t\Gamma)} \tag{6.52}$$

Recently, a particularly useful method of writing and solving the Eulerian version of Liouville's equation (6.36) has been proposed. Writing (6.36) in the following form:

$$\frac{\partial f}{\partial t} = -f\left[\Lambda + G\cdot\frac{1}{f}\nabla_\Gamma f\right] = f(\Gamma)\,\Omega^f(\Gamma) \tag{6.53}$$

which introduces the *dissipation function* Ω^f: a scalar function of phase for the given probability density f that is defined by

$$\Omega^f(\Gamma) = -\left[\Lambda(\Gamma) + G(\Gamma)\cdot\nabla_\Gamma \ln f(\Gamma)\right] \tag{6.54}$$

The solution of Liouville's equation can then written as

$$f_t(\Gamma) = \exp\left\{\int_{-t}^{0} \Omega^{f_0}(\Phi^s \Gamma)\, \mathrm{d}s\right\} f_0(\Gamma) \tag{6.55}$$

see Appendix H for some properties of Ω^f. In the case the dynamics represent an equilibrium system with equilibrium probability density f_0, so that f_0 does not change in time, the dissipation function identically vanishes and, conversely, $\Omega^{f_0} \equiv 0$ implies that f_0 is an invariant density. Note that the solution (6.55) of the Eulerian version of Liouville's equation is mathematically much handier than Eq. (6.50). However, its real importance rests on the fact that the dissipation function, under usual conditions, represents currents measurable on a single concrete object. In contrast, the phase space volume variation rate Λ is an abstract phase space quantity concerning ensembles of systems. This will be explained in Chapter 9, where one will also see that the evolution of phase functions, Eq. (6.31), takes a suggestive form in terms of Ω^f:

$$\mathbb{E}_t(\mathscr{O}) = \mathbb{E}_0(\mathscr{O}) + \int_0^t \mathbb{E}_0\left[(\mathscr{O} \circ \Phi^\tau)\Omega^{f_0}\right] d\tau \tag{6.56}$$

6.3 Ergodicity and Mixing

The investigation of the validity of Eq. (6.6) has led to the development of ergodic theory, concerning the properties of dynamical systems and their invariant measures. Born to solve the physical problem of linking the atomic description to the measurable thermodynamic properties of macroscopic objects in equilibrium, or evolving toward equilibrium, ergodic theory developed independently of physics as a most elegant mathematical theory. In its standard formulation, this mathematical theory states that there are four equivalent properties called *ergodicity*, including the validity of Eq. (6.6) Collet and Eckmann (2006). Let the triple $(\mathscr{M}, \Phi, \mu)$ be a dynamical system with μ an invariant measure under the dynamics Φ defined on the phase space $\mathscr{M}$, the following properties are equivalent:

(E1) For every integrable function $\mathscr{O} : \mathscr{M} \mapsto \mathbb{R}$, time average and phase space average coincide for μ-almost every x, which means for all $x \in \mathscr{M}$ apart from a set E with $\mu(E) = 0$, this writes

$$\overline{\mathscr{O}}(x) = \mathbb{E}_\mu(\mathscr{O}), \quad \mu - \text{a.e. } x \in \mathscr{M} \tag{6.57}$$

(E2) For every measurable $A \subset \mathscr{M}$ and for μ-a.e. $x \in \mathscr{M}$, the fraction of time a trajectory spends in A is given by

$$\tau_A(x) = \mu(A), \quad \mu - \text{a.e. } x \in \mathscr{M} \tag{6.58}$$

where the fraction of time is defined by

$$\tau_A(x) = \lim_{t\to\infty} \frac{1}{t} \int_0^t \chi_A(\Phi^s x)\, ds \tag{6.59}$$

(E3) There are no non-trivial integrals of motion. In other words, let $\mathcal{O}$ be a function of phase, such that $\mathcal{O}(\Phi^t x) = \mathcal{O}(x)$, for all t and for μ-a.e. $x \in \mathcal{M}$, and let $\mathcal{O}$ be integrable. Then,

$$\mathcal{O}(x) = K_{\mathcal{O}} = \text{constant}\,, \quad \mu - \text{a.e. } x \in \mathcal{M} \tag{6.60}$$

(E4) The dynamical system is metrically indecomposable. In other words, let A be an invariant measurable set, i.e., $\Phi^t A = A$ for all t. Then, either $\mu(A) = 1$ or $\mu(A) = 0$. When that is the case, the expression $\mathcal{M} = A \cup (\mathcal{M}\backslash A)$ is called metrically trivial decomposition of $\mathcal{M}$.

A detailed derivation of Birkhoff's ergodic theorem is given in Appendix B. For our purposes, the four above conditions can essentially be described as follows:

- E1 addresses the issue of physical measurements as time averages, Eq. (6.6). Its validity avoids recourse to the (not revealing, and for macroscopic objects even impossible) integration of the microscopic dynamics, thanks to phase space integration with respect to a supposedly simple probability distribution. Thus, the main purpose of the theory is accomplished.
- E2 expresses Boltzmann's expectation that one observation finds the (single) system of interest in a given region $A \subset \mathcal{M}$ with frequency proportional to the volume of that region. Boltzmann considered indeed Hamiltonian particle systems, whose natural invariant measure is the normalized phase space volume, which may also be understood as equal *a priori* probability. This has an immediate remarkable consequence: every measurable open set in $\mathcal{M}$ having positive volume will be visited with positive frequency. In other words, trajectories of dynamical systems obeying E2 will densely explore the whole phase space in time. For the very high-dimensional phase space of many particle systems, this gives a glimpse of how strong (hence restrictive) the ergodicity condition is and, in any case, how exceedingly long such an exploration would take.
- E3 implies the uniqueness of the equilibrium state for Hamiltonian particle systems if that is expressed by a probability density in the sense of the classical ensembles. Let ρ_t be such a density at time t. Our assumption of conservation of probability and the conservation of phase space volumes implies that $\rho_t(x) = \rho_0(\Phi^{-t}x)$, cf. Eq. (6.29). Then, to have $\rho_t = \rho_0$ for every t, ρ_t must be uniform in $\mathcal{M}$. Note: This uniqueness does not imply that there are no other invariant probability distributions; it merely says that they cannot be absolutely continuous with respect to the Lebesgue measure. Moreover, the uniqueness does not imply convergence (relaxation) to it should the initial distribution be non-uniform. As explained in Sect. 6.1.1, convergence of measures, like the measures

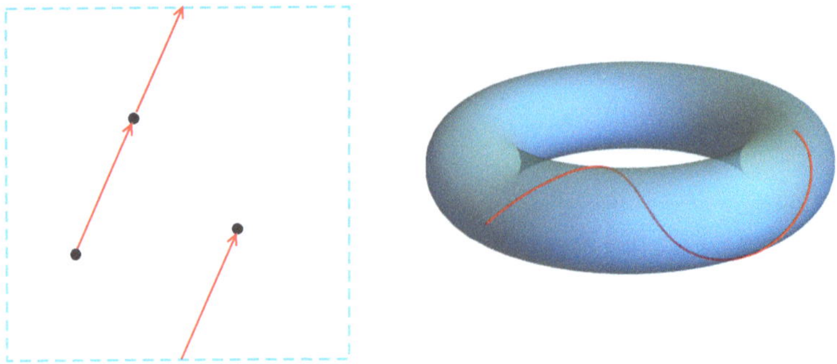

Fig. 6.2 Constant velocity translation on the square with periodic boundary conditions, $\Gamma = (x, y) \in \mathscr{M} = [0, 1] \times [0, 1]$; $\Phi^t \Gamma = (x + v_x t, y + v_y t)$ mod 1. If v_x/v_y is irrational, each trajectory explores densely $\mathscr{M}$. The unique invariant density is $\rho(\Gamma) = 1$, i.e., the uniform distribution, $d\mu = dxdy$. Ensembles of Γ's, here represented by small disks, do not spread to cover uniformly $\mathscr{M}$. Regarding phase space as analogous to real space, and ensemble members (phase points) as analogous to molecules of a gas, ergodicity does not imply relaxation in time toward a uniform distribution in a container. Apparently, more than ergodicity is needed. However, phase space is not analogous to real space and phase points are not analogous to molecules: different Γ in $\mathscr{M}$ do not represent different particles, but only different phases of the same system

themselves, must be expressed in terms of observables. Thus, it only makes sense to consider the *weak* convergence of a sequence of measures, which is defined by

$$\mu_t \xrightarrow{W} \mu_\infty \quad \text{if} \quad \lim_{t\to\infty} \int_{\mathscr{M}} \mathscr{O} d\mu_t = \int_{\mathscr{M}} \mathscr{O} d\mu_\infty \,, \quad \forall \mathscr{O} \in \mathscr{F} \tag{6.61}$$

- E4 is very elegant and concisely geometrically expressed, which is useful in the rigorous investigation of dynamical systems.

It is rather simple to prove that these properties are equivalent to each other. Therefore, the validity of one of them implies the validity of all the others, Mañe (1987).

It is easy to see that ergodicity is realized in simple dynamical systems whose phase space trajectories move parallel to each other and with the same phase space "velocity," i.e., same G, in the notation of Eq. (6.35). Take, for instance, a two-dimensional torus as $\mathscr{M}$, and let trajectories be oriented so that they do not close on a periodic orbit (cf. Fig. 6.2).[3] This is but an example of ergodic system whose trajectories remain close to each other in time, while they explore the whole phase space, thus satisfying E2. Now, if the equilibrium distribution is the so-called microcanonical ensemble, which for an ideal gas is the uniform distribution in phase space, and if the initial distribution is not like that, the ergodic property does not lead in time to the equilibrium distribution. A lump of probability continues to go round and round in $\mathscr{M}$, never spreading to become uniform. This will be certified by the averages of observables that will correspondingly fluctuate in time, as the support of this evolving probability moves within $\mathscr{M}$.

[3] Cutting the torus to obtain a square, with periodic boundary conditions, it suffices that the components of G be not rational with respect to each other.

One could take this fact as indicating that ergodicity is not sufficient to describe the *irreversible* relaxation of macroscopic states to their equilibrium. Therefore, one could think that a stronger property than ergodicity is needed to describe the most common irreversible phenomena, see, e.g., Reichl (1998). In particular, the condition of *mixing* seems appropriate, because it makes the initial probability densities to spread in the phase space of the volume conserving dynamics, until they uniformly cover the phase space. It can be defined by two equivalent properties:

(M1) For every pair of measurable sets in the phase space, $A, B \subset \mathscr{M}$, the following holds:

$$\lim_{t\to\infty} \mu(\Phi^{-t} A \cap B) = \mu(A)\mu(B)\,, \quad \text{or} \quad \lim_{t\to\infty} \mu(A \cap \Phi^{t} B) = \mu(A)\mu(B) \tag{6.62}$$

(M2) For every pair of square integrable observables, with respect to $d\mu$, e.g., $\mathscr{O}, \mathscr{P} \in L^2$ the following holds:

$$\lim_{t\to\infty} \int_{\mathscr{M}} (\mathscr{O} \circ \Phi^t)\mathscr{P} d\mu = \int_{\mathscr{M}} \mathscr{O} d\mu \int_{\mathscr{M}} \mathscr{P} d\mu\,, \tag{6.63}$$

i.e. $\lim_{t\to\infty} \mathbb{E}_\mu \left[(\mathscr{O} \circ \Phi^t)\mathscr{P}\right] = \mathbb{E}_\mu(\mathscr{O})\mathbb{E}_\mu(\mathscr{P})$

The meaning of $M1$ is that in time $\Phi^t B$ will occupy a fraction of A equal to its measure in the full $\mathscr{M}$. As A can be taken in any region of phase space, in particular, within the support of the invariant measure under consideration, this means that an initial ensemble supported in B will reach in time and will stay in any other corner of the support of μ. When the microcanonical ensemble is the relevant invariant measure, this means that the probability "contained" in B will eventually be uniformly distributed over the whole $\mathscr{M}$. This can be seen observing that the invariance of μ implies $\mu(A \cap \Phi^t B) = \mu(\Phi^{-t}(A \cap \Phi^t B)) = \mu(\Phi^{-t} A \cap B)$, and $\mu(A)\mu(\Phi^t B) = \mu(A)\mu(B)$ Then, for sufficiently long times, one may approximately write

$$\frac{\mu\left(A \cap \Phi^t B\right)}{\mu(A)} \approx \mu(B)\,, \quad \text{provided} \quad \mu(A) \neq 0 \tag{6.64}$$

which becomes exact in the $t \to \infty$ limit. The fact that single trajectories of mixing systems invade the support of the invariant measure proves that mixing implies ergodicity; the fact that lumps of ergodic trajectories do not need to invade the phase space proves that ergodicity does not imply mixing. Therefore, mixing is an effectively stronger property than ergodicity. A mixing counterpart to the uniform translation on the two-dimensional torus is the so-called Arnold's CAT,[4] defined by Arnold and Avez (1968)

[4] CAT stands for Canonical (it preserves area and orientation) Automorphism of the Torus.

$$\begin{pmatrix} x' \\ y' \end{pmatrix} = \begin{pmatrix} 1 & 1 \\ 1 & 2 \end{pmatrix} \begin{pmatrix} x \\ y \end{pmatrix} \mod 1 \tag{6.65}$$

It is instructive to examine one proof of the equivalence of properties M1 and M2. Let us begin with M2 $\Rightarrow$ M1, taking the indicator functions $\mathscr{O} = \chi_A$ and $\mathscr{P} = \chi_B$, which are L^2 functions. Observing that $\chi_A \circ \Phi^t = \chi_{\Phi^{-t}A}$, one can write

$$\int_{\mathscr{M}} (\mathscr{O} \circ \Phi^t) \mathscr{P} d\mu = \int_{\mathscr{M}} \chi_{\Phi^{-t}A} \mathscr{P} d\mu = \mu(\Phi^{-t} A \cap B) \tag{6.66}$$

and

$$\int_{\mathscr{M}} \chi_A d\mu = \mu(A), \quad \int_{\mathscr{M}} \chi_B d\mu = \mu(B) \tag{6.67}$$

which directly imply the validity of Eq. (6.62). To prove M1$\Rightarrow$M2, consider first two simple functions of phase:

$$\mathscr{O} = \sum_i \mathscr{O}_i \chi_{A_i}, \quad \mathscr{P} = \sum_j \mathscr{P}_j \chi_{B_j} \tag{6.68}$$

where $\mathscr{O}_i$ and $\mathscr{P}_j$ are constant coefficients. Because $\chi_{A_i} \circ \Phi^t = \chi_{\Phi^{-t}A_i}$, one can write

$$\mathbb{E}_\mu \left[(\mathscr{O} \circ \Phi^t) \mathscr{P} \right] = \sum_{i,j} \mathscr{O}_i \mathscr{P}_j \mathbb{E}_\mu \left[\chi_{\Phi^{-t}A_i} \chi_{B_j} \right] = \sum_{i,j} \mathscr{O}_i \mathscr{P}_j \mu(\Phi^{-t} A_i \cap B_j) \tag{6.69}$$

which converges in time to

$$\sum_{i,j} \mathscr{O}_i \mathscr{P}_j \mu(A_i) \mu(B_j) = \sum_{i,j} \mathscr{O}_i \mathscr{P}_j \mathbb{E}_\mu \left(\chi_{A_i} \right) \mathbb{E}_\mu \left(\chi_{B_j} \right) = \mathbb{E}_\mu \left(\mathscr{O} \right) \mathbb{E}_\mu \left(\mathscr{P} \right) \tag{6.70}$$

Recalling that simple functions are dense in L^2 we have obtained the result.

Property M2, in particular, expresses the decay of correlations of microscopic events in phase space. When the correlation function of $\mathscr{O}$ and $\mathscr{P}$, defined by

$$G(t) = \mathbb{E}_\mu \left[(\mathscr{O} \circ \Phi^t) \mathscr{P} \right] - \mathbb{E}_\mu \left(\mathscr{O} \right) \mathbb{E}_\mu \left(\mathscr{P} \right) \tag{6.71}$$

does not vanish, the measurements of $\mathscr{O}$ and $\mathscr{P}$ are in a sense not independent, while the decay in time of such correlation indicates a sort of irreversible loss of memory about the initial state. Indeed, a well-known back of the envelope calculation appears to confirm that the goal of describing relaxation to an equilibrium state, for a Hamiltonian particle system, is achieved when mixing holds Sinai (1976). The calculation is the following:

$$\begin{aligned} \mathbb{E}_{\mu_t} \left(\mathscr{O} \right) &= \int_{\mathscr{M}} \mathscr{O} \rho_t \, \mathrm{d}x = \int_{\mathscr{M}} \mathscr{O} (\rho_0 \circ \Phi^{-t}) \, \mathrm{d}x \\ &= \int_{\mathscr{M}} (\mathscr{O} \circ \Phi^t) \rho_0 \, \mathrm{d}x \overset{t \to \infty}{\longrightarrow} \int_{\mathscr{M}} \mathscr{O} \, \mathrm{d}x \int_{\mathscr{M}} \rho_0 \, \mathrm{d}x = \mathbb{E}_{\mathrm{d}x} \left(\mathscr{O} \right) \end{aligned} \tag{6.72}$$

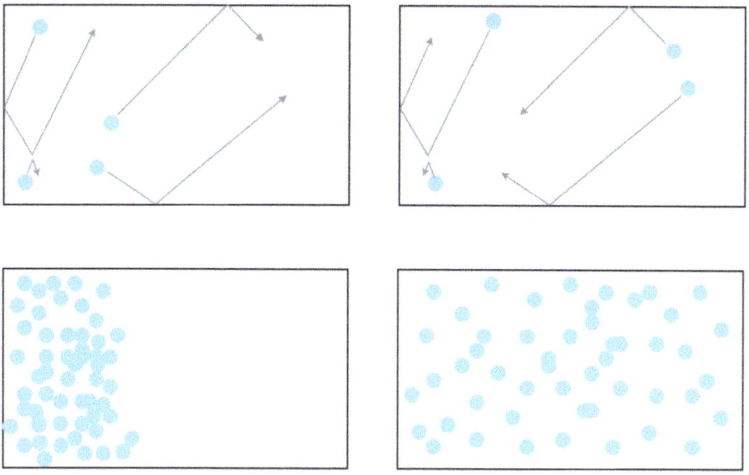

Fig. 6.3 Mixing is verified for two hard disks in a two-dimensional box, whose phase space is eight-dimensional (top panels). An initial non-uniform probability density in this phase space irreversibly converges in time to the uniform probability density. However, no direction of time is evidenced. Looking at the two figures one cannot decide which precedes the other. The two particles have no notion of uniformity in their two-dimensional real space. Different is the case of an $N = 11$ particles system (bottom panels). Not only a uniform probability distribution is established in time in its 44-dimensional (!!) phase space, but a kind of uniform mass distribution in the two-dimensional real space begins to make sense. Larger N will make the convergence to the uniform mass distribution, hence the direction of time, more evident. Mass and probability are not identified. A uniform probability distribution in phase space does not imply in general, a uniform density in real space, like non-uniform probability distributions may include a large fraction of uniform mass distributions in real space. Identification of mass and probability is not a trivial step, it requires suitable conditions

proving that μ_t weakly converges to μ_∞. Here, the dynamics is assumed to be phase space volume preserving, so that the Jacobian determinant denoted by J^{-t} equals 1. Then, $\rho_0 =$ constant is the invariant probability density (its integral equals 1), and mixing has been applied considering the probability density as a second observable $\mathscr{P} : \mathscr{M} \to \mathbb{R}$. The result is that the initial value of the observable irreversibly converges to its microcanonical value.

Despite the mathematical consistency, difficulties arise with taking probabilities in phase space as not only analogous to fluids in real space, but as if their behavior immediately implies physical consequences. Arguably the most useful mathematical tool in physics and many other fields, probabilities remain *immaterial* abstract concepts. Indeed, the ergodic construction presented above, while quite informative, allows for many paradoxes. For instance, mixing is a property of the invariant state, the one to which the time evolution is expected to converge. It cannot really speak of previous transient states: it concerns a property which is the decay of microscopic correlations *within* the steady state. Mixing also holds for systems made of a single particle confined within convex walls, which surely cannot expand to fill its container, cf. Fig. 6.3.

One difficulty with the calculation (6.72) is evidenced by the fact that the probability density ρ_t is treated like a microscopic physical property of a given object. Although it appears to be like any other function of phase, it belongs to the probability measure, and its meaning and usefulness rest on the fact that it is not an observable. One observable, the total kinetic energy say, can be defined for a single concrete object of interest, while a probability density only refers to the abstract notion of ensemble of abstract identical systems. The crowding or otherwise of phase points representing such phases in certain regions of phase space has nothing to do with the crowding of particles and the evolution of a given single system. That is determined by the equations of motion and the corresponding initial condition Γ_0, no matter how "crowded" with other initial conditions the region around Γ_0 is.

This discussion goes beyond the scope of the present book, but it allows us to introduce the question to be investigated in the next section: under which conditions does probability turn in some sense *material*? Statistical mechanics has often successfully operated identifying probability with mass or other physical quantities. In one of his letters to Lewis Campbell, J.C. Maxwell makes quite an interesting point about that:

> As it is Saturday night I will not write very much more. I was thinking today of the duties of [the] cognitive faculty. It is universally admitted that duties are voluntary, and that the will governs understanding by giving or withholding Attention. They say that Understanding ought to work by the rules of right reason. These rules are, or ought to be, contained in Logic; but the actual science of Logic is conversant at present only with things either certain, impossible, or entirely doubtful, none of which (fortunately) we have to reason on. Therefore the true Logic for this world is the Calculus of Probabilities, which takes account of the magnitude of the probability (which is, or which ought to be in a reasonable man's mind). This branch of Math., which is generally thought to favour gambling, dicing, and wagering, and therefore highly immoral, is the only 'Mathematics for Practical Men', as we ought to be Harman (1990).

However, acknowledging that we have no better tool than probability does not imply that probability is the *panacea*. Certain conditions must be satisfied. One gets a glimpse of that in Poincaré's considerations on the laws that govern the large number of molecules constituting a gas:

> It seems at first that the orderless collisions of this innumerable dust can only engender an inextricable chaos before which the analyst must retire. But the law of great numbers, that supreme law of chance, comes to our assistance. In face of a semi-disorder we should be forced to despair, but in extreme disorder this statistical law re-establishes a kind of average or mean order in which the mind can find itself again. Poincaré (1914).

Understanding pros and cons of probabilistic notions helps to proceed in uncharted territories, avoiding despair.

6.4 Boltzmann Equation and the H-Theorem

The dynamical and probabilistic notions introduced above may be used to obtain the macroscopic thermodynamic description. Because it is not possible to decide *a priori* which dynamics are better suited to the task, one almost universally begins with classical Hamiltonian mechanics, because it has proven successful in describing an incredible variety of phenomena occurring at the macroscopic scales. This was Laplace's opinion:

> One may regard the present state of the universe as the effect of its past and the cause of its future. An intellect which at a certain moment would know all forces that set nature in motion, and all positions of all items of which nature is composed, if this intellect were also vast enough to submit these data to analysis, it would embrace in a single formula the movements of the greatest bodies of the universe and those of the tiniest atom; for such an intellect nothing would be uncertain and the future just like the past would be present before its eyes.[5] Laplace (1814).

This is a very bold statement, because experience at Laplace's time was limited to systems made of relatively small numbers of large objects; nothing to do, in principle, with the exceedingly large assemblies of invisibly tiny molecules that constitute a macroscopic body. Thus it ought to be subjected to test. The outcome happened to be that statistical mechanics greatly advanced our understanding of physical phenomena, adopting Hamiltonian dynamics.[6]

Consider a system of N particles, each having d degrees of freedom, characterized by a Hamiltonian $H = H(\Gamma)$, where $\Gamma = (\mathbf{q}, \mathbf{p}) \in \mathcal{M}$, and by the following equations of motion:

$$\dot{\mathbf{q}} = \frac{\partial H}{\partial \mathbf{p}}\,, \quad \dot{\mathbf{p}} = -\frac{\partial H}{\partial \mathbf{q}} \tag{6.73}$$

As already mentioned, the connection with physical measurements begins from the observation that they take a positive time, and from the assumption that they yield the time average of one phase function. If the measurement takes a time τ, and it starts when the microstate is Γ, the result of the measurement is postulated to be

$$\overline{\mathcal{O}}_{0,\tau}(\Gamma) = \frac{1}{\tau}\mathcal{O}_{0,\tau}(\Gamma) = \frac{1}{\tau}\int_0^{\tau} \mathcal{O}(\Phi^t \Gamma)\,\mathrm{d}t \tag{6.74}$$

where by the symbol $\mathcal{O}_{s,t}$ one means the time integral of $\mathcal{O}$ from time s to time t, and a bar over a symbol denotes time averaging.

[5] Note, however, that Laplace made that statement in an essay on probabilities. Indeed, he considered such an intellect out of our reach. This was not the manifesto of reductionism, as often portrayed, but the vindication of probabilities.

[6] Alternatives, like stochastic processes, have also considerably contributed. A topical question thus concerns the link between deterministic and stochastic approaches. This section will shed some light also on this issue.

If one accepts this picture, it follows that measurements depend on τ, hence are subjective because different observers may choose different τ's, and on Γ, hence their result is a random variable, because Γ is unknown. However, this contradicts Thermodynamics, that is an objective and deterministic theory, universally confirmed by experimental tests. One possibility to accord Eq. (6.74) with experience is to assume that τ be very large, virtually infinitely larger than the microscopic scales concerning the evolution of Γ, so that

$$\frac{1}{\tau}\int_0^\tau \mathscr{O}(\Phi^t\Gamma)\,\mathrm{d}t \approx \overline{\mathscr{O}}(\Gamma) = \lim_{\tau\to\infty}\frac{1}{\tau}\int_0^\tau \mathscr{O}(\Phi^t\Gamma)\,\mathrm{d}t \tag{6.75}$$

and that during such a long time, $\mathscr{O}$ has explored many times—with proper frequencies—its range of values making the initial condition Γ irrelevant:

$$\lim_{\tau\to\infty}\frac{1}{\tau}\int_0^\tau \mathscr{O}(\Phi^t\Gamma)\,\mathrm{d}t \approx o \in I\!R \tag{6.76}$$

Mathematically the first approximate equality is due to the fact τ cannot be infinite, and the second to the fact that the range of $\mathscr{O}$ may not be perfectly explored. However, as long as the approximations fall below the scale of thermodynamic interest, equality can be used, and o represents the result of a measurement. This picture is justified by Fermi as follows:

> Studying the thermodynamical state of a homogeneous fluid of given volume at given temperature [...] we observe that there is an infinite number of states of molecular motion that correspond to it. With increasing time, the system exists successively in all the dynamical states that correspond to the given thermodynamical state. From this point of view we may say that a thermodynamical state is the ensemble of all the dynamical states through which, as a result of the molecular motion, the system is rapidly passing. Fermi (1956).

And Callen adds that, if the system passes sufficiently rapidly through representative atomic states, a measurement effectively averages over all of them, and the system is in equilibrium. If, on the other hand, the transitions from atomic state to atomic state is ineffective (e.g., the system is trapped in a small subset of states, or the transition rate is too slow) a macroscopic measurement does not yield a proper average over all possible atomic states, and the system is not in equilibrium. A different description must be adopted. However, when the microscopic values of $\mathscr{O}$ are indeed sufficiently rapidly explored, compared to macroscopic observation times, a single system of interest reveals itself through $\mathscr{O}$ as the average over the *ensemble* of such possibilities, cf. Fig. 6.4. To the observer, the system revealed by $\mathscr{O}$ *appears* like the average over the ensemble of its microscopic phases, and the result of the measurement equals an average with respect to a phase space probability distribution, called *ensemble*. This is a tenet of Statistical Mechanics.

Remark 6.4.1 This requires the macroscopic state to be stationary; if it shifts during the measurement, the observable cannot explore its range with the frequencies corresponding to that state, and different initial microstates may lead to different

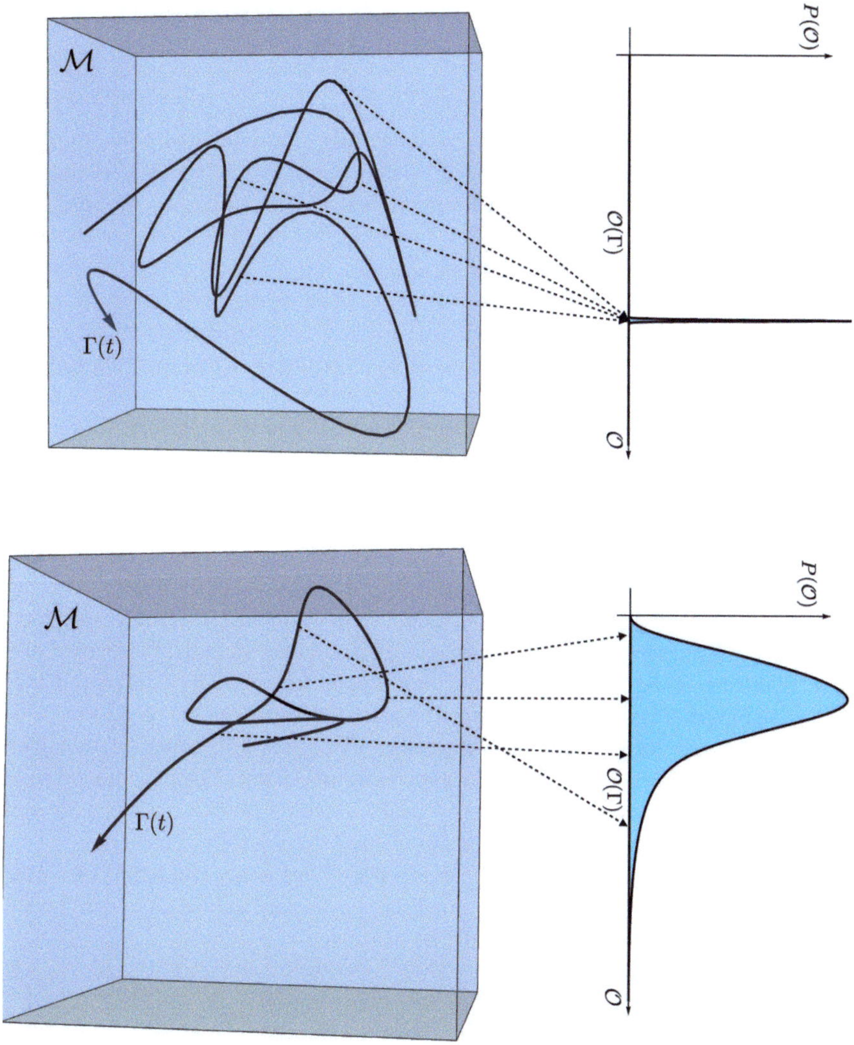

Fig. 6.4 While the phase wanders in phase space $\mathscr{M}$, the phase function $\mathscr{O}$ explores its range of values. Given the ultra-astronomically large dimensionality of $\mathscr{M}$, it is not possible for a trajectory to even coarsely explore $\mathscr{M}$. The range of $\mathscr{O}$ may nevertheless be thoroughly spanned, with proper frequencies in each of its sub-intervals. This way, the distribution of the values of $\mathscr{O}$ is built, and Eq. (6.57) may be verified in physically sensible times. The distribution of $\mathscr{O}$ must be stationary, otherwise it is not properly sampled in time, and averages depend on the initial condition. The τ required in Eq. (6.76) depends on the distribution. When Khinchin's hypotheses are verified ($N \ggg 1$, and $\mathscr{O}$ is a sum of independent molecular contributions) the range of $\mathscr{O}$ is practically a single point, and the corresponding distribution a δ-function (top panel). If the system is not large or correlations persist (e.g., nanoscopic devices or long-range interactions systems), the distribution may spread with positive variance (bottom panel)

expected values, or different observable values. In fact, this is the case of, e.g., aging systems.

Given a dynamical system like Eqs. (6.73), it is not obvious that Eq. (6.76) holds, and even if it does, the statement is so complex that it may be impossible to prove. Therefore, one introduces the so-called *Ergodic Hypothesis*, commonly stating that $\mathcal{M}$ is densely explored by almost all trajectories. The result can be formulated as follows: there is an invariant probability distribution μ on $\mathcal{M}$, such that a measurement yields Eq. (6.6) for μ-almost every $x \in \mathcal{M}$, and for every phase function $\mathcal{O}$. But while this certainly suffices for $\mathcal{O}(\Phi^t \Gamma)$ to span its range of values, it is way too strong a condition. It may hide the true reasons why experience taught us that Eq. (6.6) can be used, at least in the case of equilibrium systems, with the classical ensembles as probability distributions.

This fact can be understood as follows. Through a single variable $\mathcal{O}$, the time average may appear like a full phase space average, even if its phase space trajectory $\Phi^t \Gamma$ explores a limited region of $\mathcal{M}$. In fact, as long as the range of values of $\mathcal{O}$ has been visited with proper frequencies, the result is the same, cf. Fig. 6.4. This is of course easier for the observables whose range is narrower. In this framework, $\mathcal{O}$ is treated like a random variable with a given range of values. Different states are characterized by different values o, which may arise when the same range is explored with different frequencies, corresponding to different distributions. If one examines more than one phase variable, it may take longer for all to have explored their respective ranges. If this condition had to be verified by all possible phase functions, the time required for a system of many degrees of freedom could be so hugely long, to make the ergodic hypothesis physically irrelevant. Luckily, macroscopic observations only concern a handful of variables, which in addition are quite well behaved, allowing ergodicity to be continually and successfully accepted.

In some cases, this reasoning can be rigorously justified. The approach developed by Khinchin for rarefied gases tells that phase space subtleties are irrelevant, compared to the fact that Khinchin (2013):

(a) macroscopic systems are made of very many particles: $N \ggg 1$;
(b) only several and special phase functions are physically relevant;
(c) it does not matter if ensemble averages disagree with time averages on a limited sets of trajectories.

For rarefied gases, the relevant phase functions $\mathcal{O}$ are sums of molecular contributions o: $\mathcal{O}(\Gamma) = \sum_{n=1}^{N} o(\mathbf{q}_n, \mathbf{p}_n)$, where $(\mathbf{q}_n, \mathbf{p}_n)$ is the vector of configurations and momentum of the n-th particle. These functions are appropriate for the pressure, temperature, and density of rarefied gases, whose energy can also be expressed as $H = \sum_{n=1}^{N} h(\mathbf{q}_n, \mathbf{p}_n)$, because interactions between particles are on average negligible.[7] Then, in the microcanonical case, which here means a uniform probability distribution in $\mathscr{M}$, Khinchin proved the validity of the following relation:

$$\text{Prob}\left(\frac{|\overline{\mathscr{O}} - \mathbb{E}_{\text{mc}}(\mathscr{O})|}{|\mathbb{E}_{\text{mc}}(\mathscr{O})|} \geq K_1 N^{-1/4}\right) \leq K_2 N^{-1/4} \tag{6.77}$$

where K_1 and K_2 are positive constants, $\overline{\mathscr{O}}$ is the time average of $\mathscr{O}$, and $\mathbb{E}_{\text{mc}}(\mathscr{O})$ its phase space average in the microcanonical ensemble. In other words, the probability (in the microcanonical sense) that time averages differ by a small amount from the phase space averages is small if N is large: the larger N the smaller the probability of even smaller differences.

In this framework, the physically relevant ergodicity follows by and large from the $N \ggg 1$ condition, combined with the validity of the law of large numbers. That makes the range of sum variables of interest quite narrow, close to a single constant, when the single molecular contributions o can be considered independent, identically distributed random variables. The range of all the observables of interest does not take a long time.[8] The details of the microscopic dynamics, including transitivity or lack of transitivity, are irrelevant, while ensembles, i.e., probabilities in phase space turn considerably useful. Even though probability, *per se*, is an "immaterial" and abstract mathematical notion, it becomes in some sense "real," under the above conditions.

To understand more deeply when and why probability can be treated as real, which arguably makes it the most useful mathematical tool in Physics, let us consider evolving states, rather than stationary states, challenging Remark 6.4.1. If a stationary probability distribution correctly represents an equilibrium state, a time-dependent probability might, perhaps under specific conditions, represent an evolving material state.

6.4.1 Boltzmann Equation from the Probabilistic BBGKY Hierarchy

Consider N identical spherical hard particles of mass 1, so the momentum $\mathbf{p}$ numerically equals the velocity $\mathbf{v}$, that are subjected to no external forces, and that only interact elastically when they collide, which implies no accelerations between collisions, hence the total momentum is conserved $\dot{\mathbf{p}} = 0$. One may think of their interaction potential as being zero when the particles' centers are at a distance larger than their diameter, and infinitely high when this distance equals their diameter, so that particles are neither deformed, nor they penetrate each other, when they collide.

[7] Interactions are essential for the condition of LTE, hence for the applicability of thermodynamic notions. However, their time averages can be small compared to the kinetic energy, since they may last very short times.

[8] See, e.g., Ref. Baldovin et al. (2021) for an interesting analysis of the observables that relax to an equilibrium state, while others do not relax, because the underlying Toda dynamics are integrable, hence not ergodic.

Denote a probability density in the phase space $\mathscr{M}$ by

$$f_N^{(N)}(\Gamma; t) = f_N^{(N)}(\mathbf{q}_1, \mathbf{p}_1; ...; \mathbf{q}_N, \mathbf{p}_N; t) \tag{6.78}$$

with notation indicating the joint probability of N particles out of N. Liouville's equation can be written as

$$\frac{\partial f_N^{(N)}}{\partial t} + \left(\mathrm{div}_\Gamma \dot{\Gamma} + \dot{\Gamma} \cdot \nabla_\Gamma\right) f_N^{(N)} = \frac{\partial f_N^{(N)}}{\partial t} + \sum_{i=1}^{N} \mathbf{p}_i \cdot \nabla_{\mathbf{q}_i} f_N^{(N)} = 0 \tag{6.79}$$

where we have used the fact that the dynamics is Hamiltonian, hence the first term in parenthesis is null, and that $\dot{\Gamma} = (\dot{\mathbf{q}}_1, 0; ...; \dot{\mathbf{q}}_N, 0)$. If we introduce the s-particle distribution function:

$$f_N^{(s)}(\mathbf{q}_1, \mathbf{p}_1; ...; \mathbf{q}_s, \mathbf{p}_s; t) = \int \mathrm{d}\mathbf{q}_{s+1}\, \mathrm{d}\mathbf{p}_{s+1} ... \mathrm{d}\mathbf{q}_N\, \mathrm{d}\mathbf{p}_N\, f_N^{(N)}(\mathbf{q}_1, \mathbf{p}_1; ...; \mathbf{q}_N, \mathbf{p}_N; t) \tag{6.80}$$

and we integrate Liouville's equation over the variables $\mathbf{q}_{s+1}, \mathbf{p}_{s+1} ... \mathbf{q}_N, \mathbf{p}_N$, we obtain Cercignani (2006)

$$\frac{\partial f_N^{(s)}}{\partial t} + \sum_{i=1}^{s} \mathbf{p}_i \cdot \nabla_{\mathbf{q}_i} f_N^{(s)} = F^{(s)}\left(f_N^{(s+1)}, f_N^{(s+1)'}\right) \tag{6.81}$$

where F is a function of the joint $(s+1)$ particles' distribution before collision $f_N^{(s+1)\prime}$, and of the corresponding distribution after collision $f_N^{(s+1)}$. The set of distributions $f_N^{(s)}$ and their evolution equations, Eq. (6.81), constitute the *BBGKY hierarchy*. The important fact is that (6.81) is not a closed equation: computing $f_N^{(s)}$ requires knowledge of $f_N^{(s+1)}$. In particular, taking $s = 1$ and making $F^{(1)}$ explicit by solving the elastic collision dynamics, one obtains

$$\frac{\partial f_N^{(1)}}{\partial t} + \mathbf{p} \cdot \nabla_{\mathbf{q}} f_N^{(1)} = (N-1)\sigma^2 \int_{\mathbb{R}^3} \int_{S_-} [f_N^{(2)'} - f_N^{(2)}]\, |(\mathbf{p} - \mathbf{p}_*) \cdot \mathbf{n}|\, \mathrm{d}\mathbf{p}_* \mathrm{d}\mathbf{n} \tag{6.82}$$

where S_- is the hemisphere in $\mathbf{n}$ corresponding to particles moving one toward the other before the collision, and $\mathbb{R}^3$ the space of the velocities p_*.[9]

This is what one obtains for the 1-particle probability distribution function, integrating out all particles but one, from the phase space probability distribution function $f_N^{(N)}$. Here, σ is the diameter of one particle, hence σ^2 is the collision cross section; $\mathbf{n}$ is the unit vector in the direction going from the center of the test particle of position $\mathbf{q}$ and momentum $\mathbf{p}$, labeled by 1, to that of another particle of momentum $\mathbf{p}_*$; thus

[9] In dimensions lower than $d = 3$, fluids are in general untypical due to the fact that correlations do not decay fast enough, yielding convergence problems.

S_- is the hemisphere within which $(\mathbf{p} - \mathbf{p}_*) \cdot \mathbf{n} < 0$, i.e., particles directed toward each other before collision, and $(N - 1)$ is the number of particles with which the test particle can collide, assuming they all have the same probability of doing it, having integrated over all their momenta $\mathbf{p}_*$. Moreover, having denoted by $f_N^{(2)}$ the 2-particles distribution function, whose expression immediately before and after the collision, respectively, take the form:

$$f_N^{(2)'} = f_N^{(2)}(\mathbf{q}, \mathbf{p}'; \mathbf{q}_*, \mathbf{p}'_*; t) \quad \text{and} \quad f_N^{(2)} = f_N^{(2)}(\mathbf{q}, \mathbf{p}; \mathbf{q}_*, \mathbf{p}_*; t) \tag{6.83}$$

where $\mathbf{q}_* = \mathbf{q} + \sigma \mathbf{n}$ gives the position of the second particle at the moment of collision, and

$$\mathbf{p}' = \mathbf{p} - \left[\mathbf{n} \cdot (\mathbf{p} - \mathbf{p}_*)\right] \mathbf{n} \quad \text{and} \quad \mathbf{p}'_* = \mathbf{p}_* + \left[\mathbf{n} \cdot (\mathbf{p} - \mathbf{p}_*)\right] \mathbf{n} \tag{6.84}$$

and where $\mathbf{p}'$ and $\mathbf{p}'_*$ are, respectively, the momenta of the test particle and of the other colliding particle before collision, while $\mathbf{p}$ and $\mathbf{p}_*$ denote the corresponding momenta after collision.

To solve this equation, one needs a closure assumption on $f_N^{(2)}$. The following

$$f_N^{(2)}(\mathbf{q}, \mathbf{p}; \mathbf{q}_*, \mathbf{p}_*; t) = f_N^{(1)}(\mathbf{q}, \mathbf{p}; t) f_N^{(1)}(\mathbf{q}_*, \mathbf{p}_*; t) \tag{6.85}$$

is known in the kinetic theory of gases as the *Stosszahlansatz*, or hypothesis of molecular chaos. It amounts to assuming that when two particles collide, they are independent. As the dynamics of particles is deterministic, this statement may only have a statistical meaning, justified by some kind of dynamical *randomness*, that can be legitimately called *chaos*.[10] Then, writing f in place of $f_N^{(1)}$ to simplify the notation, and recalling that $\mathbf{v} = \mathbf{p}$ because we assumed unit masses, Eq. (6.82) takes the form:

$$\frac{\partial f}{\partial t} + \mathbf{v} \cdot \nabla_{\mathbf{q}} f = (N - 1)\sigma^2 \int_{\mathbb{R}^3} \int_{S_-} [f' f'_* - f\, f_*]\, |(\mathbf{v} - \mathbf{v}_*) \cdot \mathbf{n}|\, d\mathbf{v}_* d\mathbf{n} \tag{6.86}$$

where primes mean before collision, $*$ second particle, and small σ implies $\mathbf{q}_* \approx \mathbf{q}$.[11]

Equation (6.86) is the celebrated *Boltzmann equation*. Given its probabilistic derivation, which begins with the distribution $f_N^{(N)}$ in phase space, $f(\mathbf{q}, \mathbf{p}, t)\, d\mathbf{q}\, d\mathbf{p}$ represents the probability that particle 1 be found in the volume element $d\mathbf{q}$ around

[10] In the theory of dynamical systems, the term chaos is often used to indicate systems that have at least one positive Lyapunov exponent. This notion was not available to Boltzmann and, indeed, he did not need this kind of chaos.

[11] The gain and loss terms within square brackets in Eq. (6.86) strictly pertain to the two different hemispheres, S_- and S_+, for particles moving in opposite directions with respect to each other. However, they can be combined in a single integral by flipping the sign of $\mathbf{n}$, which does not alter the value of the integral, and they can even be written as an integral over the whole sphere S, Cercignani et al. (1994) pp. 29–31.

$\mathbf{q}$, with momentum within the set $d\mathbf{p}$ around $\mathbf{p}$, at time t. The classical interpretation of probability we have so far adopted tells us that: given a very large collection of independent macroscopically equal systems, each made of N hard spheres, the fraction of them having particle 1 within $d\mathbf{q}\,d\mathbf{p}$ equals f. If one assumes, in addition, that particles are indistinguishable, a simple combinatorial factor turns f into the fraction of ensemble members having one particle, whichever it is, in $d\mathbf{q}\,d\mathbf{p}$ at time t.[12] If the frequentist notion of probability is adopted, the reasoning is totally equivalent, with the fraction of ensemble members replaced by the fraction of time, in a microscopically long interval, a particle spends in $d\mathbf{q}\,d\mathbf{p}$ Cercignani (2006) p. 86. It remains to connect this probability to material properties, like the density of the gas in $d\mathbf{q}\,d\mathbf{p}$. This will be done in the next section.

6.4.2 Physical Derivation in Real "μ-Space"

The probability distribution f appearing in Eq. (6.86) is obtained projecting $f_N^{(N)}$ defined on the corresponding $2dN$-dimensional phase space, down to the $2d$-dimensional 1-particle space. This space of the positions and velocities of a single particle is called μ-space, to distinguish it from the $2dN$-dimensional Γ-space, or phase space, here denoted by $\mathscr{M}$. Also, given a single system of interest and $f_N^{(1)}(\mathbf{q}, \mathbf{p}; t) > 0$, there is no need for its particle 1 to actually lie within the volume $d\mathbf{q}\,d\mathbf{p}$, since only a fraction of the systems of the ensemble has particle 1 in $d\mathbf{q}\,d\mathbf{p}$: particle 1 of the complementary fraction does not lie in $d\mathbf{q}\,d\mathbf{p}$.[13] In particular, $d\mathbf{q}\,d\mathbf{p}$ could be empty because the BBGKY hierarchy can be developed for any $N \geq 1$,[14] and a few particles only occupy a limited region of the μ-space at any time t. Therefore, one could conclude that even the Boltzmann equation is *immaterial*, and there is no compelling reason to take it as a description of a given physical system.

However, a physical derivation of the Boltzmann equation is also available. It does not start from a probability distribution on a phase space $\mathscr{M}$, which is a continuum of geometric points, each of which represents a whole N-particles system. This derivation refers to a very large number N of small, but finite size, particles of mass 1, with positions and velocities in a given volume $\Delta\mathbf{q}\Delta\mathbf{p}$, that is tiny compared to

[12] Let us say the ensemble is made of a very large number M of systems made of N particles, and that each of the M systems corresponds to a different microscopic phase, within those of the same macroscopic state. How can an ensemble of independent material objects share the same coordinates, apart from undetectable infinitesimal displacements? Wouldn't they mess with each other? They do not because the notion of ensemble describes a single system in a given macroscopic state that can be found in many different microscopic phases at different times, and this collection of phases is then abstractly represented at once as the ensemble. If a concrete collection of objects is considered, they must be spatially separated so that they do not interact. Then, the coordinates $\mathbf{q}$ can be interpreted as referring to the center of mass of each of the different systems.

[13] If particles are indistinguishable and the 1-particle probability distribution has been correspondingly normalized, the same can be stated for any single particle of the systems of the ensemble.

[14] For $N = 1$ it trivially reduces to a single equation.

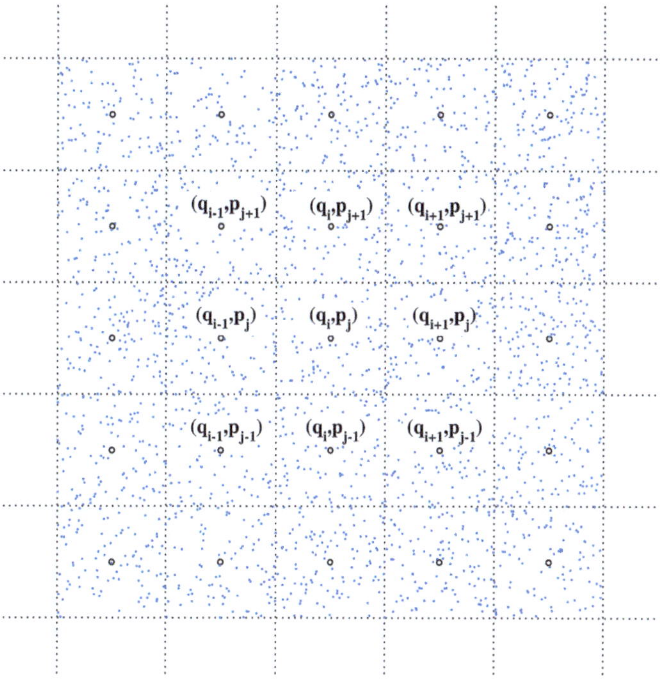

Fig. 6.5 Subdivision of available μ-space (the physical space) in discrete cells, sufficiently large to contain very many finite sized particles, but sufficiently small to look like a point to one macroscopic observer

macroscopic scales, but not infinitesimal. The particles in such a volume constitute a discrete rather than a continuous set, and can be represented by a continuous density of mass only when many of them populate $\Delta\mathbf{q}\Delta\mathbf{p}$ and a proper coarse graining is performed Falcioni et al. (2007); Kreuzer (1981), as discussed in Sect. 2.5. In fact, no more than a finite number of finite size particles can lie in $\Delta\mathbf{q}$.[15] Now, discretize the container and the space of velocities in cells C_{ij} of size $\Delta\mathbf{q}\Delta\mathbf{p}$ centered around a discrete set of lattice points $(\mathbf{q}_i, \mathbf{p}_j)$, cf. Fig. 6.5. The mass density within them is

[15] This reveals one essential difference of probability and mass densities, despite they may superficially look similar. While the density of mass is derived from a finite number of particles occupying a fraction of the volume of interest, the points concerning a probability density are a continuum occupying the whole volume, because they represent independent systems that do not overlap in space, however small $\Delta\mathbf{q}$ may be. A single system made up of N particles hardly has any similarity with a collection of M identical systems or with probability distributions on abstract spaces.

defined as the mass of one particle times the number density[16] $\rho(\mathbf{q}_i, \mathbf{p}_j; t)$, around $(\mathbf{q}_i, \mathbf{p}_j)$. Letting n_{ij} be the number of particles in cell C_{ij}, at time t one may thus write

$$n_{ij}(t) = \rho(\mathbf{q}_i, \mathbf{p}_j; t)\Delta\mathbf{q}\Delta\mathbf{p}\,; \quad \sum_{i,j} n_{ij}(t) = N \tag{6.87}$$

and may try to approximate this set of discrete mass densities using a continuous function ρ, so that

$$N = \int_{V_\mu} \rho(\mathbf{q}, \mathbf{p}; t)\, \mathrm{d}\mathbf{q}\, \mathrm{d}\mathbf{p} \tag{6.88}$$

where V_μ is the set of positions and momenta available to particles of the system. This makes no sense unless n_{ij} is very large for every cell C_{ij}. Therefore, cells must be relatively large. Furthermore, particles have to collide: their motion has to be random, so that a uniform distribution is rapidly achieved within each C_{ij}, and a single value of ρ represents the whole cell, as if it were a single point. Large n_{ij}, together with dynamical randomness, makes the particles entering or leaving a cell just a negligible number, compared to those staying inside. At the same time, the cells must be sufficiently small compared to the observation scale, that they appear like a point. For the above to make sense, a wide separation of scales, possible only if N is very large, is required, something that is not necessary in the derivation of, e.g., Eq. (6.82).

Let us now follow the motion of one set of particles lying in one cell $\Delta\mathbf{q}\Delta\mathbf{p}$ at time t, as they stroll in V_μ. Let us assume that ρ accurately describes the mass density in μ-space as a continuum. In the absence of external forces, collisions are the only mechanism that allows velocities to change. If in a time $\mathrm{d}t$ the particles in $\Delta\mathbf{q}\Delta\mathbf{p}$ undergo no collisions, the position $\mathbf{q}$ of one of them turns $\mathbf{q} + \mathbf{v}\,\mathrm{d}t$, while its momentum $\mathbf{p}$ is unchanged, and all the particles in the cell $\Delta\mathbf{q}\Delta\mathbf{p}$ centered about $(\mathbf{q}, \mathbf{p})$ will be found in the cell $\Delta\widehat{\mathbf{q}}\Delta\widehat{\mathbf{p}}$ centered about $(\mathbf{q} + \mathbf{v}\,\mathrm{d}t, \mathbf{p})$, which is obtained by evolving in time all points of the original cell with the same dynamics of the particles in it. Therefore, in the absence of collisions, one has

$$\rho(\mathbf{q} + \mathbf{v}\,\mathrm{d}t, \mathbf{p}, t + \mathrm{d}t)\Delta\widehat{\mathbf{q}}\Delta\widehat{\mathbf{p}} = \rho(\mathbf{q}, \mathbf{p}, t)\Delta\mathbf{q}\Delta\mathbf{p} \tag{6.89}$$

As $\mathbf{q}$ and $\mathbf{p}$ are canonically conjugate variables, and each particle can be considered as a whole system (hence Hamiltonian and phase volumes are preserved) between consecutive collisions, one obtains $\Delta\widehat{\mathbf{q}}\Delta\widehat{\mathbf{p}} = \Delta\mathbf{q}\Delta\mathbf{p}$, and

$$\rho(\mathbf{q} + \mathbf{v}\,\mathrm{d}t, \mathbf{p}, t + \mathrm{d}t) = \rho(\mathbf{q}, \mathbf{p}, t) \tag{6.90}$$

On the other hand, if particles collide, one additional term arises and one must introduce a new term:

$$\rho(\mathbf{q} + \mathbf{v}\,\mathrm{d}t, \mathbf{p}, t + \mathrm{d}t) = \rho(\mathbf{q}, \mathbf{p}, t) + \left(\frac{\mathrm{d}\rho}{\mathrm{d}t}\right)_{coll} \mathrm{d}t \tag{6.91}$$

[16] This is the number of particles per unit volume, at a given point.

thus defining the collision contribution to the variation of ρ. Finally, expanding the left-hand side of Eq. (6.91) to first order in $\mathrm{d}t$, one obtains

$$\rho(\mathbf{q},\mathbf{p},t)+\frac{\partial\rho}{\partial t}\,\mathrm{d}t+\mathbf{v}\cdot\nabla_{\mathbf{q}}\rho\,\mathrm{d}t=\rho(\mathbf{q},\mathbf{p},t)+\left(\frac{\mathrm{d}\rho}{\mathrm{d}t}\right)_{coll}\mathrm{d}t \tag{6.92}$$

because ρ depends on t both explicitly and implicitly through the position coordinates. Dividing by $\mathrm{d}t$ yields

$$\frac{\partial\rho}{\partial t}+\mathbf{v}\cdot\nabla_{\mathbf{q}}\rho=\left(\frac{\mathrm{d}\rho}{\mathrm{d}t}\right)_{coll} \tag{6.93}$$

where $\mathbf{v}\cdot\nabla_{\mathbf{q}}\rho$ represents streaming in position space. There is no streaming in momentum space, because no field acts on the particles, but particles may enter or exit $\Delta\mathbf{p}$ thanks to collisions.

Relation (6.93) expresses the conservation of mass, and does not require any interpretation of probabilistic nature: it is based on the deterministic Hamiltonian dynamics and on objective counting of particles (or measurement of mass).[17] Assuming that instantaneous collisions of three or more particles are negligible in a gas of hard spheres, the collision term should be expressed as a function of the number of pairs of particles located inside $\Delta\mathbf{q}$ that may either enter (gain term) or leave (loss term) $\Delta\mathbf{p}$ because of collisions. Let $\rho^{(2)}$ be the density of such pairs. As the initial condition of the N particles is not known, going beyond Eq. (6.93) requires some assumption on the collision term. One may then rely on the large value of N, on the smallness of the particles, and on the disorder produced in configuration and velocity space by the collisions. That allows each of the $n_{\Delta\mathbf{q}}$ particles in an elementary volume $\Delta\mathbf{q}$ to collide with any of the remaining $n_{\Delta\mathbf{q}}-1\approx n_{\Delta\mathbf{q}}$, because not screened by each other. Therefore, the factorization of $\rho^{(2)}$ for large numbers of particles results in a sensible possibility that is expressed by

$$\rho^{(2)}\left(\mathbf{q},\mathbf{p};\mathbf{q}+\sigma\mathbf{n},\mathbf{p}_{*};t\right)=\rho^{(1)}\left(\mathbf{q},\mathbf{p};t\right)\rho^{(1)}\left(\mathbf{q}+\sigma\mathbf{n},\mathbf{p}_{*};t\right) \tag{6.94}$$

Then, integrating over all possible velocities yields an expression like Eq. (6.86), which makes sense if particles are many and small, i.e., for a dilute gas. Introducing $\rho=\rho^{(1)}/N$, the equation for the mass distribution turns formally identical to Eq. (6.86), and its solutions enjoy the same properties that f enjoys.

At the same time, the conceptual difference between ρ and f could not be greater. In the first case, one considers the fraction of identical abstract systems that enjoy certain properties out of the total of an ideal hypothetical continuum of identical systems; in the second case, one considers the number of particles of a single concrete, experimentally observable system. Probabilistically, all calculations can be performed for a large collection of identical systems made of any number $N>1$ of particles. Physically, a single systems is considered and N, like each n_{ij}, must be

[17] Unlike probability, the number of particles or, equivalently, the mass contained in a volume of real space is an objective measurable quantity.

very large. There is no stringent reason for one quite sophisticated abstract notion to behave like a material tangible object. Nevertheless, this is precisely what happens if probability is required to evolve like a fluid made of geometric points obeying the equations of motion of the system that each of them separately represents, and the assumptions made in the derivation of the Boltzmann equation are valid. That may well be mathematically too hard to prove within a generic Hamiltonian framework, but experience has demonstrated that it does hold in very many situations; situations in which *probability* turns *material*: in some sense, probability can be identified with mass. At which cost? When can the Stosszahlansatz be justified?

In the first place, one would think that large N makes particles independent before colliding, because they may have hardly met before, but this is not enough. Particles are not independent after having collided with each other. Moreover, the factorization condition

$$f_N^{(s)}(q_1, p_1, ..., q_s, p_s; t) = \prod_{j=1}^{s} f_N^{(1)}(q_j, p_j; t) \tag{6.95}$$

is hard to obtain because particles do not overlap, and s of them occupy a volume of order $O(s\sigma^3)$. Thus, large s makes large the region to be excluded from the free motion of particles, resulting in correlations among them. This suggests that $s = 2$ is the best candidate for the postulated independence. Also, large N is needed to make rare the encounters between the same pair of particles, if the density is small, otherwise packed particles may collide forever with the same small group of companions. Then, σ better be small. On the other hand, collisions of hard spheres are defocusing, and have a randomizing effect on positions and velocities, which is beneficial for the decay of correlations. Therefore, N and σ should be tuned so that the total cross section is not negligible. Grad identified the scaling regime in which all the above is realized, now known as Boltzmann-Grad limit Grad (1949). It consists in keeping $N\sigma^2$ positive and finite, while N grows

$$N\sigma^2 = \text{constant} > 0 \quad \text{which implies} \quad \sigma^2 \sim \frac{1}{N}, \quad N\sigma^3 \sim N^{-\frac{1}{2}} \tag{6.96}$$

In this limit, the volume occupied by particles is negligible, while the total cross section is finite. More precisely, let λ denote the mean free path and τ the mean free time as, respectively, the typical microscopic length scale of the system, which is the distance traveled on average by a particle between two collisions, and the corresponding mean time. The first is defined by the ratio of the volume to the total cross section of the particles, the second by the ratio between this distance and the typical speed. Then, denoting by $|\Omega|$ the available volume, and letting $\beta = 1/k_B T$, so that the typical speed for particles of mass 1 is $\sqrt{2/\beta}$, one has

$$\lambda = \frac{|\Omega|}{N\pi\sigma^2}, \quad \tau = \sqrt{\frac{\beta}{2}} \frac{|\Omega|}{N\pi\sigma^2} \tag{6.97}$$

which remain finite in the Boltzmann-Grad limit. This implies a net effect produced by particles collisions, which is necessary to randomize their motion. Together with the vanishing of the excluded volume, the limit of Eq. (6.97) is also necessary to make correlations rapidly decay, when N is large. Physically, this picture corresponds to a rarefied gas, in which particles collide, but their interaction energy is negligible compared to their kinetic energy. Whether the assumptions apply or not, only experience can tell.[18] An important result implied by the Boltzmann-Grad limit is that $f_N^{(2)}$ stays factorized in time, if the solution of the sth equation of the BBGKY hierarchy $f_N^{(s)}$, exists is unique and is initially factorized as s factors $f_N^{(1)}$ for all s Cercignani (2006).

Before concluding this section, we note that there exists quantum mechanical and relativistic generalizations of the Boltzmann equation, cf., e.g., Ref. Ashcroft and Mermin (1976); Cercignani and Kremer (2002), which prove successful in an exceedingly wide range of applications, including transport of electrons in solid conductors and in astrophysical plasmas. The most direct extensions are the classical ones, concerning particles which get accelerated by external fields, such as electric fields. In this case, a streaming in velocity space is added to the streaming in position space. Denoting by $\mathbf{F}$ the external force and by m the mass of the particles, its effect on $\mathbf{v}$ appears as the acceleration $\mathbf{F}/m$, which smoothly brings a particle outside or inside an element of velocity space (streaming in velocity space), analogously to the velocity that takes a particle inside or outside an element of position space. Consequently, the Boltzmann equation may be rewritten as

$$\frac{\partial \rho}{\partial t} + \mathbf{v} \cdot \nabla_{\mathbf{q}} \rho + \frac{\mathbf{F}}{m} \cdot \nabla_{\mathbf{v}} \rho = \left(\frac{\mathrm{d}\rho}{\mathrm{d}t} \right)_{coll} \tag{6.98}$$

In sufficiently hot plasmas made of particles of electric charge c, collisions can be neglected. Therefore, denoting by $\mathbf{E}$ and $\mathbf{B}$, respectively, the electric and the magnetic fields acting on those particles, Eq. (6.98) can be rewritten as

$$\frac{\partial \rho}{\partial t} + \mathbf{v} \cdot \nabla_{\mathbf{q}} \rho + \frac{c}{m} (\mathbf{E} + \mathbf{v} \times \mathbf{B}) \cdot \nabla_{\mathbf{v}} \rho = 0 \tag{6.99}$$

which is known as the Vlasov equation.

6.4.3 The $\mathscr{H}$-Theorem

Assuming there are no external forces acting on the system, for instance, the effect of gravity is negligible, the stationary state is space homogeneous, and one may consider that it is homogeneous at start. Therefore, the solutions of the Boltzmann equation obey

$$\frac{\partial f}{\partial t}(\mathbf{v}, t) = (N-1)\sigma^2 \int \left[f' f'_* - f f_* \right] |(\mathbf{v} - \mathbf{v}_*) \cdot \mathbf{n}| \, \mathrm{d}\mathbf{v}_* \mathrm{d}\mathbf{n} = 0 \tag{6.100}$$

[18] To get an idea of how the Boltzmann-Grad limit complies with reality, think of a volume of 1 cm^3, at room temperature and atmospheric pressure, containing $N = 10^{20}$ molecules, that make a rarefied gas with $\sigma = 10^{-8}$. Then, $N\sigma^2 \approx 1$ m^2, while $N\sigma^3 \approx 10^{-4}$ cm^3.

where there is no dependence on the position $\mathbf{q}$. A sufficient condition for such an equilibrium is then expressed by

$$f_{eq}(\mathbf{v}')f_{eq}(\mathbf{v}'_*) - f_{eq}(\mathbf{v})f_{eq}(\mathbf{v}_*) = 0 \tag{6.101}$$

for any collision in which the velocities $\mathbf{v}'$ and $\mathbf{v}'_*$ of two particles turn, respectively, into $\mathbf{v}$ and $\mathbf{v}_*$. As a matter of fact, condition (6.101) is also necessary, as can be seen introducing the $\mathscr{H}$-functional:

$$\mathscr{H}(t) = \int f(\mathbf{v}, t) \log f(\mathbf{v}, t) \, \mathrm{d}\mathbf{v} \tag{6.102}$$

Differentiating $\mathscr{H}$ with respect to time, one gets

$$\frac{\mathrm{d}\mathscr{H}}{\mathrm{d}t} = \int \frac{\partial f}{\partial t}(\mathbf{v}, t)\Big[1 + \log f(\mathbf{v}, t)\Big]\,\mathrm{d}\mathbf{v} \tag{6.103}$$

which implies that $\mathrm{d}\mathscr{H}/\mathrm{d}t$ necessarily vanishes if $\partial f/\partial t$ does. When f is a solution of the Boltzmann equation in a volume delimited by regular elastically reflecting (hence adiabatic) walls, the first equality of Eq. (6.100) for the time derivative of f yields

$$\frac{\mathrm{d}\mathscr{H}}{\mathrm{d}t} = (N-1)\sigma^2 \int \big[f' f'_* - f\, f_*\big]\big[1 + \log f\big]\,|(\mathbf{v} - \mathbf{v}_*) \cdot \mathbf{n}|\; \mathrm{d}\mathbf{v}_*\mathrm{d}\mathbf{n}\mathrm{d}\mathbf{v} \tag{6.104}$$

Here, the role of test particle with velocity $\mathbf{v}$ and colliding particle with velocity $\mathbf{v}_*$ can be exchanged, since they are equal particles, and integration runs over all velocities of both. Then, $(1 + \log f)$ can be replaced by $(1 + \log f_*)$, without modifying the right-hand side of Eq. (6.104), and one obtains

$$\frac{\mathrm{d}\mathscr{H}}{\mathrm{d}t} = (N-1)\sigma^2 \int \big[f' f'_* - f\, f_*\big]\big[1 + \log f_*\big]\,|(\mathbf{v} - \mathbf{v}_*) \cdot \mathbf{n}|\; \mathrm{d}\mathbf{v}_*\mathrm{d}\mathbf{n}\mathrm{d}\mathbf{v} \tag{6.105}$$

since the other terms in the integral are invariant under the exchange of particles. Adding the result to Eq. (6.104), one can write

$$2\frac{\mathrm{d}\mathscr{H}}{\mathrm{d}t} = (N-1)\sigma^2 \int \big[f' f'_* - f\, f_*\big]\big[2 + \log f f_*\big]\,|(\mathbf{v} - \mathbf{v}_*) \cdot \mathbf{n}|\; \mathrm{d}\mathbf{v}_*\mathrm{d}\mathbf{n}\mathrm{d}\mathbf{v} \tag{6.106}$$

Also, the time reversibility allows the velocities before collision $(\mathbf{v}', \mathbf{v}'_*)$ to be exchanged with those after collision $(\mathbf{v}, \mathbf{v}_*)$, because for every collision path there is a path that exactly traces the first in reversed motion, and integration over all velocities allows again the exchange of roles of particles. Performing this exchange produces

$$2\frac{\mathrm{d}\mathscr{H}}{\mathrm{d}t} = (N-1)\sigma^2 \int \big[f f_* - f'\, f'_*\big]\big[2 + \log f' f'_*\big]\,|(\mathbf{v} - \mathbf{v}_*) \cdot \mathbf{n}|\; \mathrm{d}\mathbf{v}_*\mathrm{d}\mathbf{n}\mathrm{d}\mathbf{v} \tag{6.107}$$

hence, summing Eqs.(6.106) and (6.107), and dividing by 4, one obtains

$$\frac{\mathrm{d}\mathcal{H}}{\mathrm{d}t} = \frac{(N-1)}{4}\sigma^2 \int \underbrace{\left[f' f'_* - f f_*\right]\left[\log f f_* - \log f' f'_*\right]}_{\leq 0} |(\mathbf{v}-\mathbf{v}_*)\cdot\mathbf{n}| \, \mathrm{d}\mathbf{v}_*\mathrm{d}\mathbf{n}\mathrm{d}\mathbf{v} \tag{6.108}$$

The integrand of Eq. (6.108) is never positive,[19] so one can write

$$\frac{\mathrm{d}}{\mathrm{d}t}\mathcal{H} \leq 0 \tag{6.109}$$

and equality holds only when $f f_* = f' f'_*$, which is to say:

$$\log f_{eq}(\mathbf{v}) + \log f_{eq}(\mathbf{v}_*) = \log f_{eq}(\mathbf{v}') + \log f_{eq}(\mathbf{v}'_*) \tag{6.110}$$

This condition is thus both sufficient and necessary for the equilibrium state, which is stationary and homogeneous. It only remains to find the form of the equilibrium 1-particle distribution. For that, observe first that Eq. (6.110) is a conservation law, associated to a collision, since the left-hand side refers to post-collision molecular quantities, while the right-hand side refers to the pre-collision values of the same molecular quantities. Denote by $\{G_\ell\}_{\ell=1,2,...}$ the set of independent molecular quantities conserved in a collision between two molecules. Then, the most general solution of Eq. (6.110) takes the form:

$$\log f_{eq}(\mathbf{v}) = G_1(\mathbf{v}) + G_2(\mathbf{v}) + ... \tag{6.111}$$

where the sum could be finite or infinite, depending on the case. In our case of classical particles, the sum only includes the energy, the momentum, and a constant for mass. One may thus write

$$\log f_{eq}(\mathbf{v}) = c_0 + c_1\mathbf{v} + c_2\mathbf{v}^2 \tag{6.112}$$

Exponentiating this expression, one may eventually write

$$f_{eq}(\mathbf{v}) = Ce^{-c(\mathbf{v}-\mathbf{v}_0)^2} \tag{6.113}$$

where the normalization C, the factor c, and the reference velocity $\mathbf{v}_0$ must be identified. The result is the Maxwell-Boltzmann distribution.[20]

---

[19] The products of the probability densities are positive, and for positive quantities a and $b = a + h$ one has

$$(a-b)\log\left(\frac{b}{a}\right) = -h\log\left(1+\frac{h}{a}\right) < 0$$

because negative h implies negative logarithm, while positive h implies positive logarithm.

[20] The details of this calculation can be found in many textbooks, see, e.g., Ref. Huang (1987).

For isolated dilute gases, $\mathscr{H}$ reduces to minus the celebrated Boltzmann entropy

$$S_B = k_B \log W \tag{6.114}$$

where W is the number of different discrete ways in which a given thermodynamic state can be microscopically realized. If this (objective) number is (subjectively) identified with a normalized volume in phase space, W is replaced by a probability, and one concludes that equal volumes in phase space correspond to events of equal probability. Equation (6.114) is also called *bridge law*, since it connects the microscopic description afforded by W with the macroscopic thermodynamic description provided by the entropy and its derivatives, via the mind-boggling Boltzmann constant k_B. The Boltzmann postulate (6.114) was rewritten by Einstein as

$$\Pr(\text{state}) \propto e^{\Delta S/K_B} \tag{6.115}$$

meaning that a fluctuation away from the equilibrium state, characterized by a (negative) variation of entropy ΔS, is exponentially unlikely. This step began the industry of fluctuation theories, and imparts momentum to stochastic modeling in Physics.

6.4.4 Thermodynamics from Hamiltonian Mechanics?

One noted that the structure of Eqs. (6.81) preserves the factorization in s single particle densities of the multiparticle densities, which is essential for the validity of the Boltzmann equation. Why should the distribution of pairs of particles be factorized to start with?

First of all, note that the construction of the Boltzmann equation proceeds from expressing the evolution of the density due to the collisions among particles. If this density is smooth at some point before two molecules collide, the question is whether collisions preserve smoothness or produce wrinkles and eventually singularities. In other words, whether collisions tend to maximize or reduce the distances among particles in $(\mathbf{q},\mathbf{p})$ space. Smoothness is indeed required for the equation to continue to hold in time, since derivatives have to exist. Therefore, because there are TRI dynamics that produce singularities as well as TRI dynamics that smooth out the distributions of particles both in phase space and in real space, we conclude that the Boltzmann equation is suitable only for the second case Cercignani (2006). Then, repeated smoothing of the distribution gradually produces a homogeneous state, which means maximum microscopic *disorder*. In phase space, this corresponds to the uniform distribution, known as the microcanonical ensemble, which amounts to lack of correlations and ultimately to the factorization of the distributions.

Collisions of hard spheres are defocusing, which seems to imply a wider volume occupied by the colliding particles, hence higher smoothness, after a collision. However, the situation is not that simple: Hamiltonian dynamics are TRI, with reversal operator $\mathscr{T}$ defined by $\mathscr{T}(\mathbf{q}, \mathbf{p}) = (\mathbf{q}, -\mathbf{p})$, cf. Fig. 6.1. Therefore, each defocusing collision is associated with a symmetric focusing collision. This is the basis of the so-called *reversibility objection* or *Loschmidt paradox*, which states that the $\mathcal{H}$-theorem cannot be a consequence of reversible microscopic dynamics, such as that of hard spheres. In fact, Loschmidt correctly argued that if $\mathcal{H}$ decreases in a time interval for a given choice of initial conditions, there is another initial condition that leads $\mathcal{H}$ to increase in the same time interval: it *"suffices"* to reverse the velocities of the second path.

The $\mathcal{H}$-theorem was declared impossible also on the grounds of the Poincaré recurrence theorem, that had been proven shortly earlier. This objection, known as the *recurrence objection* or *Zermelo paradox*, notices that given any finite tolerance, the phase space trajectories of mechanical systems with bounded phase space take a finite time T_R to return to their initial condition within that tolerance. Therefore, $\mathcal{H}$ may initially decrease but, being a continuous function of phase, sooner or later it will return as close as one wishes to its initial value, hence will sooner or later violate the monotonic behavior implied by the Boltzmann equation.

Indeed, both objections are mathematically well motivated, but they do not consider a fact, which Boltzmann himself pointed out: a gas is not any kind of dynamical system, it is one with very many particles! While obvious, this fact has far-reaching consequences that are not immediately clear in abstract mathematical terms, but that become evident when concrete numbers are given. For instance, the molecules of air at usual temperature and pressure in a volume of 1 cm^3 can be estimated to have a typical T_R of order $O(10^{10^{19}})$ years! It is also obvious that the difficulty in aiming the velocities of colliding spheres, so that they move from a disordered distribution to an ordered one increases exceedingly rapidly with the number N of balls. But even granting that, such an effort is totally pointless, because, even if successful for some time, after balls have grouped in a region of the container, their motion will move them again apart of each other.

How can the formal correctness of Loschmidt and Zermelo's reasonings be reconciled with the most convincing and verifiable Boltzmann's theory?

Take $\mathcal{H}$ for a system made of a number N of finite size balls in a given container with straight reflecting walls. In order to compute $\mathcal{H}$, the density ρ of the balls can be computed subdividing the container into volumes that are not too small otherwise, from a thermodynamic point of view the quantity ρ turns nonsense, neither too large, that means inaccurate resolution. Now, N not sufficiently large yields a fluctuating $\mathcal{H}$, like Fig. 6.6. In fact, T_R will also be within reach and $\mathcal{H}$ will return in time to its initial value. However, taking larger N, the fluctuations decrease compared to the values of $\mathcal{H}$, and T_R increases. Therefore, $\mathcal{H}$ for a system made of a macroscopic number of particles experiences neither deviations from the monotonic decreasing curve predicted by Boltzmann's theory nor the increase due to recurrence. The Boltzmann theory is vindicated well beyond the needs of thermodynamics.

A more detailed derivation of the Boltzmann equation and the $\mathcal{H}$-theorem is given in Appendix E.

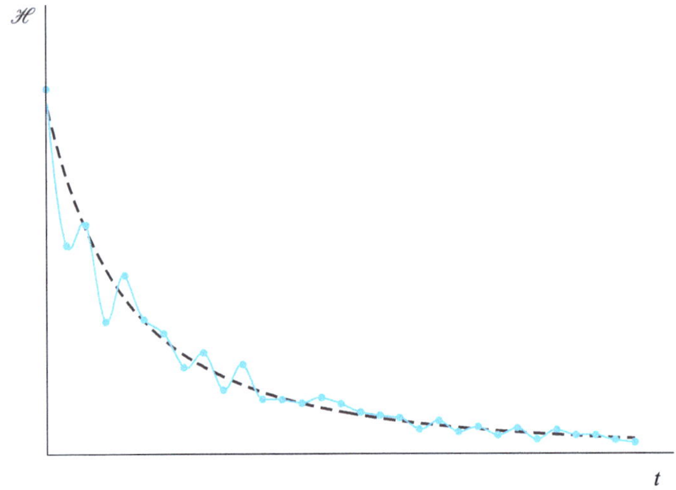

Fig. 6.6 Evolution of Boltzmann $\mathscr{H}$-functional for a finite number N of particles, as obtained from their number density ρ. It only approximately follows the monotonic decrease it would enjoy, were ρ replaced by the solution f of the corresponding Boltzmann equation. Moreover, on a much longer scale $O(T_R)$ not represented here, $\mathscr{H}$ climbs up again to higher values, because of Poincaré's recurrences. But larger N implies smaller fluctuations about the smooth curve, and longer times T_R

6.5 From Ergodic Theory to Big Data

An interesting application of ergodic theory concerns the modern subject of big data and machine learning, which are based on the idea that knowledge of the past (data) suffices to predict the future. Given a time series of some kind of signal, one may think that there is an unknown underlying dynamical system, Φ, that generates such signal. For simplicity, take the dynamics to be discrete in time:

$$\mathbf{x}_1 = \Phi \mathbf{x}_0 \tag{6.116}$$
$$\mathbf{x}_2 = \Phi \mathbf{x}_1 = \Phi^2 \mathbf{x}_0 \tag{6.117}$$
$$\vdots$$
$$\mathbf{x}_k = \Phi \mathbf{x}_{k-1} = \Phi^k \mathbf{x}_0 \tag{6.118}$$

Then, let $A \subset \mathscr{M}$ be a measurable set in phase space, and denoted by $\tau_A(\mathbf{x})$ the recurrence time in that set, i.e., the shortest time after which a point of A returns to A:

$$\tau_A(\mathbf{x}) = \inf\{k \geq 1 : \mathbf{x} \in A \text{ and } \Phi^k \mathbf{x} \in A\} \tag{6.119}$$

If μ denotes the invariant measure, which is the stationary probability distribution on $\mathcal{M}$, a theorem by Kac proves that the average recurrence time for ergodic systems is given by Kac (1957)

$$\mathbb{E}_{\mu}(\tau_A) = \frac{1}{\mu(A)} \tag{6.120}$$

For systems with N degrees of freedom, whose range is $O(L)$ for each component of $\mathbf{x}$, and for A of linear size ε, one then obtains

$$\mathbb{E}_{\mu}(\tau_A) \sim \left(\frac{L}{\varepsilon}\right)^N \tag{6.121}$$

or

$$\mathbb{E}_{\mu}(\tau_A) \sim \left(\frac{L}{\varepsilon}\right)^D \tag{6.122}$$

if the steady state lives on a set of fractal dimension D. These times grow *exponentially* with N. In fact, this expresses in mathematical terms Boltzmann's answer to Zermelo paradox: for a system of a macroscopic number of degrees of freedom, recurrence takes far longer than any physically imaginable time.

To build a model from known data, one looks for past states similar to the present one, and assumes that the future will approximately repeat what happened after the selected past state. This is the ancient method of analogues that is translated in a modern mathematical language thanks to the notion of recurrence. In modern times, Maxwell noted that:

> It is a metaphysical doctrine that from the same antecedents follow the same consequents. No one can gainsay this. But it is not of much use in a world like this, in which the same antecedents never again concur, and nothing ever happens twice. Indeed, for aught we know, one of the antecedents might be the precise date and place of the event, in which case experience would go for nothing. The metaphysical axiom would be of use only to a being possessed of the knowledge of contingent events, scintia simplicis intelligentiœ,—a degree of knowledge to which mere omniscience of all facts, sctentiia visionis, is but ignorance. The physical axiom which has a somewhat similar aspect is "from like antecedents follow like consequents." But here we have passed from sameness to likeness, from absolute accuracy to a more or less rough approximation. There are certain classes of phenomena, as I have said, in which a small error in the data only introduces a small error in the result. Such are, among others, the larger phenomena of the Solar System, and those in which the more elementary laws in Dynamics contribute the greater part of the result. The course of events in these cases is stable Campbell and Garnett (1882).

Although he immediately added:

> There are other classes of phenomena which are more complicated, and in which cases of instability may occur, the number of such cases increasing, in an exceedingly rapid manner, as the number of variables increases Campbell and Garnett (1882).

In practice, the method of analogues assumes that the phenomena of interest are stable in Maxwell's sense, and proceeds as follows. Given a sequence of past events

$$\mathbf{x}_1, \ \mathbf{x}_2, \ \ldots, \ \mathbf{x}_M \tag{6.123}$$

they are called *"analogues"* if they resemble each other in pairs with a certain degree of accuracy ε, i.e., if:

$$\left|\mathbf{x}_i - \mathbf{x}_j\right| \leq \varepsilon \tag{6.124}$$

Then, the approximate prediction that can be made is expressed by

$$\mathbf{x}_{M+1} = \mathbf{x}_{k+1} \ , \quad \text{if } \mathbf{x}_k \text{ is an analogue of } \mathbf{x}_M \tag{6.125}$$

One then models the phenomenon defining a function $\mathbf{f}$ such that sequences of states are approximated by $\mathbf{x}_{k+1} = \mathbf{f}(\mathbf{x}_k)$ within the chosen tolerance. Now, to find analogues one needs long series of data, as shown by Kac theorem: longer for higher accuracy or for larger space, and exponentially long in the number of degrees of freedom. For instance, even a relatively low accuracy such as $\varepsilon = L/10$ makes the task problematic, because of the large dimensionality of the problems for which this approach is usually invoked, i.e., those that are too complex to be tackled in a theory that starts from an understanding of the fundamental mechanisms. The problem is enhanced in the case of chaotic dynamics that exponentially magnify in time the deviations from the analog initial conditions Collet and Eckmann (2006). In principle, D equal to 6 or 7 could already be out of reach in those cases.

One could add further considerations to refute the idea that the scientific method is made obsolete by the deluge of data. For instance, the tale attributed to Bertrand Russell of a turkey before Thanksgiving day shows how naive induction is doomed to failure. The story goes like this: a turkey is fed every day. Every single feeding will firm up the bird's belief that it is the general rule of life to be fed every day by friendly members of the human race "looking out for its best interests," as a politician would say. On the afternoon of the Wednesday before Thanksgiving, something unexpected will happen to the turkey. It will incur a revision of belief.

Nevertheless, there are successes in using data that cannot be denied, and that can certainly be integrated within a scientific framework, rather than opposing it; there still is a lot to understand about what can be extracted from data, and proceeding by examples is a most promising way to go. In fact, once again, it may be beneficial to distinguish sufficient from necessary conditions, noting that negative results, such as those about recurrence mentioned above do not prevent the method to succeed. For instance, in big data practice there might be implicit use of a priori knowledge that helps forecasting.

6.6 Current Research

Research in probability and its applications is such a wide field that remains impossible to summarize. Therefore, here we only mention a few aspects of that, referring to well-established publications. The statistical theory of dynamical systems is illustrated, for instance, in the following books Ruelle (2014) and Viana and Oliveira (2016). For applications to physics, including turbulence and transport phenomena, see, e.g., Cencini et al. (2009). The mathematical theory of the Boltzmann equation, especially its use in the derivation of the hydrodynamic equations is illustrated in the following books: Villani (2002) and Saint-Raymond (2009). In reality, variations and applications of the Boltzmann equations are countless. The kinetic theory of wave turbulence Nazarenko (2011) finds applications from quantum to astrophysical phenomena, including anomalous transport. There are Boltzmann equations for thermostated rather than Hamiltonian dynamics Bonetto et al. (2014), and for a wide spectrum of variously interacting systems, such as the components of blood, or the individuals in a social network, etc., see, e.g., Cristiani et al. (2014). It is impossible to list all of them. Of course, there is a difference between these variations and the original Boltzmann equation: the interaction rules are often rather loosely known, which makes the fundamental hypothesis of molecular chaos hard to justify. Nevertheless, they allow useful insight in non-standard and even non-physical phenomena. The interested readers may easily find up-to-date literature.

Linear Response Theory 7

"Watchman, what is left of the night?" The watchman replies, "Morning is coming, but it's still night"

Isaiah 21:11-12

The kinetic theory of gases has accomplished a sophisticated and hard task: to describe the response of a *single* physical system to a perturbation that pushed it away from its natural equilibrium state. This has been possible because the behavior of the projection to the μ-space of the probability density in phase space (representing an *ensemble* of systems) can be identified with the behavior of the density of mass of a single system, under proper conditions. Apart from technicalities, it suffices that: 1. probability is transported in an exceedingly high dimensional very abstract phase space, as matter is transported in the concrete three-dimensional real space, and that 2. the Boltzmann-Grad limit, referring to rarefied gases, correctly represents the system of interest. However, numerous variations of the Boltzmann equation are formulated and satisfactorily applied in a very wide spectrum of phenomena that do not strictly correspond to rarefied gases. Identifying the necessary, rather than the sufficient conditions for the validity of Boltzmann-like equations is an interesting, non-trivial task that is related to what happens when a generic system in an equilibrium or nonequilibrium state is perturbed by external actions. The simplest situation one likes to investigate is that corresponding to responses proportional to the perturbations generating them, i.e., the realm of linear response theory.

7.1 Time-Independent Perturbations

Let us begin considering a Hamiltonian particle system, whose energy is initially expressed by the Hamiltonian $\mathscr{H}_0(\mathbf{Q}, \mathbf{P})$, where $(\mathbf{Q}, \mathbf{P}) = \Gamma$ are its canonical vari-

C. Mejía-Monasterio and L. Rondoni, *Nonequilibrium Statistical Mechanics I*,
Theoretical and Mathematical Physics,
https://doi.org/10.1007/978-3-031-98249-1_7

ables, that constitute one point in the phase space, $\Gamma \in \mathcal{M}$. Assume that this system is initially in equilibrium at temperature T, and that its phases are distributed canonically, i.e., with the following density in phase space:

$$f_0(\Gamma) = \frac{\exp(-\beta \mathcal{H}_0(\Gamma))}{\int \mathrm{d}\Gamma \exp(-\beta \mathcal{H}_0(\Gamma))} = \frac{1}{Z_0} e^{-\beta \mathcal{H}_0(\Gamma)}, \quad \beta = \frac{1}{k_B T} \tag{7.1}$$

Thus, as long as our *single* system does not evolve in time, and remains in this thermodynamic equilibrium, its macroscopically relevant microscopic variables are thought to explore their range of values with frequency corresponding to the statistics of f_0. For instance, letting the kinetic energy of the system $K(\Gamma)$ be one such variable, the quantity

$$\nu(K_1, K_2) = \int_{\{\Gamma \in \mathcal{M} : K(\Gamma) \in (K_1, K_2)\}} f_0(\Gamma)\, \mathrm{d}\Gamma \tag{7.2}$$

is taken, in this formalism, to represent the frequency with which the time-dependent kinetic energy of the single system takes values within the interval (K_1, K_2). No abstract notion of a collection of identical objects is needed here, neither any dense exploration of $\mathcal{M}$: it suffices that the range of values of K is explored with proper frequencies.[1]

Remark 7.1.1 The range of K in the canonical ensemble is $[0, \infty)$. In reality, the kinetic energy of a finite system, in equilibrium with another (even if much larger) finite system, is bounded. Assuming that it is unbounded not only substantially simplifies all calculations but also yields practically exact estimates if the systems of interest are macroscopic. Within the scope of equilibrium thermodynamics, hence *for a special set of observables*, averaging with respect to the canonical ensemble is proven totally justified.

Starting at time 0, let the system be perturbed by a Hamiltonian interaction that contributes the extra energy $\lambda\, A(\Gamma)$, with $\lambda \in \mathbb{R}$. From time 0, then, the system evolves as prescribed by the new Hamiltonian $\mathcal{H}(\Gamma) = \mathcal{H}_0(\Gamma) + \lambda A(\Gamma)$. If the perturbed system relaxes in time to the thermodynamic equilibrium at inverse temperature $\beta = 1/kT$, the final state is expressed by the new canonical ensemble:

$$f = \frac{\exp(-\beta \mathcal{H})}{\int \mathrm{d}\Gamma\ \exp(-\beta \mathcal{H})} = \frac{1}{Z} e^{-\beta \mathcal{H}} \tag{7.3}$$

[1] Whether the identification of the frequencies concerning the evolution of a single system and the probabilities expressed by Eq. (7.2) is correct should be checked case by case: no general proof starting from the details of the dynamics is available. Nevertheless, the hypothesis that the canonical ensemble describes a macroscopic system in equilibrium with a heat bath has proven so successful that most textbooks and research papers don't even mention it as an hypothesis: they merely use it. In the case of novel phenomena, such as those concerning many nonequilibrium states or small systems, care should be taken.

That the perturbation is followed by relaxation to the new canonical equilibrium with Hamiltonian $\mathscr{H}$ and temperature T is far from a trivial question. Nevertheless, one may consider two systems: the first in the unperturbed equilibrium, the other in the perturbed equilibrium, and one may estimate what difference that makes for the observables of interest. In case λ is small, the first-order approximation in λ constitutes a good approximation of f that yields

$$
\begin{aligned}
f(\Gamma) &= \frac{e^{-\beta\mathscr{H}_0(\Gamma)}e^{-\beta\lambda A(\Gamma)}}{\int d\Gamma e^{-\beta\mathscr{H}_0(\Gamma)}e^{-\beta\lambda A(\Gamma)}} = \\
&= \frac{e^{-\beta\mathscr{H}_0(\Gamma)}[1-\beta\lambda A(\Gamma)+O(\lambda^2)]}{\int d\Gamma e^{-\beta\mathscr{H}_0(\Gamma)}[1-\beta\lambda A(\Gamma)+O(\lambda^2)]} = \\
&= \frac{e^{-\beta\mathscr{H}_0(\Gamma)}}{\int d\Gamma e^{-\beta\mathscr{H}_0(\Gamma)}}\frac{1-\lambda\beta A(\Gamma)}{1-\lambda\beta\mathbb{E}_0(A)} + O(\lambda^2) \\
&= f_0(\Gamma)\,[1-\lambda\beta A(\Gamma)]\,[1+\lambda\beta\mathbb{E}_0(A)] + O(\lambda^2)
\end{aligned}
$$

i.e., for small λ, one can write

$$
f(\Gamma) \approx f_0(\Gamma)\Big\{1-\lambda\beta\,[A(\Gamma)-\mathbb{E}_0(A)]\Big\}
$$

where $\mathbb{E}_0(\cdot)$ means average with respect to the unperturbed ensemble f_0. This becomes the long time response of the system to the perturbation with small λ, if the state actually evolves from the equilibrium characterized by f_0 to the equilibrium characterized by f. In that case, the variation of a generic observable $\mathscr{O}$, due to the small perturbation, is given by

$$
\Delta\mathbb{E}(\mathscr{O}) = \int \mathrm{d}\Gamma\ \mathscr{O}(\Gamma)\big[f(\Gamma)-f_0(\Gamma)\big] \tag{7.4}
$$

$$
\approx -\lambda\beta\big[\mathbb{E}_0(\mathscr{O}A)-\mathbb{E}_0(\mathscr{O})\mathbb{E}_0(A)\big] \tag{7.5}
$$

which means that the response of the variable $\mathscr{O}$ is determined by the equilibrium correlation function of $\mathscr{O}$ and A. In the case that $\mathscr{O} = A = \mathscr{H}_0$, one obtains

$$
\frac{\Delta\mathbb{E}(\mathscr{H}_0)}{\lambda\beta} \approx -\big[\mathbb{E}_0(\mathscr{H}_0^2)-\mathbb{E}_0(\mathscr{H}_0)^2\big] = -k_B T^2 C_V \tag{7.6}
$$

where

$$
\mathbb{E}_0(\mathscr{H}_0^n) = \int \mathscr{H}_0(\Gamma)^n f_0(\Gamma)\,\mathrm{d}\Gamma \tag{7.7}
$$

Here, C_V can be interpreted as the heat capacity of the system at constant volume, because the equilibrium average of $\mathscr{H}_0$ can be interpreted as the unperturbed internal energy U. Then, by definition, one has

$$
C_V = \left.\frac{\partial U}{\partial T}\right|_V = \frac{\partial\mathbb{E}_0(\mathscr{H}_0)}{\partial T} = \frac{\partial\mathbb{E}_0(\mathscr{H}_0)}{\partial\beta}\frac{\mathrm{d}\beta}{\mathrm{d}T} = -\frac{1}{k_B T^2}\frac{\partial\mathbb{E}_0(\mathscr{H}_0)}{\partial\beta} \tag{7.8}
$$

where

$$\frac{\partial \mathbb{E}_0(\mathcal{H}_0)}{\partial \beta} = \frac{\partial}{\partial \beta}\left[Z_0(\beta)^{-1}\int \mathcal{H}_0\, e^{-\beta \mathcal{H}_0}\, \mathrm{d}\Gamma\right], \quad \text{with} \quad Z_0(\beta) = \int e^{-\beta \mathcal{H}_0}\, \mathrm{d}\Gamma \tag{7.9}$$

and

$$\begin{aligned} \frac{\partial}{\partial \beta} Z_0(\beta)^{-1} &= -\frac{1}{Z_0(\beta)^2}\frac{\partial}{\partial \beta}\int e^{-\beta \mathcal{H}_0}\, \mathrm{d}\Gamma = -\frac{1}{Z_0(\beta)^2}\int \frac{\mathrm{d}}{\mathrm{d}\beta} e^{-\beta \mathcal{H}_0}\, \mathrm{d}\Gamma \\ &= \frac{1}{Z_0(\beta)^2}\int \mathcal{H}_0\, e^{-\beta \mathcal{H}_0}\, \mathrm{d}\Gamma = \frac{1}{Z_0(\beta)}\mathbb{E}_0(\mathcal{H}_0) \end{aligned} \tag{7.10}$$

hence

$$\int \mathcal{H}_0 e^{-\beta \mathcal{H}_0}\, \mathrm{d}\Gamma \left[\frac{\partial}{\partial \beta} Z_0(\beta)^{-1}\right] = \int \mathcal{H}_0 e^{-\beta \mathcal{H}_0}\, \mathrm{d}\Gamma\, \frac{1}{Z_0(\beta)}\mathbb{E}_0(\mathcal{H}_0) = \mathbb{E}_0(\mathcal{H}_0)^2 \tag{7.11}$$

while

$$\frac{1}{Z_0(\beta)}\frac{\partial}{\partial \beta}\int \mathcal{H}_0 e^{-\beta \mathcal{H}_0}\, \mathrm{d}\Gamma = -\frac{1}{Z_0(\beta)}\int \mathcal{H}_0^2\, e^{-\beta \mathcal{H}_0}\, \mathrm{d}\Gamma = -\mathbb{E}_0(\mathcal{H}_0^2)$$

Therefore,

$$\frac{\partial \mathbb{E}_0(\mathcal{H}_0)}{\partial \beta} = \mathbb{E}_0(\mathcal{H}_0)^2 - \mathbb{E}_0(\mathcal{H}_0^2) \tag{7.12}$$

which leads to

$$C_V = \frac{1}{k_B T^2}\left[\mathbb{E}_0(\mathcal{H}_0^2) - \mathbb{E}_0(\mathcal{H}_0)^2\right] \tag{7.13}$$

That we are dealing with an isochoric process is implicit in the fact that no other energy sources are considered, while a variation of the volume implies work, which is a source or a sink for the internal energy. At constant pressure P, for example, one would have to add a $P\mathrm{d}V$ contribution to the heat needed to change the temperature of the object of interest.

This result is quite impressive: It links a macroscopic measurable quantity, such as the heat capacity, to (invisible) equilibrium fluctuations. In other words, it allows us to assess the size of microscopic fluctuations, *via* macroscopic observations.

That a system will actually relax moving from one canonical equilibrium to another canonical equilibrium is not guaranteed. It is thus interesting to study the time evolution described by the present framework, in which a stationary state of a single object is described by the set of frequencies of the values of its observables, and these frequencies are represented by a probability and hence by an ensemble. In accordance with Fermi's and Callen's reasoning, Sect. 6.4, the single system can be considered the ensemble of its microscopic phases, because its macroscopic state does not change in time.

In the next section, we investigate the problem of relaxation, pretending that an evolving probability in phase space still describes the state of the system of interest as it changes in time. We know that this is possible under the conditions of the Boltzmann equation, but we will try this in more general cases, including those of time-dependent perturbations. At the end, we will discuss why this apparently paradoxical framework is, in fact, an excellent microscopic model of many macroscopic phenomena.

7.2 Time-Dependent Perturbations

The relaxation of the state of a given system to a given equilibrium state is the consequence of a number of facts, which is hard to exhaustively link to the dynamics of its elementary constituents. Furthermore, it depends on the observables that are assumed to characterize the state. Indeed, it is usually possible to identify quantities that never settle about an asymptotic value, whether averaged in time or not.

Forget for a while that—strictly speaking or apart notable exceptions—a single thermodynamic system can be described by an ensemble in phase space only if its state is stationary and daringly take the evolution of such an ensemble as the evolution of the state of the system. In other words, think of the evolution of the ensemble averages of the observables of interest as the real evolution of the state, cf. Chap. 6. For instance, while the rule that sets of phase points carry a probability in $\mathcal{M}$ makes it impossible, except in uninteresting situations, for the probability to come to rest, cf. Eqs. (6.50) and (6.55), phase averages of observables can settle about certain asymptotic values. Inspired by the Boltzmann equation, which successfully combines the notions of mass and probability, let us adopt this point of view.

Given the unperturbed Hamiltonian system of the previous section, whose energy is expressed by $\mathcal{H}_0$, introduce a time-dependent perturbation, of the form $-\mathcal{F}(t)A(\Gamma)$, where $\mathcal{F}$ is the intensity of the perturbation, a pure number, and A has the dimensions of energy. The perturbed Hamiltonian is given by

$$\mathcal{H}(\Gamma, t) = \mathcal{H}_0(\Gamma) - \mathcal{F}(t)A(\Gamma) \, ; \quad \Gamma \in \mathcal{M} \tag{7.14}$$

and the equations of motion by

$$\frac{dq_j}{dt} = \frac{\partial \mathcal{H}_0}{\partial p_j} - \mathcal{F}(t)\frac{\partial A}{\partial p_j} = \frac{\partial \mathcal{H}_0}{\partial p_j} - \mathcal{F}(t)\, K_j^q \, ; \quad \text{with } K_j^q = \frac{\partial A}{\partial p_j} \tag{7.15}$$

$$\frac{dp_j}{dt} = -\frac{\partial \mathcal{H}_0}{\partial q_j} + \mathcal{F}(t)\frac{\partial A}{\partial q_j} = -\frac{\partial \mathcal{H}_0}{\partial q_j} - \mathcal{F}(t)\, K_j^p \, ; \quad \text{with } K_j^p = -\frac{\partial A}{\partial q_j} \tag{7.16}$$

where K_j^q has the dimensions of velocity and K_j^p of force. Denoting by G the corresponding vector field, which is a sum of an unperturbed part and of a perturbation:

$$G(\Gamma; t) = \begin{pmatrix} \partial \mathcal{H}_0/\partial p_1 \\ \vdots \\ -\partial \mathcal{H}_0/\partial q_N \end{pmatrix} - \mathcal{F}(t) \begin{pmatrix} K_1^q \\ \vdots \\ K_N^p \end{pmatrix} = G_0(\Gamma) + G_{ext}(\Gamma; t) \tag{7.17}$$

we have the following Liouville operator:

$$
\begin{aligned}
-i\mathcal{L} &= -\Big(G \cdot \nabla_\Gamma + \operatorname{div}_\Gamma G\Big) && (7.18)\\
&= -\Big(G_0 \cdot \nabla_\Gamma + \operatorname{div}_\Gamma G_0\Big) - \Big(G_{ext} \cdot \nabla_\Gamma + \operatorname{div}_\Gamma G_{ext}\Big) = -i\Big[\mathcal{L}_0 + \mathcal{L}_{ext}\Big]
\end{aligned}
$$

where letting $\{\cdot, \cdot\}$ denote the Poisson brackets, we have

$$
i\mathcal{L}_0 f = \{f, \mathcal{H}_0\} \ ; \quad i\mathcal{L}_{ext} f = -\mathcal{F}(t)\{f, A\} \tag{7.19}
$$

The evolution of the probabilities in phase space can then be derived by solving Liouville's equation:

$$
\frac{\partial f}{\partial t} = -i\mathcal{L} f = -i\Big(\mathcal{L}_0 + \mathcal{L}_{ext}\Big) f \tag{7.20}
$$

This is the Eulerian version of Liouville's equation, hence the position $\Gamma \in \mathcal{M}$ is fixed. Let the initial probability density in $\mathcal{M}$ be canonical:

$$
f_0(\Gamma) = \frac{1}{Z_0} e^{-\beta \mathcal{H}_0} \tag{7.21}
$$

By definition, the density f_0 is invariant under the unperturbed evolution, $i\mathcal{L}_0 f_0 = 0$, while it is not invariant under the perturbed evolution: $i\mathcal{L}_{ext} f_0$ does not vanish. The solution may be iteratively constructed as follows. When solutions exist, they are differentiable and, for a small time step Δt and any $\Gamma \in \mathcal{M}$, one may, for instance, write[2]:

$$
\begin{aligned}
f_{\Delta t}(\Gamma) &= f_0(\Gamma) - i\Big(\mathcal{L}_0 + \mathcal{L}_{ext}(0)\Big) f_0(\Gamma)\,\Delta t + O\left(\Delta t^2\right) && (7.22)\\
&= f_0(\Gamma) - i\mathcal{L}_{ext}(0)\, f_0(\Gamma)\,\Delta t + O\left(\Delta t^2\right) = e^{-i\Delta t \mathcal{L}_{ext}(0)} f_0(\Gamma) + O(\Delta t^2) && (7.23)
\end{aligned}
$$

The second time step yields

$$
\begin{aligned}
f_{2\Delta t}(\Gamma) &= f_{\Delta t}(\Gamma) - i\mathcal{L}_0 \Delta t f_{\Delta t}(\Gamma) - i\mathcal{L}_{ext}(\Delta t)\,\Delta t f_{\Delta t}(\Gamma) + O\left(\Delta t^2\right) && (7.24)\\
&= [1 - i\mathcal{L}_0 \Delta t]\, f_{\Delta t}(\Gamma) - i\mathcal{L}_{ext}(\Delta t)\,\Delta t f_{\Delta t}(\Gamma) + O\left(\Delta t^2\right) && (7.25)\\
&= e^{-i\Delta t \mathcal{L}_0} f_{\Delta t}(\Gamma) - i\mathcal{L}_{ext}(\Delta t)\,\Delta t f_{\Delta t}(\Gamma) + O\left(\Delta t^2\right) && (7.26)
\end{aligned}
$$

[2] In fact, rather than $O(\Delta t^2)$, less stringent conditions about the higher order corrections suffice. For instance, it suffices that they are of order $O(\Delta t^{1+\varepsilon})$, with any $\varepsilon > 0$.

At the last time step, that leads from time $t - \Delta t = (n-1)\Delta t$ to time $t = n\Delta t$, we have

$$f_t(\Gamma) = e^{-i\Delta t \mathscr{L}_0} f_{t-\Delta t}(\Gamma) - i\mathscr{L}_{ext}(t-\Delta t) f_{t-\Delta t}(\Gamma)\Delta t + O\left(\Delta t^2\right) \quad (7.27)$$
$$= e^{-i\Delta t \mathscr{L}_0} f_{t-\Delta t}(\Gamma) - i\Delta t e^{-i\,0\mathscr{L}_0}\mathscr{L}_{ext}(t-\Delta t) f_{t-\Delta t}(\Gamma) + O\left(\Delta t^2\right)$$

where, for notation consistency, we have expressed the unit operator acting on $\mathscr{L}_{ext}(t-\Delta t) f_{t-\Delta t}(\Gamma)$ as $i\exp(-i\,0\mathscr{L}_0)$. To construct the solution with initial condition f_0, it is convenient to proceed backward in time from f_t, iteratively rewriting Eq. (7.27). Replace the first term $f_{t-\Delta t}$ on the right-hand side of the equation with its approximation in terms of $f_{t-2\Delta t}$, and do not change the second, which is subject to the action of $\mathscr{L}_{ext}(t-\Delta t)$. This operation introduces a term $f_{t-2\Delta t}$ subject to $\mathscr{L}_{ext}(t-2\Delta t)$. Continuing, we obtain the exponential of $-it\mathscr{L}_0$ applied to f_0, which is f_0, plus a sum over all values of $\mathscr{L}_{ext}(k\Delta t) f_{k\Delta t}$, in reverse order along the trajectory. But sums commute, and we obtain an integral with memory kernel, when Δt decreases. Explicitly, we write

$$f_t(\Gamma) = e^{-i\Delta t\mathscr{L}_0}\left[e^{-i\Delta t\mathscr{L}_0} f_{t-2\Delta t}(\Gamma) - i\Delta t e^{-i0\mathscr{L}_0}\mathscr{L}_{ext}(t-2\Delta t) f_{t-2\Delta t}(\Gamma) + O\left(\Delta t^2\right)\right]$$
$$-i\Delta t e^{-i0\mathscr{L}_0}\mathscr{L}_{ext}(t-\Delta t) f_{t-\Delta t}(\Gamma) + O\left(\Delta t^2\right) \quad (7.28)$$

$$= e^{-i2\Delta t\mathscr{L}_0} f_{t-2\Delta t}(\Gamma) - i\Delta t e^{-i\Delta t\mathscr{L}_0}\mathscr{L}_{ext}(t-2\Delta t) f_{t-2\Delta t}(\Gamma)$$
$$-i\Delta t\, e^{-i0\mathscr{L}_0}\mathscr{L}_{ext}(t-\Delta t) f_{t-\Delta t}(\Gamma) + O\left(2\Delta t^2\right) \quad (7.29)$$

$$= e^{-i2\Delta t\mathscr{L}_0}\left[e^{-i\Delta t\mathscr{L}_0} f_{t-3\Delta t}(\Gamma) - ie^{-i0\mathscr{L}_0}\mathscr{L}_{ext}(t-3\Delta t) f_{t-3\Delta t}(\Gamma)\Delta t + O\left(\Delta t^2\right)\right]$$
$$-i\Delta t e^{-i\Delta t\mathscr{L}_0}\mathscr{L}_{ext}(t-2\Delta t) f_{t-2\Delta t}(\Gamma)$$
$$-i\Delta t\, e^{-i0\mathscr{L}_0}\mathscr{L}_{ext}(t-\Delta t) f_{t-\Delta t}(\Gamma) + O\left(2\Delta t^2\right) \quad (7.30)$$

$$= e^{-i3\Delta t\mathscr{L}_0} f_{t-3\Delta t}(\Gamma) - i\Delta t e^{-i2\Delta t\mathscr{L}_0}\mathscr{L}_{ext}(t-3\Delta t) f_{t-3\Delta t}(\Gamma)$$
$$-i\Delta t e^{-i\Delta t\mathscr{L}_0}\mathscr{L}_{ext}(t-2\Delta t) f_{t-2\Delta t}(\Gamma)$$
$$-i\Delta t\, e^{-i0\mathscr{L}_0}\mathscr{L}_{ext}(t-\Delta t) f_{t-\Delta t}(\Gamma) + O\left(3\Delta t^2\right) \quad (7.31)$$

$$\vdots \qquad \vdots \qquad \vdots \qquad \vdots$$

$$= e^{-in\Delta t\mathscr{L}_0} f_{t-n\Delta t}(\Gamma) - i\Delta t\sum_{k=1}^{n} e^{-i(k-1)\Delta t\mathscr{L}_0}\mathscr{L}_{ext}(t-k\Delta t) f_{t-k\Delta t}(\Gamma) + O\left(n\Delta t^2\right)$$

$$= e^{-it\mathscr{L}_0} f_0(\Gamma) - i\Delta t\sum_{k=1}^{n} e^{-i(k-1)\Delta t\mathscr{L}_0}\mathscr{L}_{ext}(t-k\Delta t) f_{t-k\Delta t}(\Gamma) + O\left(n\Delta t^2\right) \quad (7.32)$$

Then, $O\left(n\Delta t^2\right) = O(t^2/n) \to 0$ in the $\Delta t \to 0$ limit, for every finite t. Introducing $t' = t - k\Delta t$, so that $k\Delta t = t - t'$ and $(k-1)\Delta t = t - t' - \Delta t$, the $\Delta t \to 0$ limit finally yields

$$f_t(\Gamma) = f_0(\Gamma) - i\int_0^t dt' e^{-i(t-t')\mathscr{L}_0}\mathscr{L}_{ext}\left(t'\right) f_{t'}(\Gamma) \quad (7.33)$$

This equation is exact, but not particularly useful. However, it can be rewritten as

$$f_t(\Gamma) = f_0(\Gamma) - i\int_0^t dt' e^{-i(t-t')\mathscr{L}_0}\mathscr{L}_{ext}(t')\Big[f_0(\Gamma) + \\ -i\int_0^{t'} dt'' e^{-i(t'-t'')\mathscr{L}_0}\mathscr{L}_{ext}(t'')\, f_{t''}(\Gamma)\Big] \qquad (7.34)$$

where the same expression (7.33) has been used, replacing t with t' and t' with t'', to express $f_{t'}$. In this way, f_t is expressed as a sum of one term that is linear in $\mathscr{L}_{ext}$, i.e., in the intensity of the perturbation $\mathscr{F}$, Eq. (7.19), and one term that is quadratic in $\mathscr{F}$ that is negligible for sufficiently small perturbations. In that case, the following linear approximation of the solution of Liouville's equation (7.20) becomes relevant:

$$f_t(\Gamma) = f_0(\Gamma) - i\int_0^t dt' e^{-i(t-t')\mathscr{L}_0}\mathscr{L}_{ext}(t')\, f_0(\Gamma) + \text{H.O.} \qquad (7.35)$$

where H.O. stands for the negligible higher order terms in $\mathscr{F}$.

However, what is relevant are not the probabilities but the values of the observables. Therefore, let us compute the time-dependent behavior of the ensemble average of a given phase function $\mathscr{O} : \mathscr{M} \to I\!R$, assuming that it bears physical content. Denoting the time-dependent average by

$$\mathbb{E}_t(\mathscr{O}) = \int_{\mathscr{M}} \mathscr{O}(\Gamma)\, f_t(\Gamma) \mathrm{d}\Gamma \qquad (7.36)$$

one obtains

$$\begin{aligned} \Delta_t\mathbb{E}(\mathscr{O}) &= \mathbb{E}_t(\mathscr{O}) - \mathbb{E}_0(\mathscr{O}) \\ &= -i\int_{\mathscr{M}} \mathscr{O}(\Gamma)\int_0^t dt' e^{-i(t-t')\mathscr{L}_0}\mathscr{L}_{ext}(t')\, f_0(\Gamma)\mathrm{d}\Gamma + \text{H.O.} \\ &= \int_{\mathscr{M}} \mathscr{O}(\Gamma)\int_0^t dt' e^{-i(t-t')\mathscr{L}_0}\mathscr{F}(t')\{f_0, A\}(\Gamma)\,\mathrm{d}\Gamma + \text{H.O.} \end{aligned} \qquad (7.37)$$

where, Eq. (7.19) has been used. Observing that $\partial f_0/\partial\mathscr{H}_0 = -\beta f_0$, one also gets

$$\begin{aligned} \{f_0, A\} &= \frac{\partial f_0}{\partial Q}\frac{\partial A}{\partial P} - \frac{\partial f_0}{\partial P}\frac{\partial A}{\partial Q} = \frac{\partial f_0}{\partial \mathscr{H}_0}\frac{\partial \mathscr{H}_0}{\partial Q}\frac{\partial A}{\partial P} - \frac{\partial f_0}{\partial \mathscr{H}_0}\frac{\partial \mathscr{H}_0}{\partial P}\frac{\partial A}{\partial Q} \\ &= \frac{\partial f_0}{\partial \mathscr{H}_0}\left[\frac{\partial \mathscr{H}_0}{\partial Q}\frac{\partial A}{\partial P} - \frac{\partial \mathscr{H}_0}{\partial P}\frac{\partial A}{\partial Q}\right] = -\beta f_0\{\mathscr{H}_0, A\} = \beta f_0\{A, \mathscr{H}_0\} \end{aligned} \qquad (7.38)$$

which yields

$$\Delta_t\mathbb{E}(\mathscr{O}) = \beta\int_0^t \mathrm{d}t'\,\mathscr{F}(t')\int_{\mathscr{M}} \mathrm{d}\Gamma\,\mathscr{O}(\Gamma)e^{-i(t-t')\mathscr{L}_0}\{A, \mathscr{H}_0\}(\Gamma) f_0(\Gamma) + \text{H.O.} \qquad (7.39)$$

when the order of time and phase space integrals can be exchanged. Now, the perturbed evolution of the quantity A, expressed by $\dot{A}_{pert} = \{A, \mathscr{H}_0 - F(t)A\}$, equals the unperturbed evolution $\dot{A}_{unpert} = \{A, \mathscr{H}_0\}$. Indeed:

$$\underbrace{\{A, \mathscr{H}_0\}}_{\dot{A}_{unpert}} = \{A, \mathscr{H}_0 - F(t)A\} = \underbrace{\{A, \mathscr{H}\}}_{\dot{A}_{pert}} \tag{7.40}$$

This makes particularly striking the response formula (7.39), since it obtains nonequilibrium properties from equilibrium dynamics and equilibrium ensembles! analogously to the theory of the Brownian motion, but in a wider framework.[3] Considering that $\mathscr{L}_0$, hence $\exp(-i(t-t')\mathscr{L}_0)$, does not act on $\mathscr{F}$, and neglecting the higher order terms, the response formula may thus be rewritten as

$$\Delta_t \mathbb{E}(\mathscr{O}) = \beta \int_0^t \mathrm{d}t' \, \mathscr{F}(t') \int_{\mathscr{M}} \mathrm{d}\Gamma \, \mathscr{O}(\Gamma) e^{-i(t-t')\mathscr{L}_0} \dot{A}(\Gamma) f_0(\Gamma) \tag{7.41}$$

To obtain a mathematically handier and physically meaningful expression of the time-dependent response, we would like to swap the time evolution operator acting on $\dot{A} f_0$, with the evolution operator acting on the observable $\mathscr{O}$. That way, f_0 preserves its meaning as (equilibrium) ensemble of microscopic states of a single object, and the evolution concerns only the measurable quantity $\mathscr{O}$. We obtain

$$\Delta_t \mathbb{E}(\mathscr{O}) = \beta \int_0^t \mathrm{d}t' \, \mathscr{F}(t') \int_{\mathscr{M}} \mathrm{d}\Gamma \, \dot{A}(\Gamma) f_0(\Gamma) \, e^{i(t-t')\mathscr{L}_0} \mathscr{O}(\Gamma) \tag{7.42}$$

where the change in sign in the argument of the exponential produces, as is common, the adjoint of the evolution operator,[4] and the "−" sign is eliminated by the integration by parts. Introducing the *response function* R, defined by

$$R(t) = \beta \int_{\mathscr{M}} \mathrm{d}\Gamma \, f_0(\Gamma) \, \dot{A}(\Gamma) e^{it\mathscr{L}_0} \mathscr{O}(\Gamma) = \beta \mathbb{E}_0 \left[\left(\mathscr{O} \circ \Phi_0^t \right) \dot{A} \right] \tag{7.43}$$

and the subscript 0 in Φ_0^t stresses that the unperturbed evolution is used. We finally obtain

$$\Delta_t \mathbb{E}(\mathscr{O}) = \int_0^t \mathrm{d}t' \, R(t-t') \mathscr{F}(t') \tag{7.44}$$

[3] Note: $\dot{A}$ is a phase function independent of time like any other, although it appears as a time derivative. It depends on time only through its variable Γ: given Γ its value is fixed, and time does not matter.

[4] Assume that the functions to integrate belong to a Hilbert space $\mathscr{B}$, such as L^2, which is more than enough for our purpose, and denote by $< \cdot, \cdot >$ the scalar product. Suppose the operator T maps $\mathscr{B}$ to itself. Its adjoint T^* maps $\mathscr{B}$ to $\mathscr{B}$, and is defined by the equality $< g, Th > = < T^* g, h >$ for all functions $g, h \in \mathscr{B}$. Given the differential nature of the operator $\mathscr{L}_0$, integration by parts with vanishing conditions at the boundary of $\mathscr{M}$ yields the equality between Eqs. (7.41) and (7.42), where the integral represents the scalar product, $g = \mathscr{O}$, $h = f_0 \dot{A}$, $T = \exp[-i(t-t')\mathscr{L}_0]$, and $T^* = \exp[i(t-t')\mathscr{L}_0]$.

It is worth stressing that the response to a small perturbation is determined by the *equilibrium dynamics*, by the time correlation function of the perturbation and the observable of interest computed with respect to the (known) *equilibrium ensemble*!

Assuming the system of interest has N degrees of freedom, consider the quantity

$$\{A, \mathscr{H}_0\} = \sum_{j=1}^{N} \left[\frac{\partial A}{\partial q_j} \frac{\partial \mathscr{H}_0}{\partial p_j} - \frac{\partial A}{\partial p_j} \frac{\partial \mathscr{H}_0}{\partial q_j} \right] = - \sum_{j=1}^{N} \left[\frac{\partial \mathscr{H}_0}{\partial p_j} K_j^p - F_j^{(0)} K_j^q \right] \quad (7.45)$$

where $F_j^{(0)} = -\partial \mathscr{H}_0 / \partial q_j$ can be interpreted as the internal force acting on particle j due to the other particles of the system, while $K_j^p = -\partial A / \partial q_j$ can be interpreted as the force due to an external agent. The dimensions of $\{A, \mathscr{H}_0\}$ are those of $\dot{A}$, i.e., of energy flow: energy per unit time. Thus, the quantity

$$J(\Gamma) = -\{A, \mathscr{H}_0\}(\Gamma) = -\frac{dA}{dt}(\Gamma) \quad (7.46)$$

is called *dissipative flux*, because its average in a stationary state represents the power of the perturbing agent, which produces work and is eventually dissipated in the environment. With this notation, the response function writes

$$R(t) = -\beta \mathbb{E}_0 \left[(\mathscr{O} \circ \Phi_0^t)\, J \right] \quad (7.47)$$

The "−" sign is conventional, but sensible. For instance, take a constant intensity $\mathscr{F}(t) = F$, and $\mathscr{F}(t)A(\Gamma) = -F \sum_1^N q_i$, so that particles are pushed by a constant external force in the positive direction. Then, $K_j^q = 0$, $K_j^p = 1$, and

$$\{A, \mathscr{H}_0\} = -\sum_{j=1}^{N} \frac{\partial \mathscr{H}_0}{\partial p_j}$$

which is opposite to the current (cf. Example 7.2.1.1 below). The "−" sign in Eq. (7.46) leads to a positive dissipation for positive currents.

7.2.1 Examples

The above relations are universally confirmed, proving the validity of the linear response theory. In fact, linear response theory is extremely successful and can be derived analogously also in stochastic frameworks, replacing the phase space with the state space, and Liouville's equation with, e.g., the Fokker-Planck equation (Marconi et al. 2008).

7.2.1.1 Diffusion in Presence of a Constant Force

Consider N particles subject to a spatially uniform force, along the x-direction. In this case, rather than taking $\mathscr{F}$ as a pure number, it is handier to give it the dimensions of force. Nothing changes in practice. Thus we take

$$\mathscr{H}_0 = \sum_{i=1}^{N} \frac{p_i^2}{2m} + V(Q) \, ; \quad \mathscr{F}(t)A(\Gamma) = -F \sum_{j=1}^{N} q_j^x$$

where q_j^x is the x-component of the position of particle j. Then, the energy dissipation writes

$$\mathscr{F}\, J(\Gamma) = -\left\{ -F \sum_{j=1}^{N} q_j^x \, , \, \mathscr{H}_0 \right\} = \sum_{j=1}^{N} F \frac{\partial \mathscr{H}_0}{\partial p_j} = \frac{F}{m} \sum_{j=1}^{N} p_j^x$$

where p_j^x is the momentum in the x-direction of particle j. As an observable, consider

$$\mathscr{O}(\Gamma) = -\frac{1}{m} \sum_{j=1}^{N} p_j^x = -\, J(\Gamma)$$

whose equilibrium average computed with f_0, the unperturbed canonical ensemble (7.1), vanishes, $\mathbb{E}_0(\mathscr{O}) = 0$, because it is odd in the momenta, while f_0 is even in the momenta, and momenta take values in $(-\infty, \infty)$. Then

$$\begin{aligned}
\Delta_t \mathbb{E}(\mathscr{O}) &= \beta \int_0^t \mathrm{d}t' F \int \mathrm{d}\Gamma f_0(\Gamma) \{A, \mathscr{H}_0\}(\Gamma) e^{i(t-t')\mathscr{L}_0} \mathscr{O}(\Gamma) \\
&= \beta \int_0^t \mathrm{d}t' \mathbb{E}_0 \left[\dot{A} \left(\mathscr{O} \circ \Phi_0^{t-t'} \right) \right] = \beta \int_0^t \mathrm{d}t' \, \mathbb{E}_0 \left[F \sum_{j=1}^{N} \dot{q}_j^x \sum_{k=1}^{N} \frac{1}{m} p_k^x \circ \Phi_0^{t-t'} \right] \\
&= \frac{\beta F}{m^2} \int_0^t \mathrm{d}t' \sum_{j,k=1}^{N} \mathbb{E}_0 \left[p_j^x \, p_k^x \circ \Phi_0^{t-t'} \right] = \beta F \int_0^t \mathrm{d}t' \, \mathbb{E}_0 \left[\mathscr{O} \left(\mathscr{O} \circ \Phi_0^{t-t'} \right) \right]
\end{aligned}$$

The behavior of the response of $\mathscr{O}$ now depends on the specific properties of the microscopic equilibrium dynamics. If it is such that particles velocities are not correlated, the sum of the mean of the cross products vanishes, and one obtains

$$\Delta_t \mathbb{E}(\mathscr{O}) = \frac{\beta F}{m^2} \int_0^t \mathrm{d}t' \sum_{j=1}^{N} \mathbb{E}_0 \left[p_j^x \left(p_j^x \circ \Phi_0^{t-t'} \right) \right] \tag{7.48}$$

and, setting $s = t - t'$, $\mathrm{d}s = -\,\mathrm{d}t'$, one gets

$$\Delta_t \mathbb{E}(\mathscr{O}) = \frac{\beta F}{m^2} \int_t^0 (-\,\mathrm{d}s) \sum_{j=1}^{N} \mathbb{E}_0 \left[p_j^x \left(p_j^x \circ \Phi_0^{s} \right) \right] = \frac{\beta F}{m^2} \int_0^t \mathrm{d}s \sum_{j=1}^{N} \mathbb{E}_0 \left[p_j^x \left(p_j^x \circ \Phi_0^{s} \right) \right]$$

Recalling the mobility is defined by $\mu = \sum_i \mathbb{E}_\infty(\dot{q}_i^x)/F = \mathbb{E}_\infty(\mathscr{O})/F$, we finally obtain

$$\mu = \frac{\beta}{m^2} \int_0^\infty \mathrm{d}t \sum_{j=1}^{N} \mathbb{E}_0 \left[p_j^x \left(p_j^x \circ \Phi_0^s \right) \right] \tag{7.49}$$

This expression is called *Green-Kubo integral* for the (linear) transport coefficient μ. It reveals a striking link between equilibrium properties (the 2-time correlation function) and nonequilibrium ones (μ), which correspond to a net dissipation, when work is done. However, as with Einstein's relation (3.28), there is no need to do any work: using Eq. (7.49) it suffices to run equilibrium simulations or observe spontaneous equilibrium fluctuations to obtain the nonequilibrium property.

7.2.1.2 Shear Flow in 2 Dimensions

As in the previous example, here $\mathscr{F}$ is not dimensionless; it is more convenient to take it as a rate, i.e., inverse of time:

$$\mathscr{F} = \frac{\gamma}{2}; \quad A(\Gamma) = \sum_{j=1}^{N} \left[q_j^y p_j^x + q_j^x p_j^y \right]$$

where $\gamma = \partial v_x / \partial y$ is the imposed shear rate. This leads to the so-called *SLLOD* equations of motion:

$$\frac{\mathrm{d}q_j^x}{\mathrm{d}t} = \frac{p_j^x}{m} + \gamma q_j^y \;, \qquad \frac{\mathrm{d}q_j^y}{\mathrm{d}t} = \frac{p_j^y}{m} + \gamma q_j^x$$
$$\frac{\mathrm{d}p_j^x}{\mathrm{d}t} = F_j^x - \gamma p_j^y \;, \qquad \frac{\mathrm{d}p_j^y}{\mathrm{d}t} = F_j^y - \gamma p_j^x$$

where (F_j^x, F_j^y) is the internal force on particle j determined by $\mathscr{H}_0$, and the rest is due to the external shearing forces. The energy dissipation takes the form:

$$\mathscr{F} J(\Gamma) = -\frac{\gamma}{2} \{A, \mathscr{H}_0\} \tag{7.50}$$

$$= -\frac{\gamma}{2} \sum_{j=1}^{N} \left[p_j^x \frac{\partial \mathscr{H}_0}{\partial p_j^y} + p_j^y \frac{\partial \mathscr{H}_0}{\partial p_j^x} - q_j^y \frac{\partial \mathscr{H}_0}{\partial q_j^x} - q_j^x \frac{\partial \mathscr{H}_0}{\partial q_j^y} \right] \tag{7.51}$$

$$= -\frac{\gamma}{2} \sum_{j=1}^{N} \left[2\frac{1}{m} p_j^x p_j^y + q_j^y F_j^x + q_j^x F_j^y \right] \tag{7.52}$$

i.e., it is proportional to the off diagonal elements of the pressure tensor, for a two-dimensional fluid in a volume V:

$$P^{xy} = \frac{1}{2V} \sum_{j=1}^{N} \left[\frac{2}{m} p_j^x p_j^y + q_j^y F_j^x + q_j^x F_j^y \right] = -\frac{1}{\gamma V} J$$

Then, the response of the observable $\mathscr{O} = -P^{xy}$ to the shear rate is given by

$$\Delta_t \mathbb{E}(-P^{xy}) = -\beta \gamma V \int_0^t \mathbb{E}_0 \left[P^{xy} \left(P^{xy} \circ \Phi_0^s \right) \right] \mathrm{d}s \tag{7.53}$$

where $\mathbb{E}_0(-P^{xy}) = 0$ results from the (quite generally valid) hypothesis that p_j^x and p_j^y are independent at equilibrium, like positions and internal forces are. The shear viscosity consequently takes the form

$$\eta = -\lim_{t\to\infty} \frac{\Delta_t \mathbb{E}(-P^{xy})}{\gamma} = \beta V \int_0^\infty \mathbb{E}_0 \left[P^{xy} \left(P^{xy} \circ \Phi_0^s \right) \right] \mathrm{d}s \tag{7.54}$$

which is the Green-Kubo integral for the (linear) transport coefficient η. Again, a quantity determining the dissipation can be obtained from equilibrium ensembles and equilibrium dynamics.

7.2.1.3 Thermal Conductivity

Thermal properties are among the most sophisticated. For instance, concerning thermal conductivity λ, one considers the following observable:

$$\mathscr{O} = \sum_{j=1}^{N} \left\{ \left(\frac{p_j^2}{2m} - \mathbb{E}_0(\varepsilon_j) \right) \cdot \frac{p_j}{m} + \frac{1}{2} \sum_{j \neq i=1}^{N} \left[(q_i - q_j) \cdot F_{ij} + U(q_i - q_j) \right] \right\} \cdot \frac{p_j}{m} \tag{7.55}$$

where $\mathbb{E}_0(\varepsilon_j)$ is the enthalpy per particle , U is the prior particle potential and F_{ij} the force of particle j on particle i. This, eventually leads to (McQuarrie 1975):

$$\lambda = \frac{1}{3\, V k_B T^2} \int_0^\infty \mathbb{E}_0 \left[\mathscr{O} \left(\mathscr{O} \circ \Phi_0^t \right) \right] \mathrm{d}t \tag{7.56}$$

another Green-Kubo integral. Like the previous ones, it consists of the time integral of the equilibrium two-times autocorrelation function of a suitable flux, within which the equilibrium dynamics is used.

7.2.2 Remarks on Linear Response Theory

Various comments are in order.

- Strictly speaking, the thermodynamic relevance of this approach is questionable. It produces formulae for abstract *ensembles* of identical objects, rather than for a single macroscopic object of interest. Single systems, starting from different initial microscopic phases, although sampled from the same equilibrium distribution, may behave differently, making the ensemble averages thermodynamically scarcely meaningful. This difficulty is overcome if all systems, or the vast majority of them, behave in the same way, so that averages coincide with the single system behavior. The conditions for the validity of the Boltzmann equation allow that.

- As discussed in Sect. 6.4 ensembles are required to be stationary, in order to properly represent the macroscopic states of a given system. The reason is that the ranges of the observables must be explored with given frequencies, during the microscopic evolution. A modification of the state corresponds to different frequencies, hence one may obtain a mixture of states, rather than representing a state. Nevertheless, given a phase function $\mathscr{O} : \mathscr{M} \to R_{\mathscr{O}}$, the evolution of $\mathbb{E}_t(\mathscr{O})$ may be slow compared to the times required for a thorough exploration of the range $R_{\mathscr{O}}$ with proper frequencies, and then the ensemble average at a specific time characterizes the macroscopic state of one system of interest at that macroscopic time.
- What happens asymptotically in time if the perturbation represents an external field indefinitely accelerating the particles of the system? How can f_0 remain the relevant probability distribution in $\mathscr{M}$? There are two issues linked to these questions. First, the use of the canonical ensemble indicates an energy exchange between system and environment, which may lead to a nonequilibrium steady state, because the perturbation energy input is balanced by the energy dissipated in the environment. Then, a small perturbation of the dynamics may result in a small change of state, in the sense that the values of the observables of interest differ little from their equilibrium values. This may be obtained with unnoticeable deformations of the equilibrium distribution. Moreover, the smaller the size of the perturbation, the longer the linear approximation holds. For sufficiently small $\mathscr{F}$, these times may well be substantially longer than physically required, allowing us to consider the systems in a steady state. Both conditions can hardly be deduced from the equations of motion and must be checked on a case by case basis.
- Equation (7.42) corresponds to the presence of memory effects. In principle, this makes all nonequilibrium phenomena described by the theory substantially different from equilibrium phenomena, no matter how small the perturbations are. However, the separation of scales required for the thermodynamic relevance of the microscopic theory helps also in this respect. Memory kernels in the mathematical representation of the microscopic dynamics do not exclude *per se* the validity of Markov processes. If correlations decay sufficiently fast in time, they can be macroscopically neglected, as in the case of the Brownian motion. In fact, viscoelastic behaviors are observed only in complex fluids or under relatively extreme conditions, i.e., far from equilibrium.
- The most famous criticism of the linear response theory was formulated by van Kampen. He maintained that while there is no doubt that linear response physically holds, the theory we have summarized cannot be the right explanation: *"Linear response theory does provide expressions for the phenomenological coefficients, but I assert that it arrives at these expressions by a mathematical exercise, rather than by describing the actual mechanism which is responsible for the response"* (van Kampen 1971). As a matter of fact, van Kampen was right, in principle,

since the Green-Kubo calculations invoke non-trivial dynamical ingredients. For instance, an electric field **E** perturbs the trajectories of electrons, resulting in deviations that are expected to be of order $O(t^2e\mathbf{E}/2m)$, where t is the time since **E** has been switched on, m the mass of one electron, and e its charge. Then, van Kampen concluded that the induced current is linear in the magnitude E of the field only if $t^2e\mathbf{E}/2m$ is much smaller than the mean spacing of impurities d. For $d = 1$ Å and $t = 1$ s, he then concluded that the field has to be smaller than 10^{-18} V/cm, which would make the linear regime useless. This led van Kampen to believe that the theory outlined above does not explain linear response. The point was reasonable, but it was later realized that particles interacting with short range repulsion tame themselves under quite variegated conditions, making results robust for macroscopic observables of thermodynamic (hence large) systems. Indeed, a current is a phenomenon due to a large assembly of trajectories, and for this assembly, it only requires a sufficiently fast decay of correlations. This collective property is favored rather than impaired by chaos.

- One important aspect of linear response theory is that it does not require any knowledge of the origin of the perturbation. Is that an obvious fact? In general, it is not, that is why Onsager introduced the following (Onsager 1931b):

  Regression hypothesis: *The average regression of fluctuations will obey the same laws as the corresponding macroscopic irreversible process.*

  If a system finds itself in a state other than the steady state, Onsager stated that on average it will relax to the steady state in a way that does not depend on how it got away from it. Relaxation after a spontaneous fluctuation, or from an artificially created state, on average follows the same path. For equilibrium states, this is commonly accepted; for nonequilibrium states, it must be seen case by case.
- When the approach is physically relevant, the response formula (7.42) and the Green-Kubo integrals are quite striking: they obtain nonequilibrium properties from equilibrium ensembles *and* equilibrium dynamics: only f_0 and $\mathscr{L}_0$ play a role. The result is secured by the equilibrium correlation function, that relates equilibrium microscopic dynamics to nonequilibrium macroscopic ones, yielding the transport coefficients.

Linear response theory is therefore practically complete[5] As noted numerous times, a number of assumptions must be verified. Ensemble averages must coincide with time-dependent single systems behaviors, which requires a separation of scales that allows observables to swiftly span their range. The good news is that macroscopic objects moderately perturbed away from their equilibrium behave in accord with these requirements, which amount to a kind of *physical* ergodicity, cf. Sect. 6.4. Further good news is that perturbations do not need to be particularly small compared to our daily life scales. Ohm's law, Fourier's law, Fick's law, the Navier-Stokes

[5] Its numerical implementation may meet some difficulty. In nonequilibrium molecular dynamics, the problem is successfully tackled in many situations using *thermostatting* terms in the equations of motion, Evans and Morriss (2021).

equations, and many others relying on linear transport coefficients work beautifully under a huge range of situations. This is due to the fact that the perturbation per degree of freedom is typically small if a macroscopic number of particles is involved. Linear response is vindicated by experience.

7.3 Linear Response in Magnetic Field

The Onsager reciprocal relations constitute one of the major successes in deriving linear nonequilibrium thermodynamics relations from a statistical mechanics point of view. They are based on the property of local thermodynamic equilibrium and of time reversal invariance of the dynamics of atoms and molecules. Provided a proper choice is made for the "forces" X_j, with $j = 1, ..., n$, each of which separately induces properly defined "fluxes" J_j, the linear regime is characterized by constant matrices $L = (L_{jk})_{j,k=1}^n$ and $R = (R_{jk})_{j,k=1}^n$, such that

$$J_i = \sum_{j=1}^{n} L_{ij} X_j \, , \qquad X_i = \sum_{j=1}^{n} R_{ij} J_j \, ; \quad i = 1, ..., n \tag{7.57}$$

Then, Onsager's result is that the entries of L, known as linear transport coefficients, enjoy a special symmetry, that depends on whether the fluxes are even or odd functions of the particles velocities. Denoting by the superscript $(\alpha\alpha)$ the coefficients linking the forces of two even fluxes, by the superscripts $(\alpha\beta)$ and $(\beta\alpha)$ the coefficients linking one odd force to one even flux and vice versa, and by the superscript $(\beta\beta)$ the coefficients linking the forces of two odd fluxes, the Onsager reciprocal relations take the form:

$$L_{ij}^{\alpha\alpha} = L_{ji}^{\alpha\alpha} \, , \qquad L_{ij}^{\alpha\beta} = -L_{ji}^{\beta\alpha} \, , \qquad L_{ij}^{\beta\beta} = L_{ji}^{\beta\beta} \tag{7.58}$$

The set of Eq. (7.58) constitutes the theoretical expression of phenomena that had been previously observed, such as the Soret-Dufour effect, concerning a mixture in which neither temperature nor concentration are uniform. In this case, the heat flow gets a contribution from the thermodynamic force due to the concentration gradient, with proportionality constant L_{qm}, and the diffusion of mass gets a contribution from the thermodynamic force associated with the temperature gardient, with proportionality constant $L_{mq} = L_{qm}$. Similarly, the thermo-electric effect, or the Peltier-Seebeck-Thomson effect, couple heat flows with electric currents.

Onsager relations introduce constraints on the conversion of heat to work, like other thermodynamic relations. Therefore, their violation may in principle allow non-dissipative thermodynamic currents. At the same time, it was argued that the Onsager reciprocal relations do not hold in systems subjected to a magnetic field, or in a rotating reference frame, because their dynamics were thought to violate time reversal invariance. Onsager himself stated: *"The reciprocal relations (2.4) and their equivalent, the principle of the least dissipation of energy (5.11) are derived from the assumption of microscopic reversibility. We may expect that these relations will break down in cases where magnetic or Coriolis forces are acting, and they do.*" (Onsager 1931a). Casimir then developed Onsager's argument, stating that the reciprocal relations should be replaced by the following Casimir (1945): The Onsager-Casimir relations

$$L_{ij}(\mathbf{B}) = \pm L_{ji}(-\mathbf{B}), \quad i, j = 1, ..., n \tag{7.59}$$

where the "+" or "−" sign is determined as in Eq. (7.58), if **B** is the magnetic field in which the system of interest is immersed. Analogous modification concerns the inversion of the angular velocity of rotating frames. While this relation is definitely correct, it defeats the purpose of describing the transport coefficient for a given system: two systems, one in the magnetic field **B** and the other in the magnetic field—**B**, must be considered. Then, the L_{ij} coefficient of one system cannot be inferred from the L_{ji} coefficient of that same system: a second experiment, with a different system, must be performed.

However, violations of the Onsager relations[6] or non-dissipative currents have not been observed, so far, hence the question whether magnetic fields really break the Onsager symmetry has remained open.

This question has received an answer, in favor of the validity of the Onsager relations even in the presence of magnetic fields (Luo et al. 2020), or rotating frames, cf. Ref. Bonella et al. (2015). There are two aspects to be considered: while magnetic fields and rotating frames break the usual time reversal symmetry, based on the inversion of momenta, $(Q, P) \to (Q, -P)$, it does not break other possible time reversal symmetries, in the sense of Eq. (6.3). This has direct application on the correlation functions from which the transport coefficients can be computed, via Green-Kubo integrals. For instance, the operator $\mathcal{T} : \mathcal{M} \to \mathcal{M}$ defined by

$$\mathcal{T}(x, y, z, p_x, p_y, p_z) = (x, -y, z, -p_x, p_y, -p_z) \tag{7.60}$$

previously used in the case of shearing fluids (Evans and Morriss 2021), preserves the form of the equations of motion in presence of a magnetic field, if applied together with time inversion $t \to -t$:

$$\begin{aligned}
\dot{x}_i &= \frac{p_{iX}}{m_i} + \omega_i y_i & &\longmapsto & -\dot{x}_i &= -\frac{p_{iX}}{m_i} - \omega_i y_i \\
\dot{y}_i &= \frac{p_{iY}}{m_i} - \omega_i x_i & &\longmapsto & \dot{y}_i &= \frac{p_{iY}}{m_i} - \omega_i x_i \\
& & &\longmapsto & & \\
\dot{z}_i &= \frac{p_{iZ}}{m_i} & & & -\dot{z}_i &= -\frac{p_{iZ}}{m_i} \\
& & &\longmapsto & & \\
\dot{p}_{iX} &= F_{iX} + \omega_i (p_{iY} - m_i \omega_i x_i) & &\longmapsto & \dot{p}_{iX} &= F_{iX} + \omega_i (p_{iY} - m_i \omega_i x_i) \\
\dot{p}_{iY} &= F_{iY} - \omega_i (p_{iX} + m_i \omega_i y_i) & &\longmapsto & -\dot{p}_{iY} &= -F_{iY} + \omega_i (p_{iX} + m_i \omega_i y_i) \\
\dot{p}_{iZ} &= F_{iZ} & & & \dot{p}_{iZ} &= F_{iZ}
\end{aligned}$$

[6] Up to the present time the only experimental reports on the validity of Eq. 7.59 we are aware of, are studies on the thermopower of Bismuth crystals (Smith and Wolfe 1966).

Interestingly, also electric fields break time reversal symmetries, but that is not noted, usually, because they preserve the standard one which involves the inversion of momenta. Now, assuming that one involution operator $\mathscr{I}$ preserves the equations of motion under time reversal, the calculation of the equilibrium time correlation functions proceeds as follows. The equilibrium probability density in phase space by definition obeys $f_{\mathbf{B}}(\mathscr{I}\Gamma) = f_{\mathbf{B}}(\Gamma)$. Take two observables, $\mathscr{O}$ and $\mathscr{P}$, such that $\mathscr{O}(\mathscr{I}\Gamma) = \eta_{\mathscr{O}}\,\mathscr{O}(\Gamma)$ and $\mathscr{P}(\mathscr{I}\Gamma) = \eta_{\mathscr{P}}\,\mathscr{P}(\Gamma)$, where $\eta_{\mathscr{Q}} = \pm 1$ is the signature of $\mathscr{Q}$ under the time reversal operation $\mathscr{T}$. Then, one can write

$$\mathbb{E}_{\mathbf{B}}\left[\mathscr{O}\left(\mathscr{P}\circ\Phi^t_{\mathbf{B}}\right)\right] = \int dX\, f_{\mathbf{B}}(X)\,\mathscr{O}(X)\,\mathscr{P}(\Phi^t_{\mathbf{B}}X) \tag{7.61}$$

$$= \int dY\, f_{\mathbf{B}}(\mathscr{I}Y)\,\mathscr{O}(\mathscr{I}Y)\,\mathscr{P}(\Phi^t_{\mathbf{B}}\mathscr{T}Y) \tag{7.62}$$

$$= \eta_{\mathscr{O}} \int \mathrm{d}Y\, f_{\mathbf{B}}(Y)\,\mathscr{O}(Y)\,\mathscr{P}(\mathscr{I}\Phi^{-t}_{\mathbf{B}}Y) \tag{7.63}$$

$$= \eta_{\mathscr{O}}\,\eta_{\mathscr{P}} \int \mathrm{d}Y\, f_{\mathbf{B}}(Y)\,\mathscr{O}(Y)\,\mathscr{P}(\Phi^{-t}_{\mathbf{B}}Y)$$

$$= \eta_{\mathscr{O}}\,\eta_{\mathscr{P}}\,\mathbb{E}_{\mathbf{B}}\left[\mathscr{O}\left(\mathscr{P}\circ\Phi^{-t}_{\mathbf{B}}\right)\right] \tag{7.64}$$

where $\Phi^t_{\mathbf{B}}$ is the time evolution in presence of the magnetic field $\mathbf{B}$. This does not only show that Onsager reciprocal relations also hold in the presence of a constant magnetic field, but adds predictive power to the theory. For instance, given a second admissible time reversal operation, $\tilde{\mathscr{I}}$, with the corresponding signatures be $\tilde{\eta}_{\mathscr{P}}$ and $\tilde{\eta}_{\mathscr{O}}$. Then, in a stationary state, in which the origin of times can be shifted forward without changing the result, both relations

$$\mathbb{E}_{\mathbf{B}}\left[\mathscr{O}\left(\mathscr{P}\circ\Phi^t_{\mathbf{B}}\right)\right] = \eta_{\mathscr{O}}\,\eta_{\mathscr{P}}\,\mathbb{E}_{\mathbf{B}}\left[\mathscr{O}\left(\mathscr{P}\circ\Phi^{-t}_{\mathbf{B}}\right)\right] = \eta_{\mathscr{O}}\,\eta_{\mathscr{P}}\,\mathbb{E}_{\mathbf{B}}\left[\left(\mathscr{O}\circ\Phi^t_{\mathbf{B}}\right)\mathscr{P}\right]$$

$$\mathbb{E}_{\mathbf{B}}\left[\mathscr{O}\left(\mathscr{P}\circ\Phi^t_{\mathbf{B}}\right)\right] = \tilde{\eta}_{\mathscr{O}}\,\tilde{\eta}_{\mathscr{P}}\,\mathbb{E}_{\mathbf{B}}\left[\mathscr{O}\left(\mathscr{P}\circ\Phi^{-t}_{\mathbf{B}}\right)\right] = \tilde{\eta}_{\mathscr{O}}\,\tilde{\eta}_{\mathscr{P}}\,\mathbb{E}_{\mathbf{B}}\left[\left(\mathscr{O}\circ\Phi^t_{\mathbf{B}}\right)\mathscr{P}\right]$$

hold, and either $\eta_{\mathscr{O}}\,\eta_{\mathscr{P}} = \tilde{\eta}_{\mathscr{O}}\,\tilde{\eta}_{\mathscr{P}}$, or

$$\mathbb{E}_{\mathbf{B}}\left[\mathscr{O}\left(\mathscr{P}\circ\Phi^t_{\mathbf{B}}\right)\right] = 0 \tag{7.65}$$

Implying, e.g., that certain transport coefficients vanish, i.e., that certain thermodynamic forces do not contribute to certain currents.

The number of time reversal symmetries in the presence of magnetic fields and rotating frames, which have quantum counterparts, turns out to be quite large (Carbone et al. 2022). This substantially extends the validity of the Onsager reciprocal relations, although it is not clear how far. In addition to that, the statistical effects may extend even more their range of validity. Indeed, calculations of averages may cancel the effects of intrinsic irreversibilities, still resulting in the validity of the Onsager reciprocal relations. Nevertheless, the above setting, which corresponds to that of Onsager, Casimir, and textbooks such as Landau and Lifshitz (1980), De Groot and Mazur (1984) etc., does not exhaust all possibilities and violations of Onsager Reciprocal Relations in magnetic fields are not excluded.

7.3.1 Linear Response About Nonequilibrium Steady States

The above formalism can be extended to perturbations on nonequilibrium steady states. For instance, to steady states whose dynamics is dissipative, hence not Hamiltonian, as in the important case of viscous hydrodynamics. Recently, it has been shown that the approach we have outlined does indeed apply, if the steady state is represented by a regular probability density, as commonly happens when the dynamical equations contain noise terms, cf. Marconi et al. (2008). But this is not the case if the process in addition to be dissipative is also deterministic, which means $\dot{\Gamma} = G(\Gamma)$ and $\text{div}\, G < 0$ everywhere in $\mathscr{M}$ or at least on average. The invariant distribution μ of such a dynamical system is typically singular and supported on a fractal attractor (Collet and Eckmann 2006). Consequently, it is no longer obvious that the effect of a perturbation can be related to the unperturbed statistics, because even very small perturbations may lead to microscopic states whose probability vanishes with respect to the unperturbed measure μ, making it irrelevant.

In certain cases[7] a perturbation $\delta\Gamma$ about a microstate Γ and its evolution $\Phi^t\delta\Gamma$ can be decomposed in two parts, $(\Phi^t\delta\Gamma)_\parallel$ and $(\Phi^t\delta\Gamma)_\perp$, respectively perpendicular and parallel to the fibers of the attractor (Ruelle 1998):

$$\Phi^t\delta\Gamma = (\Phi^t\delta\Gamma)_\parallel + (\Phi^t\delta\Gamma)_\perp$$

The first addend can be related to the dynamics on the attractor, while the second may not. In particular, Ruelle showed that the effect of a perturbation $\delta F(t) = \delta F_\parallel(t) + \delta F_\perp(t)$ on the response of a generic (smooth enough) observable $\mathscr{O}$ is given by

$$\mathbb{E}_t(\mathscr{O}) - \mathbb{E}_0(\mathscr{O}) = \int_0^t R_\parallel^{(\mathscr{O})}(t-\tau)\delta F_\parallel(\tau)d\tau + \int_0^t R_\perp^{(\mathscr{O})}(t-\tau)\delta F_\perp(\tau)d\tau \quad (7.66)$$

where $\mathbb{E}_0$ denotes averaging with respect to the initial non-invariant distribution μ_0 obtained by perturbing the invariant distribution μ, $R_\parallel^{(\mathscr{O})}$ may be expressed in terms of correlation functions evaluated with respect to μ, while $R_\perp^{(\mathscr{O})}$ depends on the dynamics along the stable manifold, hence it may not.

Remark 7.3.1 This is not the situation of the Green-Kubo theory summarized above. There, the initial state is the equilibrium one, and the perturbation gradually drives it away from equilibrium. That favors the interpretation of the initial ensemble as the

[7] Smooth, uniformly hyperbolic dynamical systems, such as axiom-A systems (Ruelle 2014).

initial state of the system. Here, the system is initially represented by a transient probability distribution, μ_0, which is close to that of a stationary state, and the response consists in the consequent evolution of probabilities μ_t. The physical interpretation of this process is more delicate.

While Ruelle's approach is mathematically relevant for hyperbolic low-dimensional dynamical systems, it has been pointed out (Colangeli et al. 2012) that the difficulty due to $R_{\perp}^{(\mathscr{O})}$ should not concern the systems of many interacting particles, which are of statistical mechanical interest. In those cases, rather than the full phase space, one considers the much lower dimensional projections, corresponding to a few physically relevant observables. Hence, one is typically interested in the marginals of singular phase space measures on spaces of low dimension, and these marginals are usually regular, especially if they result from a very large number of integrations.

Therefore, one may proceed as follows. Take a d-dimensional dissipative dynamical system, and perform an impulsive perturbation $\Gamma \to \Gamma + \delta\Gamma$ of the initial conditions.[8] This produces a perturbed initial distribution that differs from the stationary one and that represents the effect of some, at the moment unspecified, external action. For simplicity, let all components of $\delta\Gamma$ vanish except the i-th, $\delta\Gamma_i$.

The singular invariant probability distribution μ is correspondingly deformed, and turns into a non-invariant distribution μ_0, whose evolution μ_t is expected to return to μ in the $t \to \infty$ limit. Whether it really does, and how, is a non-trivial question as that is related to the phase space dynamics. In this framework, for every measurable set $E \subset \mathscr{M}$, one takes $\mu_0(E) = \mu(E - \delta\Gamma)$, with $E - \delta\Gamma = \{\Gamma \in \mathscr{M} : \Gamma + \delta\Gamma \in E\}$. Then, for the observable $\mathscr{O}(\Gamma) = \Gamma_i$, one writes

$$\mathbb{E}_t(\Gamma_i) - \mathbb{E}_0(\Gamma_i) = \int \Gamma_i \, \mathrm{d}\mu_t(\Gamma) - \int \Gamma_i \, \mathrm{d}\mu(\Gamma) \tag{7.67}$$

which may be approximated by substituting μ with a piecewise constant distribution, on the cells of an ε-partition made of a finite set of d-dimensional hypercubes $\Lambda_k(\varepsilon)$ of side ε and centers Γ_k. With no approximations, the probabilities under μ and μ_t of the hypercubes $\Lambda_k(\varepsilon)$ are given by

$$P_k(\varepsilon) = \int_{\Lambda_k(\varepsilon)} \mathrm{d}\mu(\Gamma) \,, \quad P_{t,k}(\varepsilon) = \int_{\Lambda_k(\varepsilon)} \mathrm{d}\mu_t(\Gamma) \tag{7.68}$$

[8] By the theory of Green's functions for linear operators, knowledge of the response to an impulsive perturbation leads to the response to a generic perturbation, by integration. Suppose L is a linear operator, and the response is determined by equation $Lf(x) = p(x)$, for x in some space, and suppose the solution to the equation $L\mathscr{G}(x, s) = \delta(s - x)$ is known. Then the solution of the original problem is given by $f(x) = \int \mathscr{G}(x, s)p(s) \, \mathrm{d}s$: to see that, multiply by $p(s)$ and integrate the impulsive equation, then take L outside the integral.

The corresponding coarse grained invariant density $\rho(\Gamma; \varepsilon)$ takes the form:

$$\rho(\Gamma; \varepsilon) = \sum_k \rho_k(\Gamma; \varepsilon) \ , \quad \text{with} \quad \rho_k(\Gamma; \varepsilon) = \begin{cases} P_k(\varepsilon)/\varepsilon^d & \text{if } x \in \Lambda_k(\varepsilon) \\ 0 & \text{otherwise} \end{cases} \tag{7.69}$$

For the evolving probability, the i-th direction of phase space plays a role different from the others. Therefore, we treat it separately, introducing a number Z_i of one-dimensional bins of the form:

$$\left[\Gamma_i^{(q)} - \frac{\varepsilon}{2}, \Gamma_i^{(q)} + \frac{\varepsilon}{2}\right) ; \quad q \in \{1, 2, ..., Z_i\}$$

Referring to this bin, and integrating the approximate distribution (7.69), one obtains

$$p_i^{(q)}(\varepsilon) = \int_{\Gamma_i^{(q)} - \frac{\varepsilon}{2}}^{\Gamma_i^{(q)} + \frac{\varepsilon}{2}} \left\{ \int \rho(\Gamma; \varepsilon) \prod_{j \neq i} \mathrm{d}\Gamma_j \right\} \mathrm{d}\Gamma_i \ ; \quad q \in \{1, 2, ..., Z_i\} \tag{7.70}$$

Each of these quantities represents the invariant probability that the i-th component Γ_i of Γ lies in one of the Z_i bins. The marginal of the evolving coarse grained probability, say $p_{t,i}^{(q)}(\varepsilon)$, can be analogously obtained. Both the invariant and the non-invariant probability densities, $\rho_i^{(q)}(\varepsilon)$ and $\rho_{t,i}^{(q)}(\varepsilon)$, are obtained dividing by ε, and the ε-coarse grained response function follows:

$$B_i^{(q)}(\Gamma_i, \delta\Gamma, t, \varepsilon) = \frac{1}{\varepsilon}\left[p_{t,i}^{(q)}(\varepsilon) - p_i^{(q)}(\varepsilon)\right] = \rho_{t,i}^{(q)}(\varepsilon) - \rho_i^{(q)}(\varepsilon) \tag{7.71}$$

For high-dimensional dynamical systems, the right-hand side of Eq. (7.71) is regularized by the many integrations, and tends to a smooth function of Γ_i even under the $\varepsilon \to 0$, hence $Z_i \to \infty$, limit. Consequently, $B_i^{(q)}(\Gamma_i, \delta\Gamma, t, \varepsilon)$ yields an expression similar to that of standard response theory, in the sense that it depends solely on the unperturbed state, even if that is supported on a fractal set. For systems of many interacting particles, this is the expected result because, in general, the sets on which distributions are singular are not very precisely orthogonal to the directions of projection, and even a single projection often results in a regular projected distribution. Consequently, our procedure makes unnecessary the explicit calculation of $R_\perp^{(\mathcal{O})}$ in Eq. (7.71). However, this does not mean that $R_\perp^{(\mathcal{O})}$ is negligible (Cessac and Sepulchre 2007), but it says that, apart from peculiar low-dimensional dynamics, there is no need to explicitly compute it, and that the response can be referred to the unperturbed dynamics.

7.3.1.1 Stochastic Example

While dissipative steady states of deterministic dynamics are represented by singular distributions in phase space, stochastic processes are based on probability densities in the space of states. Therefore, one proceed as follows Boffetta et al. (2003). Let x represent a state, and ρ be the invariant density characterizing the different states. A perturbation is introduced, initially shifting all states by a quantity δx so that a new non-invariant density is created $\rho'(x) = \rho(x - \delta x)$. Let the evolution be determined by the transition probability density

$$W\left(X, 0 \to X', t\right) \tag{7.72}$$

for a state X at time 0 to turn X' at time t.[9] The average of the evolution of one observable, e.g., the component x_i of the state vector x, now writes

$$\mathbb{E}_\rho\left[\left(\Phi^t x\right)_i\right] = \int x_i\, \rho\left(x^0\right) W\left(x^0, 0 \to x, t\right) \mathrm{d}x\, \mathrm{d}x^0 \tag{7.73}$$

for the unperturbed distribution, and

$$\mathbb{E}_{\rho'}\left[\left(\Phi^t x\right)_i\right] = \int x_i\, \rho'\left(x^0\right) W\left(x^0, 0 \to x, t\right) \mathrm{d}x\, \mathrm{d}x^0 \tag{7.74}$$

for the perturbed distribution. Assuming that ρ never vanishes in the state space, the response is given by

$$\begin{aligned} \mathbb{E}_{\rho'}\left[\left(\Phi^t x\right)_i\right] - \mathbb{E}_\rho\left[\left(\Phi^t x\right)_i\right] &= \int x_i \left[\rho'\left(x^0\right) - \rho\left(x^0\right)\right] W\left(x^0, 0 \to x, t\right) \mathrm{d}x\, \mathrm{d}x^0 \\ &= \int x_i \frac{\left[\rho'\left(x^0\right) - \rho\left(x^0\right)\right]}{\rho\left(x^0\right)} \rho\left(x^0\right) W\left(x^0, 0 \to x, t\right) \mathrm{d}x\, \mathrm{d}x^0 \end{aligned} \tag{7.75}$$

Introducing

$$F\left(X^0, \delta x\right) = \frac{\left[\rho'\left(X^0\right) - \rho\left(X^0\right)\right]}{\rho\left(X^0\right)} \tag{7.76}$$

and assuming that ρ can be expanded as a Taylor series:

$$\rho'\left(X^0\right) \rho\left(X^0 - \delta X\right) = \rho\left(X^0\right) + \sum_{i=1}^{n} \left.\frac{\partial \rho}{\partial X_i}\right|_{X^0} \left(-\delta X_i\right) + O\left(\delta X^2\right) \tag{7.77}$$

[9] In principle this formalism includes deterministic dynamics as well. In that case one would have $W\left(X, 0 \to X', t\right) = \delta\left(x - \Phi^t X^0\right)$, with usual notation. However, singularities may take a toll, as proved by Ruelle.

one obtains

$$F\left(X^0, \delta x\right) = \frac{1}{\rho\left(X^0\right)}\left[-\sum_{i=1}^{n} \left.\frac{\partial \rho}{\partial X_i}\right|_{X^0} \delta X_i + O\left(\delta X^2\right)\right] \tag{7.78}$$

$$= \left[-\sum_{i=1}^{n} \left.\frac{\partial \ln \rho}{\partial X_i}\right|_{X^0} \delta X_i + O\left(\delta X^2\right)\right] \tag{7.79}$$

The response then takes the form:

$$\mathbb{E}_{\rho'}\left[\left(\Phi^t x\right)_i\right] - \mathbb{E}_{\rho}\left[\left(\Phi^t x\right)_i\right] = \sum_{i=1}^{n} R_{ij}(t)\, \delta X_j \tag{7.80}$$

where analogously to the standard linear response theory, the unperturbed correlation function is used:

$$R_{ij}(t) = -\mathbb{E}_{\rho}\left[\left(\Phi^t X\right)_i \left.\frac{\partial \ln \rho}{\partial X_j}\right|_X\right] \tag{7.81}$$

More generally, for a generic observable $\mathcal{O}$, the response formula has the form

$$\mathbb{E}_{\rho'}\left[\mathcal{O} \circ \Phi^t\right] - \mathbb{E}_{\rho}\left[\mathcal{O} \circ \Phi^t\right] = -\sum_{j=1}^{n} \mathbb{E}_{\rho}\left[\left(\mathcal{O} \circ \Phi^t\right) \frac{\partial \ln \rho}{\partial X_j}\right] \delta X_j \tag{7.82}$$

If the unperturbed state were canonical, we would have

$$\ln \rho = -\beta H + \text{const} \tag{7.83}$$

and we would fall back on the traditional linear response theory.

In this example, concerning the perturbations about nonequilibrium steady states, linear response is obtained from an unperturbed correlation function, like linear response about equilibrium states. This is possible because, unlike the deterministic dissipative case, the unperturbed nonequilibrium distribution of the stochastic system is quite generally positive over the whole state space.

7.4 Current Research

Present day science and technology can go beyond the thermodynamic realm, and what is observed at our scale under special conditions is magnified in small or strongly driven systems. Think of lightnings or even the simple diving in a swimming pool. Non-linear (Ferry et al. 2012) and anomalous phenomena (Klafter and Sokolov 2005) may arise not only in special media: they depend also on how experiments are performed. Slowly entering in the water of a swimming pool is very different from hitting it flat when disorderly falling from on high. In the second case, the memory effect reflects the inertia of water, which takes longer than the first impact to make room for the falling body. Analogously, thermal treatments of alloys, plastic deformations of materials, etc. correspond to situations in which the available ranges are not properly explored within the times of interest. Perhaps, at present the best known situations in which thermodynamics has to be replaced by general theories are those concerning small systems, e.g., nanometric devices, or macromolecules such as proteins, in presence of external drivings. As in the case of no drivings, such as the pollen particles in the Brownian motion, fluctuations of observables are comparable with average signals. Moreover, for a reduced number of degrees of freedom, it is hard to apply sufficiently small forcings, that a linear response is produced. Also, large systems, such as the Earth atmosphere, can be seen on a different scale as small systems, because their relevant components are not many, although very complex (Yin et al. 2024). The classical statistical mechanical formalism needs adjustments to treat such cases, and research in that direction is very much a hot topic. Therefore, the statistical treatment seems to be the best option, and there are numerous attempts to produce a suitable general theory of response, although that remains at the moment just a goal. Among the myriad of works that are worth considering, here are a few: Refs. Hill (2001), Sekimoto (2010), Ghil and Lucarini (2020), Maffioli et al. (2022). It is also quite encouraging the fact that the rules, if not the details, of statistical mechanics apply to both large and small systems, with the difference that taking the limit of large numbers of degrees of freedom allows comprehensible analytical results, while keeping the number of degrees of freedom finite introduces finite size corrections, often quite cumbersome and system dependent. On the other hand, the direct simulation of small systems is already possible, and can be profitably combined with the indications that come from statistical mechanics (Tuckerman 2023).

8 Onsager-Machlup Theory

> *The ability to reduce everything to simple fundamental laws does not imply the ability to start from those laws and reconstruct the universe.*
>
> *More is different, Philip Warren Anderson*

The theory developed by Onsager and Machlup was meant to quantify the probability of fluctuation paths as they evolve in time, rather than the mere statistics of fluctuation values. It is a theory based on stochastic perturbations of deterministic dynamics that describe the irreversible relaxation to equilibrium of the thermodynamic observables of macroscopic objects when they are moderately shifted from their equilibrium states. The deterministic contribution represents the irreversible thermodynamic behavior, while the stochastic contribution represents the molecular chaos, assumed to be generated by time reversal invariant Hamiltonian dynamics. The treatment begins with the description of a system made of N particles, and then the $N \to \infty$ limit is taken, turning the reversible noisy evolution deterministic and irreversible. In the process, one uncovers a symmetry between the onset and the relaxation of a fluctuation, which amounts to two kinds of macroscopic evolution: the spontaneous emergence of fluctuations and the relaxation from a non-equilibrium state. This theory constitutes an early application of functional integration to physics, in addition to the application to the Brownian motion, which motivated Wiener to develop this technique (Wiener 1921), and the application to quantum mechanics developed by Feynman (1942). It constitutes the starting point of a prolific line of research, recently culminated in the *macroscopic fluctuation theory* (Bertini et al. 2002), which will be briefly summarized below.

C. Mejía-Monasterio and L. Rondoni, *Nonequilibrium Statistical Mechanics I*,
Theoretical and Mathematical Physics,
https://doi.org/10.1007/978-3-031-98249-1_8

8.1 Assumptions and Mathematical Derivations

The Onsager-Machlup theory is based on the following assumptions (Onsager and Machlup 1953; Machlup and Onsager 1953):

- **A.1** (*Onsager's Regression Hypothesis*): The decay of one observable $\mathscr{X}$ from a non-equilibrium value produced by a spontaneous fluctuation obeys *on average* the macroscopic law describing the decay from the same state, when that has been produced by a macroscopic constraint, that is suddenly removed.
- **A.2** (*Correlations decay*): The values of the observable $\mathscr{X}$ at time t_1, $\mathscr{X}_{t_1}$, are statistically independent of their values at time t_2, $\mathscr{X}_{t_2}$, if $|t_1 - t_2| > \tau_d$, where the *decorrelation time* τ_d is macroscopically negligible.
- **A.3** (*Gaussianity*): The evolution of observables is a *Gaussian stochastic process*, i.e., for each finite set of times $(t_1, t_2, ..., t_k)$, the random vector $(\mathscr{X}_{t_1}, \mathscr{X}_{t_2}, ..., \mathscr{X}_{t_k})$ of the values that the observable $\mathscr{X}$ takes at those instants of time is a multivariate normal variable. This requires the number of particles constituting the system to be very large, $N \gg 1$, so that their properties be like i.i.d. random variables.
- **A.4**: (*Bridge Law*): Let Γ be the microscopic phase of the system at a time t; its probability is given by the Boltzmann postulate:

$$k_B \log[\text{Prob}(\Gamma)] = S(\Gamma) + \text{const} \tag{8.1}$$

  independently of t, where S is the entropy of the system in the phase Γ.
- **A.5**: The underlying microscopic dynamics are *Time Reversal Invariant* (TRI), Hamiltonian in particular.
- **A.6**: Observables are sufficiently many to fully characterize the entropy S, and evolve as *Markov* processes. This also means that they cannot be too many, otherwise they describe the microscopic dynamics, that are supposed to be deterministic.
- **A.7**: The system is in *Local Thermodynamic Equilibrium* (LTE), so not too far from equilibrium, and thermodynamic properties make sense.
- **A.8**: The fluxes $\dot{\alpha_i}$ of the quantities α_i depend linearly on the thermodynamic forces X_j:

$$\dot{\alpha_i} = \sum_{j=1}^{n} L_{ij} X_j \qquad X_i = \sum_{j=1}^{n} R_{ij} \dot{\alpha_j} \tag{8.2}$$

  where $L = (L_{ij})$ and $R = (R_{ij})$ are constant matrices.

- **A.9**: The states are *stationary*, in the sense that, given values $\alpha^{(i)} \in \mathbb{R}$ and times t_i, one may write

$$\text{Probability}\left[\alpha(t_k) \leq \alpha^{(k)}, k = 1, ..., p\right] = F_p \begin{pmatrix} \alpha^{(1)} & \alpha^{(2)} & \dots & \alpha^{(p)} \\ t_1 & t_2 & \dots & t_p \end{pmatrix} \tag{8.3}$$

$$= F_p \begin{pmatrix} \alpha^{(1)} & \alpha^{(2)} & \dots & \alpha^{(p)} \\ t_1 + \tau & t_2 + \tau & \dots & t_p + \tau \end{pmatrix} \tag{8.4}$$

$$= \text{Probability}\left[\alpha(t_k + \tau) \leq \alpha^{(k)}, k = 1, ..., p\right] \tag{8.5}$$

where the probability F_p is obtained integrating the probability density f_p which is a function of the p values that the variable α takes at the times $t_1, ..., t_p$:

$$F_p \begin{pmatrix} \alpha^{(1)} & \alpha^{(2)} & \dots & \alpha^{(p)} \\ t_1 & t_2 & \dots & t_p \end{pmatrix} = \int_{-\infty}^{\alpha^{(1)}} d\alpha_1 ... \int_{-\infty}^{\alpha^{(p)}} d\alpha_p \, f_p \begin{pmatrix} \alpha_1 & \alpha_2 & \dots & \alpha_p \\ t_1 & t_2 & \dots & t_p \end{pmatrix} \tag{8.6}$$

8.2 Complete Set of Extensive Variables

To select a complete set of extensive variables that fully characterize the entropy of the macroscopic system of interest, Onsager and Machlup followed these rules:

1. The variable α must be a macroscopically measurable quantity that refers to a macroscopically small subsystem but still containing very many particles.
2. α must be an algebraic sum of independent molecular variables, so that the Central Limit Theorem (CLT) applies and their (small) fluctuations are normally distributed and centered around their equilibrium value.
3. α must be even functions of the molecular variables that are odd functions of time (as prescribed by TRI).

Before entering the calculations that arise from the above assumptions, we note that the corresponding physics may initially sound fragile and insubstantial. To dispel such thoughts, Onsager and Machlup stated:

A long time lapse makes the states statistically independent

and, in the corresponding footnote, it is stated that this may hold for:

systems which "forget" their initial states in a "reasonably" short time. It is, however, precisely the choice of time scale that matters. In a sufficiently long time, all physical systems "forget"

and again:

Our system must consist of many "sufficiently" independent particles, and equilibrium must be stable at least for times of the order of laboratory measurement times.

This is also required in order that we can consider our systems as "aged"—left alone long enough to have "forgotten" their initial states

As variously discussed in this book, thermodynamics is the realm in which quotation marks can be removed, and the exact mathematical results exactly describe observations. Their range of applicability though limited is sufficiently wider than what we experience in our daily life, that we do not easily imagine its failure. Nevertheless, current science and technology take us more and more beyond this boundary, like, *e.g.*, toward systems that operate at nanoscopic scales where molecular fluctuations are not negligible, and even comparable in magnitude to the variables of interest. Keeping these facts in mind, let us analyze the Onsager-Machlup theory.

Let $\alpha_1, \alpha_2, \ldots, \alpha_n$ be our complete set of variables, which express the entropy as $S(\alpha) = S(\alpha_1, \alpha_2, \ldots, \alpha_n)$. To investigate fluctuations about an equilibrium state, it suffices to consider the α's as the deviations of corresponding quantities from their equilibrium values. Then, the maximum of the entropy is given by

$$S_{\rm max} = S_0 = S(0, 0, \ldots, 0) = S_{\rm eq} \tag{8.7}$$

which, according to thermodynamics, occurs at equilibrium. If we further assume, as is customary, that the entropy can be expanded in a Taylor series about equilibrium, we also have

$$\left.\frac{\partial S}{\partial \alpha_i}\right|_{\alpha=0} = 0\,, \quad i = 1, \ldots, n \tag{8.8}$$

Then, for small fluctuations, one can truncate the series at second order in α, obtaining:

$$S = S_0 - \frac{1}{2}\sum_{i,j=1}^{n} s_{ij}\alpha_i\alpha_j$$

where higher order terms have been neglected, and the "−" sign reflects the fact that the equilibrium entropy is a maximum. The drive to return to equilibrium is due to the *thermodynamic forces*, say X_i, which identify with the derivatives of the entropy. Within our bounds, this means

$$X_i = \frac{\partial S}{\partial \alpha_i} = -\sum_{j=1}^{n} s_{ij}\alpha_j\,, \quad i = 1, \ldots, n \tag{8.9}$$

which act as restoring forces that depend on the variables α, away from equilibrium, and vanish at equilibrium. Because of assumption **A.8** above, we get

$$\sum_{j=1}^{n} R_{ij}\dot\alpha_j = -\sum_{j=1}^{n} s_{ij}\alpha_j\,, \quad \text{which also writes:} \quad \sum_{j=1}^{n}\Big[R_{ij}\dot\alpha_j + s_{ij}\alpha_j\Big] = 0 \tag{8.10}$$

We therefore obtain

$$\dot{S} = \frac{d}{dt}\left[S_0 - \frac{1}{2}\sum_{i,j=1}^{n} s_{ij}\alpha_i\alpha_j\right] = -\frac{1}{2}\sum_{i,j=1}^{n} s_{ij}\left[\dot{\alpha}_i\alpha_j + \alpha_i\dot{\alpha}_j\right] \tag{8.11}$$

$$= -\sum_{i,j=1}^{n} s_{ij}\dot{\alpha}_i\alpha_j = -\sum_{i=1}^{n}\dot{\alpha}_i\sum_{j=1}^{n} s_{ij}\alpha_j \tag{8.12}$$

$$= \sum_{i=1}^{n}\dot{\alpha}_i\sum_{j=1}^{n} R_{ij}\dot{\alpha}_j = \sum_{i,j=1}^{n} R_{ij}\dot{\alpha}_i\dot{\alpha}_j = 2\Phi(\dot{\alpha}, \dot{\alpha}) \tag{8.13}$$

where Φ is a function of the rate of change of state:

$$\Phi(\dot{\alpha}, \dot{\beta}) = \frac{1}{2}\sum_{i,j=1}^{n} R_{ij}\dot{\alpha}_i\dot{\beta}_j \tag{8.14}$$

with $\alpha = (\alpha_1, ..., \alpha_n)$ and $\beta = (\beta_1, ..., \beta_n)$ any two states. The derivative of S may also be expressed as

$$\dot{S} = \sum_{i=1}^{n}\frac{\partial S}{\partial \alpha_i}\dot{\alpha}_i = \sum_{i=1}^{n} X_i\sum_{j=1}^{n} L_{ij}X_j = \sum_{i,j=1}^{n} L_{ij}X_iX_j = 2\Psi(X, X) \tag{8.15}$$

where the second equality comes from assumption **A.8**, and a function of the forces $X = (X_1, ..., X_n)$ and $Y = (Y_1, ..., Y_n)$:

$$\Psi(X, Y) = \frac{1}{2}\sum_{i,j=1}^{n} L_{ij}X_iY_j \tag{8.16}$$

which is a function of state that, unlike Φ, has been introduced. Given its connection with the entropy variations, Φ is called *dissipation function*, it acts as a potential for the thermodynamic forces, and it plays a role analogous to the dissipation function defined by Eq. (6.54) in the expression of probabilities. The difference is that Φ refers to observable quantities while Ω^f to microscopic phases.

8.3 Random Perturbation of Hydrodynamic Laws

In a thermodynamic situation, the state is determined by its evolution equations and is unique. Therefore, it is characterized by a single set of parameters α or forces X. In Eqs. (8.14) and (8.15), one then has $\alpha = \beta$ and $X = Y$, respectively, and given one initial condition, the subsequent evolution draws a unique path. In that case, the macroscopic (hydrodynamic) deterministic evolution is deduced from:

$$\Phi(\dot{\alpha}, \dot{\alpha}) = \Psi(X, X) = \frac{1}{2}\dot{S} \tag{8.17}$$

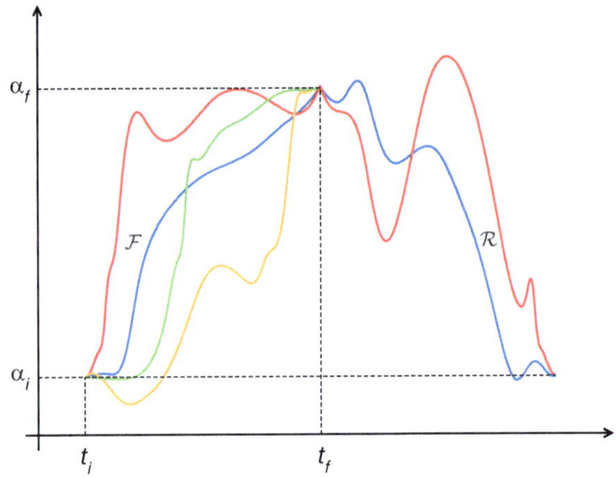

Fig. 8.1 Various non-macroscopic paths for a fluctuation, $\alpha_{\text{initial}} \to \alpha_{\text{final}}$, to arise and to decay in time. Instead, the hydrodynamic behavior is unique

with $\alpha = \alpha(t)$ solution of Eq. (8.10). Nevertheless, this formalism allows alternative paths, with $\beta \neq \alpha$ and $Y \neq X$, assuming that non-macroscopic thermodynamic conditions exist, while being unobservable (Fig. 8.1). They will then be characterized by

$$\Phi(\dot{\alpha}, \dot{\beta}) \neq \frac{1}{2}\dot{S} \neq \Psi(X, Y) \tag{8.18}$$

The idea is that, while the macroscopic dynamics admit $\alpha(t)$ as the unique solution of Eq. (8.10), and no other paths are allowed, alternative paths should exist, because of the microscopic molecular chaos. Such paths are considered equally real as the macroscopic ones, but remain invisible at macroscopic space and time scales. In principle, fluctuations could also be so large to go "upstream," every now and then. Having no other knowledge of the phenomenon, we may model it adding a *random force* ε_i to Eq. (8.10):

$$\sum_{j=1}^{n}\left[R_{i,j}\dot{\alpha}_j + s_{i,j}\alpha_j\right] = \varepsilon_i\,, \quad i = 1\ldots n \tag{8.19}$$

Then, speaking of the probability of a path, that is not the macroscopic observed path, makes sense. Let

$$f_1\begin{pmatrix}\alpha^{(1)}\\ t\end{pmatrix} \tag{8.20}$$

be the probability density for α to be close to $\alpha^{(1)}$ at time t. According to the Boltzmann principle **A.4**, this density does not depend on t, and may be written as

$$f_1\begin{pmatrix}\alpha\\ t\end{pmatrix} = f(\alpha) = \frac{e^{S/k_B}}{\mathcal{N}} \tag{8.21}$$

$$= \frac{1}{\mathcal{N}}\exp\left[S_{\text{eq}} - \frac{1}{2}\sum_{i,j=1}^{n} s_{ij}\alpha_i\alpha_j\right] = C\exp\left[-\frac{1}{2}\sum_{i,j=1}^{n} s_{ij}\alpha_i\alpha_j\right] \tag{8.22}$$

where $\mathcal{N}$ is the normalization constant, and C includes the equilibrium entropy contribution in the normalization constant. This way, the probability density of a path is seen to depend on the deviations of the path from the equilibrium state, and it decreases as this deviation increases. Denote by

$$f_1\left(\begin{array}{c|c}\alpha^{(f)} & \alpha^{(i)} \\ t_f & t_i\end{array}\right). \tag{8.23}$$

the conditional probability density for α to be close to $\alpha^{(f)}$ at time t_f, if it was $\alpha^{(i)}$ at the earlier time t_i. This can be calculated by solving Eq. (8.19), assuming that its stochastic term obeys

$$\mathbb{E}\left[\varepsilon_i\right]=0\,, \quad \mathbb{E}\left[\varepsilon_i(t)\varepsilon_j(t')\right]=q_i\delta(t-t')\delta^i_j\,, \quad i,j=1,\ldots,n \tag{8.24}$$

For a Markov process, implied by assumption **A.6**, one can write

$$f_p\left(\begin{array}{cccc}\alpha^{(1)} & \alpha^{(1)} & \ldots & \alpha^{(p)} \\ t_1 & t_2 & \ldots & t_p\end{array}\right)=f_1\left(\begin{array}{c|c}\alpha^{(p)} & \alpha^{(p-1)} \\ t_p & t_{p-1}\end{array}\right)\cdots f_1\left(\begin{array}{c|c}\alpha^{(2)} & \alpha^{(1)} \\ t_2 & t_1\end{array}\right)f\left(\alpha^{(1)}\right) \tag{8.25}$$

Assumption **A.1**, may be formulated as the steady-state conditional average of α at time $t+\tau$, given that it was α' at time t:

$$\mathbb{E}\left(\alpha,t+\tau|\alpha',t\right)=\int d\alpha_1\ldots.\int d\alpha_n\,\alpha\,f_1\left(\begin{array}{c|c}\alpha & \alpha' \\ t+\tau & t\end{array}\right) \tag{8.26}$$

As far as the average behavior is concerned, this means that it does not matter whether the starting point α' is the result of a spontaneous fluctuation, or it was artificially created by some operator. It is a consequence of the Markov assumption (Doob 1990). Finally, for a stationary process, which is the object of our investigation here, the conditional 2-times probability density

$$f_1\left(\begin{array}{c|c}\alpha^{(f)} & \alpha^{(i)} \\ t+\tau & t\end{array}\right) \tag{8.27}$$

only depends on the difference τ of the two times, not on t.

Now, for the sake of simplicity and without any real loss of generality, let us consider the case with a single extensive variable, $n=1$. The generalization to any number of variables is immediate, and will be given at the end. In this case, the evolution equation for the state is written as follows:

$$R\dot{\alpha}+s\alpha=\varepsilon \text{ with } \mathbb{E}\left[\varepsilon\right]=0\,, \quad \mathbb{E}\left[\varepsilon(t)\varepsilon(t')\right]=q\delta(t-t') \tag{8.28}$$

and averaging it one obtains

$$R\mathbb{E}\left[\dot{\alpha}\right]+s\mathbb{E}\left[\alpha\right]=0\,, \quad \text{whose solution is} \quad \mathbb{E}_\tau\left[\alpha\right]=ke^{-s\tau/R} \tag{8.29}$$

with k is a real constant. In case all paths start at time t from a given α_0, the solution takes the form:

$$\mathbb{E}_{t+\tau}[\alpha] = \mathbb{E}[\alpha, t+\tau|\alpha_0, t] = \int \alpha f_1\left(\begin{matrix}\alpha \\ t+\tau\end{matrix}\middle|\begin{matrix}\alpha_0 \\ t\end{matrix}\right) \mathrm{d}\alpha = \alpha_0 e^{-s\tau/R} \tag{8.30}$$

where the notation stresses that the process that produced the starting state and the time at which this happened are irrelevant. Furthermore, the Markov property implies that memory is lost in time, therefore α becomes independent of α_0 for $\tau \to \infty$, i.e.:

$$f_1\left(\begin{matrix}\alpha \\ t+\tau\end{matrix}\middle|\begin{matrix}\alpha_0 \\ t\end{matrix}\right) \to f_1\left(\begin{matrix}\alpha \\ t+\tau\end{matrix}\right) = \frac{1}{\mathcal{N}} e^{S(\alpha)/k_B} = C \exp\left[-\frac{s\alpha^2}{2k_B}\right] \tag{8.31}$$

in the $\tau \to \infty$ limit. The last equality comes from considering $S(\alpha) = S_0 - s\alpha^2/2$ an accurate expression of S, which is the case for near-equilibrium states. On the other hand, the $\tau \to 0$ limit yields certainty that the state is α_0, i.e.:

$$\lim_{\tau\to 0} f_1\left(\begin{matrix}\alpha \\ t+\tau\end{matrix}\middle|\begin{matrix}\alpha_0 \\ t\end{matrix}\right) = k\delta(\alpha - \alpha_0)) \tag{8.32}$$

Starting with a very peaked Gaussian distribution, practically and physically representing the Dirac δ function, the process preserves Gaussianity in time, as in the case of the Brownian motion. Then, it suffices to compute the mean and the variance of the distribution of $\alpha(t)$, at every positive time t, considering that the conditions (8.31) and (8.32) are satisfied if the variance grows from 0 to k_B/s as $\tau \to \infty$. This problem was beautifully tackled by Chandrasekhar Ref. Chandrasekhar (1943) (cf. also Ref. Risken (1989), Eq. (5.28)). The result is that the mean equals $\mathbb{E}[\alpha]_{t+\tau}$, and the variance equals $k_B\left[1 - \exp(-2s\tau/R)\right]/s$, analogously to Eq. (3.17) with $t_1 = t_2 = t$. Therefore, the distribution takes the form:

$$f_1\left(\begin{matrix}\alpha \\ t+\tau\end{matrix}\middle|\begin{matrix}\alpha_0 \\ t\end{matrix}\right) = \frac{1}{\sqrt{2\pi}}\sqrt{\frac{s}{k_B(1-e^{-2s\tau/R})}} \exp\left[-\frac{s\,(\alpha - e^{-s\tau/R}\,\alpha_0)^2}{2k_B(1-e^{-2s\tau/R})}\right] \tag{8.33}$$

where, in fact, t plays no role. Now, partitioning the time interval $(t, t+\tau)$ in p subintervals of size $\Delta\tau = \tau/p$, and introducing the discrete times $t_i = t + (i-1)\Delta\tau$ with $i = 1, ..., p+1$, we can consider a whole path, rather than a single value of α at a single time t. Denoting by

$$\alpha_0 = \alpha(t_1) = \alpha^{(1)}\,;\ \ldots;\ \alpha(t_{p+1}) = \alpha^{(p+1)} = \alpha \tag{8.34}$$

the values of $\alpha(t)$ at these time instants, we obtain

$$f_1\left(\begin{matrix}\alpha \\ t+\tau\end{matrix}\middle|\begin{matrix}\alpha_0 \\ t\end{matrix}\right) = \int \mathrm{d}\alpha^{(2)} \ldots \int \mathrm{d}\alpha^{(p)}\, f_1\left(\begin{matrix}\alpha^{(p+1)} \\ t_{p+1}\end{matrix}\middle|\begin{matrix}\alpha^{(p)} \\ t_p\end{matrix}\right) \cdots f_1\left(\begin{matrix}\alpha^{(2)} \\ t_2\end{matrix}\middle|\begin{matrix}\alpha^{(1)} \\ t_1\end{matrix}\right) \tag{8.35}$$

since we have Markovian evolution. This allows to write a discrete version of Eq. (8.28) as

$$R\frac{\alpha^{(k)} - \alpha^{(k-1)}}{\Delta\tau} + s\alpha^{(k-1)} = \varepsilon(t_k) \tag{8.36}$$

that for τ fixed becomes a better description with increasing p. The reader may have noted that, as in the paper Onsager and Machlup (1953), we have used the symbol "$=$," rather than the approximate equality symbol. Indeed, as we pointed out, the differential equation is rather an approximation of the phenomenon of physical interest: speaking of the evolution of LTE states, it amounts to considering infinitesimal what in reality are mesoscopic times. Therefore, one may better take Eq. (8.36) as the fundamental mathematical object, and later see what it becomes in a $\Delta\tau \to 0$ limit, provided such a limit suits the purpose of the study. In other words, the $\Delta\tau \to 0$ limit should be considered correct only if the mesoscopic times are negligible with respect to the characteristic times of the phenomenon one wants to describe, within the scales at which it is observed. As the stochastic term is still only known up to certain statistical properties, there is a question on how the discretization should be performed. The one of Eq. (8.36) is known as Itô discretization.[1] Multiplying by $\Delta\tau$ and dividing by R, Eq. (8.36) yields

$$\frac{\Delta\tau}{R}\varepsilon(t_k) = \alpha^{(k)} - \left(1 - \frac{s\Delta\tau}{R}\right)\alpha^{(k-1)} \approx \alpha^{(k)} - \alpha^{(k-1)}e^{-s\Delta\tau/R} \tag{8.37}$$

We may now use Chandrasekhar's expression (8.33) for every time step, given $\alpha^{(k-1)}$ and taking $\Delta\tau$ in place of τ, thus replacing the various quantities as

$$\alpha - e^{-s\tau/R}\alpha_0 \quad \longrightarrow \quad \alpha^{(k)} - e^{-s\Delta\tau/R}\alpha^{(k-1)} \approx \frac{\Delta\tau}{R}\varepsilon(t_k) \tag{8.38}$$

and

$$1 - e^{-2s\tau/R} \quad \longrightarrow \quad 1 - e^{-2s\Delta\tau/R} \approx \frac{2s\Delta\tau}{R} \tag{8.39}$$

[1] Equation (8.36) is of the kind:

$$x_k = x_{k-1} + b\left[\lambda x_k + (1-\lambda)x_{k-1}\right]\Delta\tau + \xi_k\Delta\tau$$

where x is the evolving variable of interest, ξ is a given random process affecting such evolution, k is the time index, b is a constant, and $\lambda \in [0, 1]$. Itô discretization corresponds to $\lambda = 0$, while $\lambda = 1/2$ is known as Stratonovich discretization, and it is sometimes considered the discretization of physical interest. In our case, this distinction is irrelevant.

where approximations are accurate to first order in $\Delta\tau$. This eventually yields

$$f_1\left(\begin{array}{c|c} \alpha^{(k)} & \alpha^{(k-1)} \\ t_{k-1}+\Delta\tau & t_{k-1} \end{array}\right) = \sqrt{\frac{s}{2\pi k_B\, 2s\Delta\tau/R}} \exp\left[-s\frac{(\Delta\tau\varepsilon(t_k)/R)^2}{2k_B 2s\Delta\tau/R}\right]$$

$$= \sqrt{\frac{1}{4k_B}}\sqrt{\frac{R}{\pi\Delta\tau}} \exp\left[-\frac{R}{4k_B}\left(\frac{\varepsilon(t_k)}{R}\right)^2 \Delta\tau\right] \tag{8.40}$$

Then, multiplying all the densities f_1, from time t_1 up to time t_p, and recalling Eq. (8.28), one obtains

$$f_1\left(\begin{array}{c|c} \alpha & \alpha_0 \\ t+\tau & t \end{array}\right) = \left(\frac{1}{4k_B}\right)^{p/2}\left(\frac{R}{\pi\Delta\tau}\right)^{p/2} \times \tag{8.41}$$

$$\int d\alpha^{(2)} \ldots \int d\alpha^{(p)} \exp\left[-\frac{R}{4k_B}\sum_{k=1}^{p}\left(\dot\alpha^{(k)} + \frac{s}{R}\alpha^{(k)}\right)^2 \Delta\tau\right] \tag{8.42}$$

The integrand is the exponential of a negative quantity, which can be seen as a Riemann sum approximating an integral in the $\Delta\tau \to 0$ limit:

$$\sum_{k=1}^{p}\left(\dot\alpha^{(k)} + \frac{s}{R}\alpha^{(k)}\right)^2 \Delta\tau \approx \int_t^{t+\tau} dt' \left[\dot\alpha(t') + \frac{s}{R}\alpha(t')\right]^2 \tag{8.43}$$

where the path is assigned and the integrand is a given function of time that changes when the path is changed. Therefore, the maximum of f_1 is achieved when the integral in the exponential is minimum over all possible paths $(\alpha, \dot\alpha)(t)$ connecting $\alpha^{(1)} = \alpha_0$ to $\alpha^{(p+1)} = \alpha$. Furthermore, because α and $\dot\alpha$ are extensive quantities, hence proportional to the (very large) number of particles, there is an enormous unbalance between the maximum and the nearby values. Thus, the integral essentially reduces to the maximum value of the integrand, apart from some normalization factor. Indeed, let

$$I(N) = \int_{-\infty}^{\infty} dx e^{-Nf(x)}$$

where N is large, and the analytic function $f(x)$ has a minimum in x_0, so that $f(x) = f(x_0) + g(x)$ with $g(x) > 0$. Then,

$$I(N) = e^{-Nf(x_0)} \int_{-\infty}^{\infty} e^{-Ng(x)}\, dx \approx e^{-Nf(x_0)} \int_{x_0-\varepsilon}^{x_0+\varepsilon} e^{-Ng(x)}\, dx\,, \quad \text{for small } \varepsilon$$

because $Ng(x)$ is large, hence $\exp(-Ng(x))$ practically vanishes, for large N, unless x is close to x_0. In such a small range about x_0 we may thus write

$$f(x) \approx f(x_0) + \frac{1}{2}f''(x_0)(x-x_0)^2\,, \quad f''(x_0) > 0 \tag{8.44}$$

and

$$I(N) \approx \int_{-\infty}^{\infty} e^{-Nf(x_0)} e^{-\frac{N}{2} f''(x_0)(x-x_0)^2} \, dx$$

$$= e^{-Nf(x_0)} \int_{-\infty}^{\infty} d(x-x_0) e^{-\frac{N}{2} f''(x_0)(x-x_0)^2} = \sqrt{\frac{2\pi}{Nf''(x_0)}} e^{-Nf(x_0)}$$

where the approximation becomes exact in the large N limit. Analogously, neglecting the normalization constant, we can write

$$f_1\left(\begin{matrix}\alpha \\ t+\tau\end{matrix}\middle|\begin{matrix}\alpha_0 \\ t\end{matrix}\right) \sim \exp\left(-\int_t^{t+\tau} dt' \left[\dot{\alpha}(t') + \frac{s}{R}\alpha(t')\right]^2\right)_{\min} \tag{8.45}$$

subject to the constraints $\alpha(t) = \alpha^{(1)}$ and $\alpha(t+\tau) = \alpha^{(p+1)}$. The same holds for all scalar variables.

The extension to n-dimensional vector variables follows similarly. It suffices to note that the matrices $s = (s_{ij})$ and $R = (R_{ij})$ can be diagonalized simultaneously, so that one simply deals with n independent random processes, the random terms being uncorrelated. In the diagonal form and within the regime of small fluctuations, one may then write

$$S = S_0 - \frac{1}{2}\sum_{i=1}^{n} s_i \alpha_i^2\,; \quad \frac{dS}{dt} = -\sum_{i=1}^{n} s_i \alpha_i \dot{\alpha}_i\,; \quad X_j = \frac{\partial S}{\partial \alpha_j} = -s_j \alpha_j \tag{8.46}$$

hence

$$\Phi(\dot{\alpha}, \dot{\alpha}) = \frac{1}{2}\sum_{i=1}^{n} R_i \dot{\alpha}_i^2\,; \quad \Psi(X, X) = \frac{1}{2}\sum_{i=1}^{n} \frac{1}{R_i} X_i^2 = \frac{1}{2}\sum_{i=1}^{n} \frac{1}{R_i} s_i^2 \alpha_i^2 \tag{8.47}$$

Consequently, the n-dimensional generalization of the integrand in Eq. (8.45) yields

$$\sum_{i=1}^{n} R_i \left(\dot{\alpha}_i + \frac{s_i}{R_i}\alpha_i\right)^2 = 2\Phi(\dot{\alpha}, \dot{\alpha}) - 2\dot{S}(\alpha) + 2\Psi(X(\alpha), X(\alpha)) \equiv \mathscr{L}(\alpha, \dot{\alpha}) \tag{8.48}$$

where we have introduced the *Lagrangian* $\mathscr{L}$ in which α, $\dot{\alpha}$ are independent variables, until a given thermodynamic path has been identified. This is done by obtaining the path that minimizes the integral in Eq. (8.45), i.e., solving the Lagrange equations:

$$\frac{d}{dt}\frac{\partial \mathscr{L}}{\partial \dot{\alpha}_i} - \frac{\partial \mathscr{L}}{\partial \alpha_i} = 0, \quad i = 1, \ldots, n \tag{8.49}$$

under the conditions $\alpha_i(t_1) = \alpha_i^{(1)}$ and $\alpha_i(t_2) = \alpha_i^{(2)}$, for each of the n components of α. This yields

$$\frac{d}{dt}\frac{\partial \mathscr{L}}{\partial \dot{\alpha}_j} = \frac{d}{dt}[2R_j\dot{\alpha}_j + 2s_j\alpha_j] = 2R_j\ddot{\alpha}_j + 2s_j\dot{\alpha}_j\,; \quad \frac{\partial \mathscr{L}}{\partial \alpha_j} = 2s_j\dot{\alpha}_j + 2\frac{s_j^2}{R_j}\alpha_j \tag{8.50}$$

hence the evolution equations:

$$R_j\ddot{\alpha}_j - \frac{s_j^2}{R_j}\alpha_j = 0\,, \quad j = 1, \ldots, n \tag{8.51}$$

which has solution:

$$\alpha_j(t) = c_1 e^{-s_j t/R_j} + c_2 e^{s_j t/R_j}\,, \quad j = 1, \ldots, n \tag{8.52}$$

If we are looking for solutions describing the relaxation to equilibrium in the long time limit, $c_2 = 0$ and we have

$$\alpha_j^{(1)}(t) = c_1 e^{-s_j t/R_j}\,, \quad j = 1, \ldots, n \tag{8.53}$$

which is the well-known hydrodynamic behavior. But we could as well assume that the system was in equilibrium (vanishing α) at a much earlier time ($t = -\infty$), in which case c_1 has to vanish and the solution is given by

$$\alpha_j^{(2)}(t) = c_2 e^{s_j t/R_j}\,, \quad j = 1, \ldots, n \tag{8.54}$$

The above procedure has thus revealed two possible and opposite behaviors:

- Relaxation to equilibrium: $\alpha(t \to \infty) = 0$.
- Spontaneous fluctuation from equilibrium: $\alpha(t \to -\infty) = 0$.

The second of which was not expected on hydrodynamic grounds, and makes $\alpha_j^{(2)}$ the value reached by a spontaneous fluctuation.

In fact, Eq. (8.51) is twice the order of the phenomenological laws, and contains them as special cases. Its two independent solutions can be derived from two first-order equations: the hydrodynamic law

$$\dot{\alpha}_j + \frac{s_j}{R_j}\alpha_j = 0\,, \quad j = 1, \ldots, n \tag{8.55}$$

which describes the relaxation to equilibrium, and the so-called *adjoint hydrodynamic* law

$$\dot{\alpha}_j - \frac{s_j}{R_j}\alpha_j = 0\,, \quad j = 1, \ldots, n \tag{8.56}$$

which describes the process of spontaneous fluctuation out of equilibrium. The two laws are the mirror image of each other. This is the case because the minimum of the integral (8.45) can be indifferently obtained by satisfying two symmetric expressions, one describing relaxation, and the other fluctuation.

Through the approach summarized above, Onsager and Machlup obtained the probability of nonequilibrium paths in the state space of spontaneously fluctuating thermodynamic systems, under the assumption that the macroscopic variables defining a state are Gaussian random variables whose average behavior is expressed by the macroscopic thermodynamic laws.

8.4 Macroscopic Fluctuation Theory

The Onsager-Machlup theory has motivated research beyond the small equilibrium fluctuations and, in particular, it has been generalized to the large deviations of spatially extended systems in nonequilibrium stationary states. In this realm, explicit mathematical results have been obtained for exactly solvable models of interacting particle systems on the lattice for which the hydrodynamic limit can be taken, e.g., simple exclusion processes, zero-range processes,, and others. An additivity principle has been proposed, which allows the analytical calculation of the density profile of one-dimensional diffusive systems, driven from their boundaries with unbalanced particle reservoirs (Bodineau and Derrida 2004). This and other results concerning exactly solvable models of driven diffusive systems, like the validity of the so-called fluctuation theorem, can be seen as part of the general theory proposed in Ref. Bertini et al. (2002) and extensively reviewed in Ref. Bertini et al. (2015), known as macroscopic fluctuation theory. It is based on the assumption that a hydrodynamic-like description of the system at hand is possible; hence, it is applicable beyond the linear regime, as long as local thermodynamic equilibrium holds. Here, it is worth summarizing a few basic ideas of this theory concerning, in particular, the temporal symmetry of the fluctuations. The approach consists of random perturbations of dynamical systems (Freidlin and Wentzell 1998).

Consider a dynamical system

$$\dot{\rho} = D(\rho)\,, \quad \rho \in \mathcal{M} \subset \mathbb{R}^n \tag{8.57}$$

where D is a (smooth) vector field with a globally attracting fixed point $\hat{\rho}$, and $\mathcal{M}$ is the corresponding n-dimensional phase space. The n components of ρ may represent, for instance, the values of a scalar quantity at n distinct sites constituting an extended system, such as a crystal. As in the Onsager-Machlup theory, we assume that the effect of mesoscopic fluctuations be represented by a Gaussian, δ-correlated noise ξ such that $\mathbb{E}[\xi] = 0$, $\mathbb{E}\left[\xi_i(t)\xi_j(t')\right] = K_{i,j}\delta(t - t')$, where $K = (K_{i,j})$ is a symmetric and positive definite $n \times n$ matrix. Then the perturbed dynamical system writes

$$\dot{\rho} = D(\rho) + \xi \tag{8.58}$$

As before, this equation could be obtained by taking the continuous limit of some discrete difference equation, such as

$$\frac{1}{\Delta t}\Big[\rho(k+1) - \rho(k)\Big] = D(\rho(k)) + \xi(k) \tag{8.59}$$

with

$$\mathbb{E}\left[\xi(k)\right] = 0 \quad \langle \xi_i(l), \xi_i(k)\rangle = \frac{1}{\Delta t}\delta_{l,k} \tag{8.60}$$

where $k = 0, 1, 2, \ldots$ labels time variable, and

$$\langle x, y\rangle = x^T K^{-1} y \tag{8.61}$$

is a scalar product on $\mathscr{M}$. Moreover, we take

$$p(\xi) = \sqrt{\frac{\Delta t}{(2\pi)^n |K|}}\, e^{-\Delta t \langle \xi, \xi\rangle/2} \tag{8.62}$$

as the known probability density characterizing the noise. From this, the conditional probability density that the process ρ is r_{k+1} at time $k+1$ given that it was r_k at time k takes the form:

$$P\Big(\rho(k+1) = r_{k+1} \big| \rho(k) = r_k\Big) = \int d\xi \; p(\xi)\delta\left(\frac{r_{k+1} - r_k}{\Delta t} - D(r_k) - \xi\right) \tag{8.63}$$

$$= \sqrt{\frac{\Delta t}{(2\pi)^n |K|}} \exp\left[-\frac{\Delta t}{2}\left\langle \frac{r_{k+1} - r_k}{\Delta t} - D(r_k), \frac{r_{k+1} - r_k}{\Delta t} - D(r_k)\right\rangle\right]$$

which is the sum over all realizations of noise that lead from r_k to r_{k+1} in a time unit. If one considers a whole path

$$\gamma = \Big\{\rho_i = r_0, \rho_1 = r_1, \ldots, \rho_m = \rho_f = r_m\Big\} \tag{8.64}$$

where ρ_i and ρ_f are the fixed initial and final states, and the Markov property is assumed, one obtains

$$P_m(\gamma) = \left(\frac{\Delta t}{(2\pi)^n |K|}\right)^{m/2} \times \tag{8.65}$$

$$\exp\left[-\frac{\Delta t}{2}\sum_{k=0}^{m-1}\left\langle \frac{r_{k+1} - r_k}{\Delta t} - D(r_k), \frac{r_{k+1} - r_k}{\Delta t} - D(r_k)\right\rangle\right] P_0(r_0)$$

where $P_0(\rho)$ is the initial probability density. In the $\Delta t \to 0$ limit, the sum turns an integral, and the exponent becomes

$$-\frac{1}{2}\int_{t_i}^{t_f} \Big\langle \dot\rho - D(\rho), \dot\rho - D(\rho)\Big\rangle \mathrm{d}t = -I_{[t_i, t_f]}(\gamma) \tag{8.66}$$

evaluated along the chosen path with given $\rho(t_i) = \rho_i$ and $\rho(t_f) = \rho_f$. The functional $I_{[t_i,t_f]}$, introduced here, will later play the role of the large deviation functional, of the fluctuations of ρ in the $n \to \infty$ limit. In fact, $I_{[t_i,t_f]}$ is extensive in the number of sites n, and the $n \to \infty$ limit leads to a probability density P that is negligible for all but the path that minimizes $I_{[t_i,t_f]}$. Therefore, one may conveniently introduce a functional J as

$$I_{[t_i,t_f]}(\gamma) = n\, J_{[t_i,t_f]}(\gamma) \tag{8.67}$$

where J is intensive, and may reach a limit as n grows.

The important point now is that the vector field $D(\rho)$ can be commonly decomposed in two orthogonal pieces[2]:

$$D(\rho) = -\frac{1}{2} K \nabla_\rho V(\rho) + \mathscr{A}(\rho)\,, \quad \big\langle K \nabla_\rho V(\rho), \mathscr{A} \big\rangle = 0 \tag{8.68}$$

where $\nabla_\rho = (\partial_{\rho_1}, \ldots, \partial_{\rho_n})$ is a gradient with respect to the components of ρ (not in space), V is a scalar function called quasi-potential with a minimum in $\hat{\rho}$, and $V(\hat{\rho}) = 0$. In this way, we identify a *dissipative* part and a *conservative* part in D, respectively, $-\frac{1}{2} K \nabla_\rho V(\rho)$ and $\mathscr{A}$. The first leads to entropy production and drives the system toward its (in general nonequilibrium) steady-state $\hat{\rho}$, the second, being orthogonal to this nonequilibrium trend, does not.

To maximize the probability of paths, we may follow two equivalent routes. First, directly from the definition of $I_{[t_i,t_f]}$, one obtains

$$I_{[t_i,t_f]}(\gamma) = \int_{t_i}^{t_f} \frac{1}{2} \left\langle \dot{\rho} + \frac{1}{2} K \nabla_\rho V - \mathscr{A}, \dot{\rho} + \frac{1}{2} K \nabla_\rho V - \mathscr{A} \right\rangle \mathrm{d}t \tag{8.69}$$

$$\begin{aligned} &= \frac{1}{2} \int_{t_i}^{t_f} \Big[\langle \dot{\rho}, \dot{\rho} \rangle + \frac{1}{2} \big\langle \dot{\rho}, K \nabla_\rho V \big\rangle - \langle \dot{\rho}, \mathscr{A} \rangle \\ &\quad + \frac{1}{2} \big\langle K \nabla_\rho V, \dot{\rho} \big\rangle + \frac{1}{4} \big\langle K \nabla_\rho V, K \nabla_\rho V \big\rangle - \frac{1}{2} \big\langle K \nabla_\rho V, \mathscr{A} \big\rangle \\ &\quad - \langle \mathscr{A}, \dot{\rho} \rangle - \frac{1}{2} \big\langle \mathscr{A}, K \nabla_\rho V \big\rangle + \langle \mathscr{A}, \mathscr{A} \rangle \Big] \mathrm{d}t \end{aligned} \tag{8.70}$$

[2] See, e.g., discussion in Sect. 3 of Chap. 4 of Ref. Freidlin and Wentzell (1998). For instance, Helmholtz decomposition of a vector field as $F = -\nabla U + \nabla \times W$, where the scalar function U and the vector W are properly related to F holds and is unique only under conditions that are often assumed but not general, e.g., smoothness in simply connected domains and fast decay to infinity of F and its derivatives

where the orthogonality condition yields $\langle K\nabla_\rho V, \mathscr{A}\rangle = \langle \mathscr{A}, K\nabla_\rho V\rangle = 0$. Then, one may note that

$$\int_{t_i}^{t_f} \frac{1}{2}\left\langle \dot{\rho} - \frac{1}{2}K\nabla_\rho V - \mathscr{A}, \dot{\rho} - \frac{1}{2}K\nabla_\rho V - \mathscr{A}\right\rangle \mathrm{d}t \tag{8.71}$$

$$\begin{aligned} &= \frac{1}{2}\int_{t_i}^{t_f} \Big[\langle\dot{\rho},\dot{\rho}\rangle - \frac{1}{2}\langle\dot{\rho}, K\nabla_\rho V\rangle - \langle\dot{\rho},\mathscr{A}\rangle \\ &\quad - \frac{1}{2}\langle K\nabla_\rho V, \dot{\rho}\rangle + \frac{1}{4}\langle K\nabla_\rho V, K\nabla_\rho V\rangle + \frac{1}{2}\langle K\nabla_\rho V, \mathscr{A}\rangle \\ &\quad - \langle\mathscr{A},\dot{\rho}\rangle + \frac{1}{2}\langle\mathscr{A}, K\nabla_\rho V\rangle + \langle\mathscr{A},\mathscr{A}\rangle\Big]\,\mathrm{d}t = \\ &= \frac{1}{2}\int_{t_i}^{t_f} \langle\dot{\rho} + \frac{1}{2}K\nabla_\rho V - \mathscr{A}, \dot{\rho} + \frac{1}{2}K\nabla_\rho V - \mathscr{A}\rangle\,\mathrm{d}t \\ &\quad - \frac{1}{2}\int_{t_i}^{t_f} [\langle\dot{\rho}, K\nabla_\rho V\rangle + \langle K\nabla_\rho V, \dot{\rho}\rangle]\,dt \end{aligned} \tag{8.72}$$

where one has

$$\langle\dot{\rho}, K\nabla_\rho V\rangle = \dot{\rho}^T K^{-1} K\nabla_\rho V = \dot{\rho}^T\nabla_\rho V = \sum_{i=1}^{n}\dot{\rho}_i\frac{\partial V}{\partial\rho_i} = \frac{\mathrm{d}V}{\mathrm{d}t} \tag{8.73}$$

and also

$$\langle K\nabla_\rho V, \dot{\rho}\rangle = \left(K\nabla_\rho V\right)^T K^{-1}\dot{\rho} = \nabla_\rho V^T K K^{-1}\dot{\rho} = \nabla_\rho V^T\dot{\rho} = \frac{\mathrm{d}V}{\mathrm{d}t} \tag{8.74}$$

which after some algebra leads to

$$\frac{1}{2}\int_{t_1}^{t_2}\left[\langle\dot{\rho}, K\nabla_\rho V\rangle + \langle K\nabla_\rho V, \dot{\rho}\rangle\right]\mathrm{d}t = \int_{t_1}^{t_2}\frac{\mathrm{d}V}{\mathrm{d}t}\,\mathrm{d}t = V(\rho_f) - V(\rho_i) \tag{8.75}$$

which is a fixed quantity, with vanishing variation. Therefore, minimizing Eq. (8.69) is the same as minimizing Eq. (8.71). The corresponding integrands are quadratic forms, as in the Onsager-Machlup case, hence the minimum of the integrals is obtained when either:

$$\dot{\rho} = D(\rho) = -\frac{1}{2}K\nabla_\rho V(\rho) + \mathscr{A}(\rho) \tag{8.76}$$

or

$$-\dot{\rho} = D^*(\rho) = -\frac{1}{2}K\nabla_\rho V(\rho) - \mathscr{A}(\rho) \tag{8.77}$$

Onsager-Machlup is a special instance of macroscopic fluctuation theory, in which $\mathscr{A} = 0$. The evolution Eq. (8.76) corresponds to the relaxation to the stationary, usually nonequilibrium state, while the evolution Eq. (8.77) corresponds to the non-thermodynamic fluctuation away from it in which currents flow against their driving forces. Being non-thermodynamic does not mean, however, being impossible. Indeed, present-day technology allows observations of mesoscopic fluctuations.

The fact that $D^* = D - 2\mathscr{A}$ breaks the temporal symmetry between fluctuation and relaxation and implies a form of irreversibility. Interestingly, this phenomenon does not depend on the dissipative part of the dynamics, but on the *non-dissipative* part, the one which does not contribute to transport or entropy production. The term V nevertheless plays the role of a Lyapunov function for the dynamics, and its time derivative is independent of $\mathscr{A}$:

$$\dot{V}(\rho) = \langle K\nabla_\rho V(\rho), D(\rho)\rangle \tag{8.78}$$

$$= -\frac{1}{2}\langle K\nabla_\rho V(\rho), K\nabla_\rho V(\rho)\rangle + \langle K\nabla_\rho V(\rho), \mathscr{A}\rangle \tag{8.79}$$

$$= -\frac{1}{2}\langle K\nabla_\rho V(\rho), K\nabla_\rho V(\rho)\rangle \leq 0 \tag{8.80}$$

The dynamics determined by the evolution equation

$$\dot{\rho} = +D^*(\rho) \tag{8.81}$$

is called adjoint hydrodynamics.

The models considered in this framework are stochastic, analogously to those of the Onsager-Machlup theory, but the restriction that the states be near equilibrium and that the fluctuations be small is relaxed. The state about which fluctuations take place will in general be a nonequilibrium state, and the fluctuations themselves can be as large as large deviation theory describes. Given the size n of the system of interest, fluctuations can be macroscopic, i.e., of order $O(n)$, rather than the microscopic order $O(\sqrt{n})$ of Ref. Onsager and Machlup (1953). If we assume that the stochastic (mesoscopic) processes arise from chaos in the deterministic, time reversal invariant microscopic dynamics, the question is whether predictions of the macroscopic fluctuation theory can be verified in such deterministic nonequilibrium systems. The question is far from trivial because even accepting that mesoscopic stochasticity is the result of microscopic chaos, the two levels of description are widely separated from each other and could have totally different phenomenology. A number of molecular dynamics simulations of dissipative, but time reversal invariant systems have nonetheless shown that the asymmetry predicted by the stochastic theory is indeed present also in them and that, in particular, fluctuations arise faster than they decay. It was then shown that the asymmetry arises in sufficiently chaotic systems, provided the source of chaos is not separate from that of transport (Giberti et al. 2007; Paneni et al. 2008). Independence of the sources of chaos and transport in certain systems yield symmetric fluctuations, like the equilibrium ones, centered about average nonequilibrium values. It is as if an equilibrium system was boarded on a train: there is transport, but it is not due to the microscopic dynamics of the object of interest (Gamba and Rondoni 2004).

8.5 Current Research

Apart from the literature cited in this chapter, the approach based on trajectory functionals described above feeds a lively and wide field of research that includes, for instance, stochastic models of nonequilibrium interacting particles (Giacomin et al. 2019). In certain approaches to thermodynamic fluctuations, it is applied even to anomalous transport (Serdyukov 2023). It is also found in machine learning (Lu et al. 2020), control theory (Chen et al. 2024), chemical kinetics (Renger 2022), and certain molecular simulations (Peredo-Ortiz et al. 2023), to mention just a few recent works.

9 Fluctuation Relations and Response Theory

> If the results agree wth a large series of facts, we must be content, even if the true nature of facts is not revealed in every respect.
>
> *On Certain Questions of the Theory of Gases,*
>
> *Ludwig Boltzmann*

As illustrated in the previous chapters, the study of fluctuations that began with Einstein's and Smoluchowski's theory of Brownian motion has been fundamental for the development of statistical mechanics and, more recently, for its applications to a range of fields that extends well beyond physics. In the first place, Einstein's relation, that connects diffusivity and mobility (Einstein 1956), was immediately found to apply to a much wider class of phenomena, than specks of dust or pollen suspended in a fluid, to include, e.g., fluctuations of electrons in an electrical circuit. Onsager's regression hypothesis and reciprocal relations may be considered as the following breakthrough (Onsager 1931a, b). Lastly, the works by Green and Kubo on a general linear response formalism and the fluctuation-dissipation relations completed what can be considered the classical response theory for thermodynamic systems, that is finely intertwined with the properties of microscopic fluctuations (Green 1952, 1954; Kubo 1957). This theory, which is limited to "near-equilibrium" states, has been developed for classical as well as quantum systems, for deterministic as well as stochastic processes (Kubo et al. 1998). We wrote near-equilibrium within quotes, because linear phenomena cover in fact a huge number of situations, which are hardly ever exceeded within the human space-time scales, for which thermodynamics has been developed. The fluctuation-dissipation relations link the mean response of particle systems to perturbations with a suitable correlation function, with respect to the statistics of the unperturbed system. As described in Chap. 7, this amazing relation yields nonequilibrium properties of a given system by means of an equilibrium

C. Mejía-Monasterio and L. Rondoni, *Nonequilibrium Statistical Mechanics I*,
Theoretical and Mathematical Physics,
https://doi.org/10.1007/978-3-031-98249-1_9

experiment. For instance, it gives the viscosity of a fluid, which is experimentally evident only when work is done to drag some object inside or at the surface of the fluid, without having to do any work.

While the theory of fluctuations around equilibrium states can be considered quite satisfactory and complete, its nonequilibrium counterpart faces greater difficulty to be developed, given the huge variety of phenomena it should encompass, and the fact that there is no obvious choice for the nonequilibrium ensembles. In 1993, an important step ahead was performed by D. J. Evans, E. G. D. Cohen and G. P. Morriss who proposed and tested the so-called *Fluctuation Relation* Ref. Evans et al. (1993). This relation appeared as one of the very few general exact results obtained on purely dynamical grounds, for systems that can be even quite far from equilibrium. Close to equilibrium, it can be used to derive the Green-Kubo and Onsager relations. The fluctuation relation states

$$\frac{\text{Prob}_\tau(A)}{\text{Prob}_\tau(-A)} = e^{\tau A} \tag{9.1}$$

where A and $-A$ are time averages computed over a long time τ of the dissipated power in a driven system, and $\text{Prob}_\tau(\pm A)$ is the steady-state probability of observing values close to $\pm A$. The relation was derived for the SLLOD isoenergetic model of a two-dimensional shearing fluid:

$$\begin{cases} \dot{\mathbf{q}}_i = \dfrac{\mathbf{p}_i}{m} + \gamma\, y_i \hat{\mathbf{x}} \\ \dot{\mathbf{p}}_i = \mathbf{F}_i(\mathbf{q}) + \gamma\, p_i^{(y)} \hat{\mathbf{x}} - \alpha_{th} \mathbf{p}_i \end{cases} \tag{9.2}$$

where $\Gamma = (\mathbf{q}_1, \mathbf{p}_1, \ldots, \mathbf{q}_N, \mathbf{p}_N)$ is the system state in phase space, γ is the shear rate in the y-direction, $\hat{\mathbf{x}}$ is the unit vector in the x-direction, and the friction term α_{th} (also called "thermostat") takes the form

$$\alpha_{th}(\Gamma) = -\frac{\gamma}{\sum_{i=1}^{N} \mathbf{p}_i^2} \sum_{i=1}^{N} p_i^{(x)} p_i^{(y)} \tag{9.3}$$

as prescribed by Gauss' principle of least constraint illustrated in Appendix I. The Fluctuation Relation followed from the observation that α_{th} is proportional to both the dissipated power, which means the entropy production rate under the condition of LTE, and the phase space volume variation rate Λ, i.e., the divergence of the vector field of the equations of motion, which plays a role in the theory of dynamical systems, cf. Chap. 6. The fluctuation relation for Λ was based on the applicability of relations enjoyed by axiom-A dynamical systems, under the hypothesis that they apply to a wider class of time reversal invariant dynamics (Gallavotti and Cohen 1995). Later, it was found that the connection between Λ and dissipation is not universal, but the fluctuation relations for the dissipated energy hold nonetheless.

The Jarzynski equality and the Crooks relation are other relations that concern the statistics of nonequilibrium systems. While the Fluctuation Relation is a large deviation steady-state result, because it considers the tails of the invariant distribution of the energy dissipation, the Jarzynski equality is a transient relation, i.e., referring to a non-invariant distribution, concerning an observable that probes the tails of that distribution. The Crooks relation was inspired by both the Fluctuation Relation and the Jarzynski equality, and also belongs to the class of transient fluctuation relations. Both the Jarzynski equality and the Crooks relation relate free energy differences between equilibrium states to nonequilibrium driving experiments. In this sense, transient relations close the loop with the Fluctuation-Dissipation relations, which obtain nonequilibrium properties from equilibrium experiments.

In this chapter, we begin with the Jarzynski equality, and show how simply it can be derived within the stochastic framework, in contrast to the more elaborate derivation concerning deterministic Hamiltonian systems. Indeed, stochastic and deterministic descriptions are complementary, referring in general to different experimental situations, even though at times they yield formally identical relations. That is reflected in the implicit or explicit assumptions that are part of the corresponding derivations. Jarzynski equality is exemplary in this sense. Then we derive the Crooks relation, which combines Jarzynski equality with the fluctuation relation. Finally, we focus on the fluctuation relations and on the response theory that emerged from one of Evans et al. (1993) within the framework of deterministic dynamics.

9.1 The Jarzynski Equality

9.1.1 Stochastic Derivation

Consider the following Langevin equation:

$$dX_t = K(X_t, \lambda(t))\,dt + \sqrt{\frac{2}{\beta}}\,dW_t, \quad X_t \in \Omega \subset \mathbb{R}^d, \quad K(x, \lambda) = -\frac{\partial U}{\partial x}(x, \lambda) \tag{9.4}$$

where U is a potential dependent on time through a parameter $\lambda(t)$, with t in an interval $[0, \tau]$,

W_t Wiener process, and $\beta = 1/k_B T$ the inverse "temperature." Define the Jarzynski work done up to time t, multiplied by β, as

$$A_t = \beta w_t = \beta \int_0^t \frac{\partial U}{\partial \lambda}(X_s, \lambda(s))\dot{\lambda}(s)\,ds \tag{9.5}$$

which is a mechanical quantity related to the variations of the energy of the system produced by the variations of λ. In the case in which λ represents a position in space, w_t becomes a mechanical work. For instance, when λ represents the position of the external agent exerting the force K on the system, w_t equals the work performed by the force $-K$ on the agent, which can be intended as the mechanical work done by the system on the external agent.

Remark. Measurements revealing this work do not necessarily correspond to the work done by the external agent on the system, because while the force may be known (although not guaranteed) the motion of the system of interest in general is not. In quasi-static thermodynamic transformations, this difficulty is overcome because all possible events are realized and the system remains homogeneous.

Let us also introduce the *intrinsic* free energy of the system at a given λ. Because in the following we are only interested in its variations. We define the free energy F as

$$F(\lambda) = -\frac{1}{\beta} \log Z(\lambda), \quad \text{where} \quad Z(\lambda) = \int_{\Omega} e^{-\beta U(x,\lambda)}\, dx \tag{9.6}$$

for a given value of λ. Here, the contribution of the kinetic part of the energy of the system is neglected, since in the present scheme it is not affected by variations of λ. This quantity is called intrinsic, because it refers solely to the properties of the system and does not consider its environment, although it strongly interacts with it, via the noise W_t.

The generator of the diffusion caused by the Langevin dynamics (9.4) is expressed by[1]:

$$L = -\frac{\partial U}{\partial x} \cdot \nabla_x + \frac{1}{\beta} \Delta_x \tag{9.7}$$

and its adjoint L^* in $L^2(\Omega)$ determines the evolution of the probability density of the variable X_t.[2] Consider the vector random variable (X_t, A_t), which obeys the following system of stochastic differential equations:

$$\begin{cases} dX_t = K\left(X_t, \lambda(t)\right) dt + \sqrt{\frac{2}{\beta}} dW_t \\ \\ dA_t = \beta\, \dot{\lambda}\, \dfrac{\partial U}{\partial \lambda}(X, \lambda)\, dt \end{cases} \tag{9.8}$$

[1] This is like the operator $\mathscr{L}_0$ in Eq. (7.18), which acts on the observables in Eq. (7.42), but there is no divergence of the vector field, because the deterministic part of the process is Hamiltonian. There is instead a diffusion contribution due to the stochastic part of the dynamics, which is absent in deterministic dynamics.

[2] Informally, given a bounded linear operator between two normed spaces, $T\colon \mathscr{M} \to \mathscr{N}$, its adjoint is a bounded linear operator $T^*\colon \mathscr{N} \to \mathscr{M}$, such that a bracket operation between elements of $\mathscr{N}$ equals a corresponding bracket operation between elements of $\mathscr{M}$: $\langle Tf, g\rangle_{\mathscr{N}} = \langle f, T^* g\rangle_{\mathscr{M}}$, where $x \in \mathscr{M}$ and $y \in \mathscr{N}$. For instance, let $\langle f, g\rangle_{\mathscr{N}} = \int f(x)g(x)\mathrm{d}x$, and $Tf = \mathrm{d}f/\mathrm{d}x$. Then, integrating by parts and assuming the functions vanish at the boundaries of the integration domain, we obtain $\langle Tf, g\rangle = -\int f(\mathrm{d}g/\mathrm{d}x)\mathrm{d}x$, which means $T^* g = -\mathrm{d}g/\mathrm{d}x$, with $\langle f, T^* g\rangle_{\mathscr{M}} = \int f(x) T^* g(x) dx$. Here, L^* is like $-\mathscr{L}_0$ in Eq. (7.18), which acts on the probability density in Eq. (7.20).

Its probability density $p_t(x, a)$ is obtained solving the following equation:

$$\frac{\partial p}{\partial t} = L^* p - \left(\beta\,\dot{\lambda}\,\frac{\partial U}{\partial \lambda}\right)\frac{\partial p}{\partial a} \tag{9.9}$$

because the non-random part of the process contributes the negative of the divergence of the vector field times the probability density, $\mathrm{d}(\dot{a}p)/\,\mathrm{d}t$, to the local variation of p, and $\dot{a}$ does not depend on a. Then, transforming p_t as

$$\widetilde{p}_t(x) = \int_{-\infty}^{\infty} da\, e^{-a}\, p_t(x, a) \tag{9.10}$$

the new quantity obeys

$$\int_{-\infty}^{\infty} da\, e^{-a}\frac{\partial p}{\partial t} = \int_{-\infty}^{\infty} da\, e^{-a} L^* p - \int_{-\infty}^{\infty} da\, e^{-a}\left(\beta\dot{\lambda}\frac{\partial U}{\partial \lambda}\right)\frac{\partial p}{\partial a} \tag{9.11}$$

which, under the hypothesis that $e^{-a}p(x, a)$ vanishes in the $a \to \pm\infty$ limits, yields

$$\frac{\partial \widetilde{p}_t}{\partial t} = L^* \widetilde{p}_t - \left(\beta\dot{\lambda}\frac{\partial U}{\partial \lambda}\right)\widetilde{p}_t \tag{9.12}$$

because neither L^* nor the term within brackets depend on a. As initial condition, we take

$$p_0(x, a) = \frac{e^{-\beta U(x,\lambda(0))}}{Z(\lambda(0))}\delta(a) \quad \text{which implies} \quad \widetilde{p}_0(x) = \frac{e^{-\beta U(x,\lambda(0))}}{Z(\lambda(0))} \tag{9.13}$$

because the state is initially in equilibrium, and the work done in a single time instant vanishes. This leads to the following solution for Eq. (9.12):

$$\widetilde{p}_t(x) = \frac{1}{Z(\lambda(0))}\exp\left[-\beta U(x, \lambda(t))\right] \tag{9.14}$$

In turn, the average of the exponential of $-A_\tau$ with respect to the initial distribution is expressed by

$$\mathbb{E}_0\left(e^{-A_\tau}\right) = \int_\Omega dx \int_{-\infty}^{\infty} da\, e^{-a}\, p_\tau(x, a) = \int_\Omega dx\, \widetilde{p}_\tau(x) \tag{9.15}$$

Therefore, taking $\lambda(0) = \alpha$ and $\lambda(\tau) = \omega$, Eqs. (9.14) and (9.15) eventually yield:

$$\mathbb{E}_0\left(e^{-A_\tau}\right) = \frac{Z(\lambda(\tau))}{Z(\lambda(0))} = \exp\{-\beta\left[F(\omega) - F(\alpha)\right]\} \tag{9.16}$$

whatever function λ connects α to ω in any given time τ. Equation (9.16) is called the Jarzynski Equality (JE). This mathematically elementary exercise constitutes a physically spectacular result. In the first place, it completes the theory of the fluctuation-dissipation relation, since it yields equilibrium properties by means of nonequilibrium experiments. Secondly, it suggests that the free energy difference corresponding to the equilibrium states at $\lambda = \alpha$ and $\lambda = \omega$ of any system in contact with a thermal bath can be obtained from the statistics of practically any kind of work, which can be arbitrarily slow or fast, since τ can be arbitrarily long or short. If taken literally, this formal result holds regardless of the space and time scales involved, which is, of course, not plausible as no physical theory extends to all possible scales. A comparison with the Hamiltonian derivation of the Jarzynski equality is thus quite useful to understand the strong hypotheses that underlie the Langevin description, as well as the definition of work adopted here. These hypotheses delimit the proper range of applicability of the theory, giving it predictive value, and hence making it physically relevant. In this respect, one may recall that the Langevin description has proven exceedingly successful in describing the fluctuations of standard observables of small systems in an equilibrium state at a given temperature T. This is surely a good reason to use it. However, that framework must be taken *cum grano salis* in the present situation. The difference between the classical and the current studies is that the system treated here is driven arbitrarily far from equilibrium, and, in addition, the observable to be measured is not standard: it is an exponential quantity. Such quantities, unlike the standard ones connected to thermodynamic properties, get a substantial contribution to their average from the tails, rather than the bulk, of their distributions. The consequences of this fact will be analyzed in Appendix F. In practice, not only do the tails of a distribution pose a statistically difficult, if not impossible, task, but may incorrectly describe the system. This results in bounds on the applicability of Eq. (9.16), that must be assessed case by case. A correct interpretation of the ensuing results requires this to be considered. Let us turn to the Hamiltonian derivation of the Jarzynski equality.

9.1.2 Hamiltonian Theory

Consider an N-particle system S immersed in an environment E made of M particles, both initially in equilibrium with a heat bath at temperature T. System and environment together are denoted by S+E, and the bath by B. We denote by $\mathscr{M}$ the phase space of S+E, and by $\Phi^t : \mathscr{M} \to \mathscr{M}$ the evolution operator of the phase points, which we assume to be determined by a Hamiltonian $\mathscr{H}$ that depends on a time dependent externally controlled parameter λ. The paradigmatic experiment one wants to describe is that in which S is a protein, E is a pool filled with water and B is the laboratory in which the pool is located, but the derivation of the Jarzynski equality does not need such an interpretation. Then, in general, the Hamiltonian of S+E is the sum of the energy of S, H_S, the energy of E, H_E, and the energy of the interaction between the two, h_{int}. When λ is modified in time by external agents, following a precise rule $\lambda(t)$, with $t \in [0, \tau]$, one expects that some work is directly or indirectly done on S and, in general, all contributions to the Hamiltonian are modified. Therefore, denoting by $\Gamma = (x, y)$ a point in the phase space of S+E, with $x = (\mathbf{q}_S, \mathbf{p}_S)$ the coordinates and momenta of the particles of S, and $y = (\mathbf{q}_E, \mathbf{p}_E)$ the coordinates and momenta of the particles of E, one may write

$$\mathscr{H}(\Gamma; \lambda) = H_S(x; \lambda) + H_E(y; \lambda) + h_{\text{int}}(x, y; \lambda) \tag{9.17}$$

In the derivation of the JE, one assumes instead that the variation of λ from its initial value, $\lambda(0) = \alpha$, to its final value, $\lambda(\tau) = \omega$, has no effect on H_E and h_{int}, so that the Jarzynski Hamiltonian takes the following simplified form:

$$\mathscr{H}(\Gamma; \lambda) = H_S(x; \lambda) + H_E(y) + h_{\text{int}}(x, y) \tag{9.18}$$

In the case of a protein in water, λ could represent the position of the optical device which drags the bead to which one end of the protein is glued, and its variation could correspond to the translation of this device. The coordinates of the system of interest are in general unknown to the operator, but may be distributed in a known fashion. If S+E is initially in thermodynamic equilibrium at temperature T, the statistics of the corresponding microscopic states, the phases Γ, are given by the canonical ensemble:

$$p_\alpha(\Gamma) = \frac{\exp\{-\beta \mathscr{H}(\Gamma; \alpha)\}}{Y_\alpha}, \quad \text{with } Y_\alpha = \int d\Gamma\, e^{-\beta \mathscr{H}(\Gamma;\alpha)} \text{ and } \beta = \frac{1}{k_B T} \tag{9.19}$$

At time $t = 0$, S+E is isolated from B, and the given protocol $\lambda : [0, \tau] \to \mathbb{R}$, with $\lambda(0) = \alpha$ and $\lambda(\tau) = \omega$, is performed in such a way that there is no exchange of energy between S+E and the rest of the universe, except that implied by the variation of λ. The dynamics of S+E is thus determined by the equations of motion derived from the time-dependent Hamiltonian $\mathscr{H}(\cdot; \lambda(t))$. Then, given the initial condition $\Gamma \in \mathscr{M}$, one considers the variation of H_S, which writes

$$H_S(x_\tau; \omega) - H_S(x_0; \alpha) = \int_0^\tau dt\, \dot{\lambda} \frac{\partial H_S}{\partial \lambda}(x_t; \lambda(t)) + \int_0^\tau dt\, \dot{x}_t \frac{\partial H_S}{\partial x}(x_t; \lambda(t)) \tag{9.20}$$

where x_t represents the part of the evolved phase point $\Phi^t \Gamma$ corresponding to the coordinates and momenta of S at time t. Drawing a parallel with thermodynamics, H_S is taken to formally represent the internal energy of S; the first integral in Eq. (9.20), the one in which the external agents controlling λ play a role, is taken to formally represent work; and the second integral to formally represent heat. Accepting the analogy one may write

$$W_J = \int_0^\tau dt\, \dot{\lambda} \frac{\partial H_S}{\partial \lambda}; \qquad Q_J = \int_0^\tau dt\, \dot{x}_t \frac{\partial H_S}{\partial x} \tag{9.21}$$

where W_J denotes the Jarzynski work, and Q_J the Jarzynski heat, with the subscripts J stressing that, in general, these quantities are only formal analogues of the corresponding thermodynamic quantities. This process is repeated many times, always beginning in the same equilibrium state, i.e., picking the initial condition Γ at random with respect to the distribution p_α. Of course, Γ is not known and varies from experiment to experiment. Consequently, the quantities computed along each repetition of the process can be treated as random variables, although deterministically fixed by the initial condition. Now, observing that

$$\frac{\mathrm{d}\mathscr{H}}{dt} = \frac{\partial \mathscr{H}}{\partial t} + \{\mathscr{H}, \mathscr{H}\} = \frac{\partial \mathscr{H}}{\partial t}, \quad \text{and that} \quad \frac{\partial H_E}{\partial \lambda} = \frac{\partial h_{int}}{\partial \lambda} = 0 \tag{9.22}$$

one obtains

$$\mathscr{H}(\Phi^\tau \Gamma; \omega) - \mathscr{H}(\Gamma, \alpha) = \int_0^\tau \mathrm{d}t \, \frac{d}{dt} \mathscr{H}(\Phi^t \Gamma; \lambda(t)) \tag{9.23}$$

$$= \int_0^\tau \mathrm{d}t \, \dot{\lambda} \frac{\partial \mathscr{H}}{\partial \lambda}(\Phi^t \Gamma; \lambda(t)) = \int_0^\tau \mathrm{d}t \, \dot{\lambda} \frac{\partial H_S}{\partial \lambda}(x_t; \lambda(t)) = W_J(\Gamma; \lambda) \tag{9.24}$$

as natural, because the full Hamiltonian $\mathscr{H}$ only changes because H_S does. In the ideal case of infinitely many repetitions of the experiment, the average of the quantities W_J is then given by

$$\lim_{n\to\infty} \frac{1}{n} \sum_{k=1}^{n} e^{-\beta w_k} = \int_{\mathscr{M}} \mathrm{d}\Gamma e^{-\beta W_J(\Gamma;\lambda)} p_\alpha(\Gamma) = \mathbb{E}_\alpha \left(e^{-\beta W_J} \right) \tag{9.25}$$

where w_k is the Jarzynski work computed along the k-th trial. Obviously only a finite number of repetitions can be done, and n is increased as much as possible, in order to sample, as thoroughly as one can, the ensemble of initial conditions. In turn, the canonical average with parameter α yields

$$\mathbb{E}_\alpha \left(e^{-\beta W_J} \right) = \int_{\mathscr{M}} \mathrm{d}\Gamma \, \frac{1}{Y_\alpha} e^{-\beta \mathscr{H}(\Gamma;\alpha)} \, e^{-\beta W_J(\Gamma;\lambda)} = \tag{9.26}$$

$$= \frac{1}{Y_\alpha} \int_{\mathscr{M}} \mathrm{d}\Gamma \; e^{-\beta \mathscr{H}(\Gamma;\alpha)} \, e^{\beta \mathscr{H}(\Gamma;\alpha)} \, e^{-\beta \mathscr{H}(\Phi^\tau \Gamma;\omega)} \tag{9.27}$$

$$= \frac{1}{Y_\alpha} \int_{\mathscr{M}} \mathrm{d}\Gamma \; e^{-\beta \mathscr{H}(\Phi^\tau \Gamma;\omega)} \tag{9.28}$$

The change of variable from Γ to $\Gamma' = \Phi_\lambda^\tau \Gamma$ yields

$$\int_{\mathscr{M}} \mathrm{d}\Gamma \; e^{-\beta \mathscr{H}(\Phi^\tau \Gamma;\omega)} = \int \mathrm{d}\Gamma' \; e^{-\beta \mathscr{H}(\Gamma';\omega)} \left| \frac{\partial \Gamma'}{\partial \Gamma} \right|^{-1} \tag{9.29}$$

where the Jacobian determinant $\left|\partial \Gamma' / \partial \Gamma\right| = 1$, because the dynamics are Hamiltonian. Then, one may write

$$\mathbb{E}_\alpha \left(e^{-\beta W_J} \right) = \int \mathrm{d}\Gamma' \; \frac{1}{Y_\alpha} e^{-\beta \mathscr{H}(\Gamma',\omega)} = \frac{Y_\omega}{Y_\alpha} = e^{-\beta [F_{\mathrm{S+E}}(\omega) - F_{\mathrm{S+E}}(\alpha)]} \tag{9.30}$$

where $F_{\mathrm{S+E}}(\lambda) = -\beta^{-1} \ln Y_\lambda$ is the free energy of S+E in the canonical equilibrium at temperature T and parameter λ. Note that only the initial and final Hamiltonians appear in this result, therefore, the deterministic protocol $\lambda(t)$ is irrelevant, as long as $\lambda(0) = \alpha$, $\lambda(\tau) = \omega$. Analogously to the stochastic derivation of the JE, it seems that the same result is obtained regardless of the physical scales characterizing the experiments.

However, one is not interested in the free energy variations of S+E, but in those of the *intrinsic* free energy of S, F_S, which is measurable when S does not observably interact with an environment, usually referred to as weak coupling . In that case, one has

$$F_{S+E}(\lambda) = F_{\mathrm{S}}(\lambda) + F_{\mathrm{E}}, \quad \text{and} \quad F_{\mathrm{S+E}}(\omega) - F_{\mathrm{S+E}}(\alpha) = F_{\mathrm{S}}(\omega) - F_{\mathrm{S}}(\alpha) \tag{9.31}$$

since in this framework the free energy of E does not depend on λ. However, this is generally not the case of, e.g., protein stretching experiments, since the proteins strongly interact with the water in the pool. To treat the cases in which h_{int} is not negligible, which is referred to as strong coupling, it becomes appropriate to consider a quantity that accounts for such interaction, and that is called *solvated* free energy: a property of S in E. One such quantity is obtained by introducing the following energy function:

$$H_{\mathrm{S}}^{*}(x;\lambda) = H_{\mathrm{S}}(x;\lambda) - \frac{1}{\beta} \ln \frac{\int \mathrm{d}y e^{-\beta[H_{\mathrm{E}}(y)+h_{\mathrm{int}}(x,y)]}}{\int \mathrm{d}y\, e^{-\beta H_{\mathrm{E}}(y)}} \tag{9.32}$$

which was proposed by Kirkwood (1935) to treat subsystems of macroscopic dense fluids in thermodynamic equilibrium. This is the energy of S with a contribution coming from the interaction energy averaged over the variables of E, which essentially amounts to a potential of mean force. Its meaning can be understood in terms of the marginal probability of the particles of S in E:

$$f_S(x;\lambda) = \int \mathrm{d}y\, p_\lambda(x,y) = \frac{e^{-\beta H_S(x;\lambda)}}{Y_{\mathrm{S+E}}(\lambda)} \int \mathrm{d}y\, e^{-\beta\{H_E(y)+h_{\mathrm{int}}(x,y)\}} \tag{9.33}$$

whose associated Landau free energy takes the form

$$F_{S+E}(x;\lambda) = -\frac{1}{\beta} \log f_S(x;\lambda) = -\frac{1}{\beta} \log \left[\frac{e^{-\beta H_S(x;\lambda)}}{Y_{\mathrm{S+E}}(\lambda)} \int \mathrm{d}y\, e^{-\beta\{H_E(y)+h_{\mathrm{int}}(x,y)\}} \right] \tag{9.34}$$

This shows that H_S^* is effectively a free energy. Indeed, we have

$$H_S^*(x;\lambda) = F_{\mathrm{S+E}}(x;\lambda) - \frac{1}{\beta} \log \frac{Y_\lambda}{Z_{\mathrm{E}}} = F_{\mathrm{S+E}}(x;\lambda) + F_{\mathrm{S+E}}(\lambda) - F_E = F_{\mathrm{S+E}}(x;\lambda) + \text{constant}, \tag{9.35}$$

where

$$Z_E = \int dy \exp\{-\beta H_E(y)\}, \quad F_E = -\beta^{-1} \ln Z_E = \text{free energy of E} \tag{9.36}$$

and the constant does not matter for thermodynamic potentials. The partition function for H_S^* takes the form:

$$
\begin{aligned}
Z_S^*(\lambda) = \int \mathrm{d}x\, e^{-\beta H_S^*(x;\lambda)} \mathrm{d}x = \int \mathrm{d}x\, e^{-\beta H_S(x;\lambda)} \frac{\int \mathrm{d}y\, e^{-\beta[H_E(y)+h_{\mathrm{int}}(x,y)]}}{\int \mathrm{d}y\, e^{-\beta H_E(y)}} \\
= \frac{\int e^{-\beta \mathscr{H}(\Gamma;\lambda)}\, \mathrm{d}\Gamma}{\int e^{-\beta H_E(y)}\, \mathrm{d}y} = \frac{Y_\lambda}{Z_E}
\end{aligned}
\tag{9.37}
$$

and the logarithm of $Z_S^*(\lambda)$ multiplied by $-\beta^{-1}$ yields

$$
F_S^*(\lambda) = F_{S+E}(\lambda) - F_E, \quad \textit{i.e.,} \quad F_{S+E}(\lambda) = F_S^*(\lambda) + F_E, \tag{9.38}
$$

where $F_S^*(\lambda) = -\beta^{-1} \ln\ Z_S^*(\lambda)$ is called *"solvated"* free energy. It can be interpreted as the free energy of a hypothetical system S* with Hamiltonian H_S^*. Now, the free energy of S+E results the sum of the two contributions, taken separately, as if S* and E were not interacting, or the interaction was negligible. It follows that

$$
\left[F_{S+E}(\omega) - F_{S+E}(\alpha)\right] = \left[F_S^*(\omega) - F_S^*(\alpha)\right] \tag{9.39}
$$

because in this investigation F_E does not depend on λ. Therefore, one can finally write

$$
\mathbb{E}_\alpha\left(e^{-\beta W_J}\right) = e^{-\beta\left[F_S^*(\omega) - F_S^*(\alpha)\right]}, \tag{9.40}
$$

which is the JE, for the solvated free energy. Note, however, that F_S^* and F_S are far from equivalent. Different environments lead to different estimates of the solvated free energy F_S^*, while the intrinsic free energy F_S does not depend on E.

The Hamiltonian derivation of the JE outlined above is formally quite general and apparently applicable to any particle system, under any kind of perturbation of an initial equilibrium state at a given temperature T. In effect, no restriction is imposed on the time dependence of λ. Various comments are in order:

- This relation connects thermodynamic quantities and external (arbitrarily far from equilibrium) work. It is thus very useful when applicable and, in such cases, it completes the theory of the fluctuation-dissipation relation;
- It is a *transient/ensemble* relation. Transient means a property of non-steady states; ensemble means that differently from thermodynamic relations it requires an ensemble of measurements;
- Because $\ln \mathbb{E}(\phi) \geq \mathbb{E}(\ln \phi)$ for positive ϕ, one has $\mathbb{E}_\alpha(W_J) \geq F_S^*(\omega) - F_S^*(\alpha)$, which, if W_J is the thermodynamic work, then is consistent with the second law of thermodynamics. In general, $\mathbb{E}_\alpha(W_J)$ is not a thermodynamic quantity, and the relation does not refer to a single system;
- The relation goes beyond thermodynamics when LTE is broken. Hence it is most interesting in small systems;

- $F_S^*(\omega)$ *is not* the free energy at time τ, but the one possibly reached if λ remains fixed at λ and relaxation to the canonical equilibrium takes place.

Many differences distinguish the Langevin and the Hamiltonian derivations. The first difference is that the Langevin case includes no environment or, equivalently, E and the bath B are considered as combined in a single term, which strongly interacts with the systems, since it is represented by the noise W_t. In any case, the Langevin derivation concerns the intrinsic free energy, while the Hamiltonian derivation in general aims at estimating the solvated free energy. The second difference is that while the work considered in the Langevin derivation is the mechanical work, i.e., a force times an elementary displacement, integrated over the whole trajectory, W_J is not mechanical work, because in general λ is not a position, and the derivative of H_S with respect to λ is not a force. Is W_J related to thermodynamic work? Again, in general it is not. A thermodynamic quantity means averaging a phase function over microscopic states, i.e., over an ensemble. Therefore, let us compute its average:

$$\mathbb{E}_\alpha(W_J) = \int \mathscr{H}(\Phi^\tau\Gamma;\omega)\frac{e^{-\beta\mathscr{H}(\Gamma;\alpha)}}{Y_\alpha}\mathrm{d}\Gamma - \int \mathscr{H}(\Gamma;\alpha)\frac{e^{-\beta\mathscr{H}(\Gamma;\alpha)}}{Y_\alpha}\mathrm{d}\Gamma \quad (9.41)$$

$$= U^*(\tau,\omega) - U(\alpha), \quad (9.42)$$

where $U(\alpha)$ is the initial internal energy of S+E, because it is the average of the initial Hamiltonian with respect to the initial canonical ensemble. On the other hand, $U^*(\tau,\omega)$ is the average with respect to p_α of the Hamiltonian of S+E with parameter ω computed in the τ-dependent final position in phase space. The integrand $\mathscr{H}(\Phi^\tau\Gamma;\omega) = [\mathscr{H}(\cdot;\omega)\circ\Phi^\tau](\Gamma)$ is not the Hamiltonian with parameter ω, but a function of functions, with $\mathscr{H}(\cdot;\omega)$ the external function, averaged with respect to the initial ensemble f_α. Hence, it does not yield the internal energy in the final equilibrium state $U(\omega)$, which does not depend on τ. Furthermore, this function is not averaged with respect to the final distribution p_ω but with respect to p_α. On the other hand, both W_J and its averages are mathematically well defined quantities, and can be computed in numerical simulations.

9.2 Crooks Relation

The Crooks Relation introduced in Ref. Crooks (2000) can be seen as a generalization of the Jarzynski equality (9.16), combined with ideas concerning the Fluctuation Relations, that will be treated below in Sect. 9.3. Two canonical equilibrium states at same temperature are considered, analogously to Jarzynski's theory, but in Crooks' theory two processes join them, not just one: the forward and the backward processes. In the first process, one equilibrium state is associated with the initial value $\lambda^F(0) = \alpha$ of the time-dependent parameter $\lambda^F = \lambda^F(t)$, and the other equilibrium with its final value $\lambda^F(\tau) = \omega$. In the second process, the time reversal $\lambda^R(t) = \lambda^F(\tau - t)$ is considered and the order of the two equilibria is correspondingly swapped, together with the values of the time-dependent parameter: $\lambda^R(0) = \omega$ and $\lambda^R(\tau) = \alpha$.

To obtain the relation, one may use the Girsanov theorem (Rogers and Williams 2000), concerning the change of probability distribution, by means of the auxiliary process defined by

$$dY_t = \sqrt{2/\beta}\, dW_t \tag{9.43}$$

Given a functional $L(X_{[0,\tau]})$ of the solution of the equation

$$dX_t = F(X_t, t) + \sqrt{2/\beta}\, dW_t, \quad \text{for } t \in [0, \tau] \tag{9.44}$$

The Girsanov theorem states that its expected value conditioned on the initial datum $X_0 = x$ is given by

$$\mathbb{E}_0^{(x)}[L(X_{[0,\tau]})] = \mathbb{E}_0^{(x)}\left[L\left(Y_{[0,\tau]}\right)\exp\left(\frac{\beta}{2}\int_0^\tau F(Y_t, t)\, dY_t - \frac{\beta}{4}\int_0^\tau |F(Y_t, t)|^2\, dt\right)\right] \tag{9.45}$$

where all realizations of the process start from x. One may then average over all initial conditions with the distribution of interest and obtain the full expectation $\mathbb{E}_0$. In our case, take $L(X_{[0,\tau]}) = \exp\left(-rA_t^F\right)$, with $r \in \mathbb{R}$ and A_t^F given by Eq. (9.5): $A_t^F = A_t$. Then, substituting this functional and the definition (9.5) in Eq. (9.45) we obtain

$$\begin{aligned}\mathbb{E}_0\left(e^{-rA_\tau^F}\right) = \mathbb{E}_0\Bigg\{\exp\Bigg[-r\beta\int_0^\tau \frac{\partial U}{\partial \lambda}(Y_t, \lambda(t))\,\dot{\lambda}(t)\, dt \\ -\frac{\beta}{2}\int_0^\tau \frac{\partial U}{\partial x}(Y_t, \lambda(t))\, dY_t - \frac{\beta}{4}\int_0^\tau \left|\frac{\partial U}{\partial x}(Y_t, \lambda(t))\right|^2 dt\Bigg]\Bigg\}\end{aligned} \tag{9.46}$$

where Y_t solves Eq. (9.4) with K set to 0. Observe that

$$dU\left(Y_t, \lambda(t)\right) = \frac{\partial U}{\partial \lambda}(Y_t, \lambda(t))\,\dot{\lambda}(t)\, dt + \frac{\partial U}{\partial x}(Y_t, \lambda(t))\, dY_t + \frac{1}{\beta}\frac{\partial^2 U}{\partial x^2}(Y_t, \lambda(t)))\ dt \tag{9.47}$$

Therefore, Eq. (9.46) can be rewritten as

$$\begin{aligned}\mathbb{E}_0\left(e^{-rA_\tau^F}\right) = \mathbb{E}_0\Bigg\{\exp\Bigg[-\left(r - \frac{1}{2}\right)\beta\int_0^\tau \frac{\partial U}{\partial \lambda}(Y_t, \lambda(t))\,\dot{\lambda}(t)\, dt \\ -\frac{1}{2}\beta\left[U\left(Y_\tau, \lambda(\tau)\right) - U\left(Y_0, \lambda(0)\right)\right] \\ +\frac{1}{2}\int_0^\tau \frac{\partial^2 U}{\partial x^2}(Y_t, \lambda(t)))\ dt - \frac{\beta}{4}\int_0^\tau \left|\frac{\partial U}{\partial x}(Y_t, \lambda(t))\right|^2 dt\Bigg]\Bigg\}\end{aligned} \tag{9.48}$$

for the forward process. Introducing the random variable:

$$A_t^R = -\beta\int_0^t \frac{\partial U}{\partial \lambda}\left(X_s^R, \lambda^R(s)\right)\dot{\lambda}^R(s)\, ds \tag{9.49}$$

with the reversed process X_t^R obeying the following Langevin equation

$$dX_t^R = -\frac{dU}{dx}\left(X_s^R, \lambda^R(s)\right) dt + \sqrt{\frac{2}{\beta}}\, dW_t \tag{9.50}$$

and $\mathbb{E}_\tau$ denoting the expectation with respect to the equilibrium for X_0^R expressed by

$$d\mu_\tau(x) = \frac{1}{Z(\omega)} \exp\left[-\beta U(x;\omega)\right] dx \tag{9.51}$$

one also obtains

$$\begin{aligned}\mathbb{E}_\tau\left(e^{-(1-r)A_\tau^R}\right) = \frac{Z(\alpha)}{Z(\omega)}\mathbb{E}_0\Bigg\{\exp\Bigg[&-\left(r-\frac{1}{2}\right)\beta\int_0^\tau \frac{\partial U}{\partial \lambda}(Y_t, \lambda(t))\,\dot{\lambda}(t)\,dt \\ &-\frac{1}{2}\beta\left[U\left(Y_\tau, \lambda(\tau)\right) - U\left(Y_0, \lambda(0)\right)\right] \\ &+\frac{1}{2}\int_0^\tau \frac{\partial^2 U}{\partial x^2}(Y_t, \lambda(t)))\,dt - \frac{\beta}{4}\int_0^\tau \left|\frac{\partial U}{\partial x}(Y_t, \lambda(t))\right|^2 dt\Bigg]\Bigg\}\end{aligned} \tag{9.52}$$

for the backward process, where a change of coordinates $t \to \tau - t$ has been used, together with the observation that Y_t and $Y_{\tau-t}$ are equal in law. This means that

$$\mathbb{E}_0\left(e^{-rA_\tau^F}\right) = e^{-\beta[F(\omega)-F(\alpha)]}\mathbb{E}_\tau\left(e^{-(1-r)A_\tau^R}\right) \tag{9.53}$$

which also yields the Jarzynski Equality for $r = 1$. Introducing ρ^F and ρ^R as

$$\mathbb{E}_0\left(e^{-rA_\tau}\right) = \int_{-\infty}^{\infty} e^{-ra}\rho^F(a)\,da; \qquad \mathbb{E}_\tau\left(e^{-rA_\tau^R}\right) = \int_{-\infty}^{\infty} e^{-ra}\rho^R(a)\,da \tag{9.54}$$

one obtains

$$\int_{-\infty}^{\infty} e^{-ra}\rho^F(a)\,da = e^{-\beta[G(\omega)-G(\alpha)]}\int_{-\infty}^{\infty} e^{-(1-r)a}\rho^R(a)\,da \tag{9.55}$$

$$= e^{-\beta[G(\omega)-G(\alpha)]}\int_{-\infty}^{\infty} e^{(1-r)a}\rho^R(-a)\,da \tag{9.56}$$

and since that holds for all $r \in \mathbb{R}$, the above eventually leads to

$$\rho^F(a) = e^{a-\beta[G(\omega)-G(\alpha)]}\rho^R(-a) \tag{9.57}$$

which is known as the Crooks Relation. It compares the statistics of the forward process starting from the equilibrium with parameter α, and the reverse process starting from the equilibrium with parameter ω.

The Jarzynski and Crooks relations belong to a set of response formulae concerning *transient* states, i.e., states not preserved by the dynamics. In particular, they estimate properties of canonical equilibrium states via nonequilibrium processes imposed by some external agent and, in this respect they do the opposite of the original fluctuation-dissipation relations. Furthermore, they are exact relations that hold at any finite time. Their interest lies by and large on the fact that the systems of interest do not need to reach the final equilibrium states. Moreover, the Jarzynski relation is currently used as an efficient sampling tool within the realm of machine learning, which does not need to rely on its physical content.

The Jarzynski and Crooks transient relations were preceded by the discovery of a set of fluctuation relations concerning the properties of nonequilibrium stationary states, which only hold approximately at finite times, and become exact under a sequence of limits, provided the dynamics are sufficiently *"chaotic"*. While studying them, exact transient relations, concerning general initial states, not just canonical ones, have been found together with novel ergodic properties and a general theory of response to perturbations. The rest of this chapter is devoted to them.

9.3 Fluctuation Relations

The so-called *Fluctuation Relation* derived in Ref. Evans et al. (1993), which gave rise to an impressive wave of research, involving physics, mathematics, chemistry, biology, etc., concerns fluctuations in the finite-time averages of the energy dissipation, for an isoenergetic model of interacting particles, known as SLLOD, simulating a shearing fluid. In such a system, the instantaneous energy dissipation equals the instantaneous phase space contraction rate, allowing, under proper conditions, an identification of dynamical and thermodynamic quantities. Informally, it can be expressed as Eq. (9.1). This relation became immediately very popular for a number of reasons. In the first place, this shows that the energy dissipation rate, which yields the entropy production rate when in local equilibrium, can only be positive. Indeed, taking A as a macroscopic quantity, hence of the order of Avogadro's number in microscopic units, and τ as a macroscopic observation time, hence of the order of $10^{12} - 10^{14}$ microscopic time units, the probability of a negative fluctuation vanishes. To the best of our knowledge, Ref. Evans et al. (1993) was the first to obtain this enormous imbalance between positive and negative fluctuations for a finite number of particles described by time reversal invariant equations of motion.[3] Moreover, Eq. (9.1) was at the time one of the very few existing general results about

[3] As discussed in the case of Boltzmann's kinetic theory, the Hamiltonian dynamics of N particles systems is reversible and conservative hence it cannot produce by itself an imbalance between positive and negative dissipative currents. To obtain irreversibility, one has to rely on additional ingredients. For instance, Boltzmann's H-theorem is based on the $N \to \infty$ limit and on the stochasticity introduced by the molecular chaos hypothesis. Prigogine gives up the reversibility of the equations of motion and advocates irreversible microscopic dynamics, in order to obtain irreversible collective behavior (Prigogine 1980).

far from equilibrium systems: being dynamical in nature, the Fluctuation Relation remains valid beyond the range of applicability of thermodynamics. Later it was found that it is particularly suited to describe small systems, that it can be obtained also as a transient relation, and eventually it paved the way for the development of an exact, non-perturbative, response theory. We begin considering the derivation of the transient and steady-state relations. Many numerical and experimental tests have been performed, especially but not exclusively on small systems, proving that the Fluctuation Relations do hold rather generally and express interesting properties of such systems (Rondoni and Mejía-Monasterio 2007).

9.3.1 Derivation of Fluctuation Relations

The original derivation of the steady-state Fluctuation Relation was given in Ref. Evans et al. (1993). It was based on the identification of the energy dissipation rate with the phase space volumes contraction rate $-\Lambda$, and on the unstable periodic orbits decomposition of the invariant probability measure of an Anosov dynamical system. These are smooth uniformly hyperbolic systems that are strongly chaotic and afford robust statistical features, as later more precisely expressed in Ref. Gallavotti and Cohen (1995). The Fluctuation Relation was, however, found to apply to numerical simulations of dynamical systems that reach well beyond the Anosov class, such as nonequilibrium molecular dynamics models of which SLLOD is one example, as well as to stochastic processes within which it emerged in quite a simple fashion. The question then arose as to what ingredients are necessary, rather than sufficient, for the relation to hold. Indeed, sufficient conditions may include more than needed, possibly hiding the relevant physical mechanisms. On the other hand, necessary conditions are bound to be at work when the relation is observed to hold, which helps to clarify the nature of the phenomenon. Moreover, the decomposition in periodic orbits is technically complex, not applicable to most systems of interest, and $-\Lambda$ does not represent in general the desired physically measurable quantity, which is the energy dissipation. Therefore, a different derivation of the Fluctuation Relation has been developed, which has led to a host of new relations, including response formulae, that are exact. The derivation based on the Anosov structure remains informative; therefore, it is summarized in the Appendix G. Here, we illustrate instead the later developments, legacy of a rich season of statistical mechanics.

Consider a generic time reversal invariant and dissipative dynamical system:

$$\dot{\Gamma} = G(\Gamma), \quad \Gamma \in \mathcal{M} \tag{9.58}$$

where $\mathcal{M}$ denotes the phase space. The concrete dynamics that one may have in mind in mind are standard Gaussian and Nosé-Hoover dynamics in the presence of constant boundary or bulk drivings, cf. Appendix I, that are chaotic[4] and lead

[4] By chaotic we mean that at least one Lyapunov exponent is positive. .

in time to nonequilibrium steady states. Denote by $\Phi^t \Gamma$ the solution at time t of Eq. (9.58) with initial condition Γ, for all times $t \in I\!R$. Let $\mathscr{O}$ be a real-valued function of phase. According to the standard formulation of statistical mechanics, an experimental measurement is expressed by a time average over the time interval $[t, t+\tau]$ of the experiment, which we express as

$$\overline{\mathscr{O}}_{t,t+\tau}(\Gamma) := \frac{1}{\tau}\mathscr{O}_{t,t+\tau}(\Gamma) := \frac{1}{\tau}\int_t^{t+\tau} \mathscr{O}(\Phi^s \Gamma) ds \tag{9.59}$$

where $\mathscr{O}$ is a real-valued microscopically defined quantity. This notation highlights the bounds of the measurement time t and $t+\tau$, as well as the dependence of the result on the initial phase Γ. Indeed, different initial phases result in different evolution times, hence in different time averages. Note that, for fixed t and τ, the quantity $\overline{\mathscr{O}}_{t,t+\tau}$ is just another real-valued function of phase.

Let us now endow the phase space $\mathscr{M}$ with a probability distribution μ_0 of density f_0, so that the probability of a phase space volume element around Γ be written as $\mathrm{d}\mu_0(\Gamma) = f_0(\Gamma)\mathrm{d}\Gamma$. The density f_0 may represent, for instance, an equilibrium state which, at some stage, is perturbed by dissipative forces. One may then compute the initial probability that $\mathscr{O}$ takes values in the interval (a, b). By definition, this is obtained by integrating $\mathrm{d}\mu_0$ over the set $\{\Gamma \in \mathscr{M} : \mathscr{O}(\Gamma) \in (a, b)\}$, denoted by $\mathscr{O}|_{(a,b)}$, which collects the phases Γ for which $\mathscr{O}(\Gamma)$ lies in (a, b):

$$\mu_0\left(\mathscr{O}|_{(a,b)}\right) = \int_{\mathscr{O}|_{(a,b)}} \mathrm{d}\mu_0(\Gamma) = \int_{\mathscr{O}|_{(a,b)}} f_0(\Gamma)\mathrm{d}\Gamma \tag{9.60}$$

Let $(A)_\delta = (A-\delta, A+\delta)$. Then, the probability $\mu_0(\mathscr{O}|_{(A)_\delta})$ that a measurement starting in the state described by f_0 yields within a time τ a value close to A with tolerance δ, is expressed by

$$\mu_0\left(\mathscr{O}_{0,\tau}|_{(A)_\delta}\right) = \int_{\mathscr{O}_{0,\tau}|_{(A)_\delta}} f_0(\Gamma)\mathrm{d}\Gamma$$

The probability density f_0 and the phase space variation rate $\Lambda = \mathrm{div}\, G$ can be combined to define the phase variable known as *Dissipation Function*:

$$\Omega^{(0)}(\Gamma) = -G(\Gamma)\cdot \nabla_\Gamma \ln f_0|_\Gamma - \Lambda(\Gamma) \tag{9.61}$$

This requires f_0 to be positive in all regions visited by the phase space trajectories $\{\Phi^t \Gamma\}_{t\in \boldsymbol{R}}$. Averaging $\Omega^{(0)}$ along the trajectory starting at Γ, during the time interval $[t, t+\tau]$ yields

$$\overline{\Omega}_{t,t+\tau}(\Gamma) = \frac{1}{\tau}\,\Omega^{(0)}_{t,t+\tau} = \frac{1}{\tau}\ln\frac{f_0(\Phi^t\Gamma)}{f_0(\Phi^{t+\tau}\Gamma)} - \overline{\Lambda}_{t,t+\tau}(\Gamma) \tag{9.62}$$

The motivation for this terminology rests on the fact that $\Omega^{(0)}$ represents the energy dissipation rate normalized to $k_B T$, provided f_0 is properly chosen, e.g., if it equals the equilibrium distribution regarding the dynamics with the dissipative fields set to zero.

9.3.2 Transient Fluctuation Relations

Consider time reversal invariant dynamics and an observable $\mathscr{O}$ such that $\mathscr{O}(i\Gamma) = -\mathscr{O}(\Gamma)$, namely odd under the time reversal involution $i : \mathscr{M} \to \mathscr{M}$. This allows $\mathscr{O}$ to take both positive and negative values, along a time evolution, and one may compute their relative probabilities. In particular, one may focus on the μ_0-probabilities that $\mathscr{O}_{0,\tau} \in (-A)_\delta$ and that $\mathscr{O}_{0,\tau} \in (A)_\delta$, whose ratio is expressed as

$$\frac{\mu_0(\mathscr{O}_{0,\tau}|_{(-A)_\delta})}{\mu_0(\mathscr{O}_{0,\tau}|_{(A)_\delta})} = \frac{\int_{\mathscr{O}_{0,\tau}|_{(-A)_\delta}} f_0(\Gamma) d\Gamma}{\int_{\mathscr{O}_{0,\tau}|_{(A)_\delta}} f_0(\Gamma) d\Gamma} \tag{9.63}$$

To compute the ratio (9.63), note that the phases in $\mathscr{O}_{0,\tau}|_{(-A)_\delta}$ are those, and only those, obtained from the points in $\mathscr{O}_{0,\tau}|_{(A)_\delta}$ as follows

$$\mathscr{O}_{0,\tau}|_{(-A)_\delta} = i\Phi^\tau \, \mathscr{O}_{0,\tau}|_{(A)_\delta} \tag{9.64}$$

Therefore, the numerator of (9.63) can be computed integrating over the set $\mathscr{O}_{0,\tau}|_{(A)_\delta}$, thanks to the coordinate transformation

$$\Gamma = i\Phi^\tau X, \qquad \text{whose Jacobian is} \quad J = \left|\frac{d\Gamma}{dX}\right| = e^{-\Lambda_{0,\tau}(X)} \tag{9.65}$$

where $\Lambda_{0,\tau}(X)$ is the integral of Λ along the trajectory from X to $\Phi^\tau X$. Then,

$$\int_{\mathscr{O}_{0,\tau}|_{(-A)_\delta}} f_0(\Gamma)\mathrm{d}\Gamma = \int_{\mathscr{O}_{0,\tau}|_{(A)_\delta}} f_0(i\Phi^\tau X)\, e^{-\Lambda_{0,\tau}(X)} \mathrm{d}X \tag{9.66}$$

If f_0 is even under i, as in equilibrium states, $f_0(i\Gamma) = f_0(\Gamma)$, we eventually get

Transient $\mathscr{O}$-FR:

$$\frac{\mu_0(\mathscr{O}_{0,\tau}|_{(-A)_\delta})}{\mu_0(\mathscr{O}_{0,\tau}|_{(A)_\delta})} = \frac{\int_{\mathscr{O}_{0,\tau}|_{(A)_\delta}} f_0(\Phi^\tau X) e^{-\Lambda_{0,\tau}(X)} \mathrm{d}X}{\int_{\mathscr{O}_{0,\tau}|_{(A)_\delta}} f_0(\Gamma)\mathrm{d}\Gamma} := \mathbb{E}_0\left[e^{-\Omega^{(0)}_{0,\tau}}\right]_{\mathscr{O}_{0,\tau} \in (A)_\delta} \tag{9.67}$$

where the last equality defines the phase space average of $\exp(-\Omega^{(0)}_{0,\tau})$ with respect to μ_0, under the condition that $\mathscr{O}_{0,\tau}(\Gamma) \in (A)_\delta$. Because the dissipation function is odd under time reversal if f_0 is even, one may take $\mathscr{O}_{0,\tau} = \overline{\Omega^{(0)}}_{0,\tau} = \Omega^{(0)}_{0,\tau}/\tau$, obtaining the elegant identity:

Transient Ω-FR:

$$\frac{\mu_0\left(\overline{\Omega^{(0)}}_{0,\tau}|_{(-A)_\delta}\right)}{\mu_0\left(\overline{\Omega^{(0)}}_{0,\tau}|_{(A)_\delta}\right)} = \mathbb{E}_0\left[\exp\left(-\Omega^{(0)}_{0,\tau}\right)\right]_{\overline{\Omega^{(0)}}_{0,\tau} \in (A)_\delta} = e^{-[A+\varepsilon(\delta,A,\tau)]\tau} \tag{9.68}$$

where we have used the time average of the dissipation function, rather than its integral, to highlight in the exponential the size of the dissipation function and that of the integration interval, which could be separately large or small. The correction term obeys $|\varepsilon(\delta, A, \tau)| \le \delta$, because the conditional average is over the phase space trajectory segments in which the average of $\Omega^{(0)}$ lies within a distance δ from A. Note also that in the exponential we have the integral of the dissipation function, rather than its average.

Remarkably, the time reversal invariance of the dynamics and of the probability density f_0 is the only ingredient of this derivation.

Note, however, that these transient relations are conceptually and also practically very different from the steady-state fluctuation relation of Evans et al. (1993). Indeed, Eq. (9.68) refers to the initial state characterized by μ_0, and not to the steady-state probability distribution which, as far as these relations are concerned, does not even need to exist. Taking larger τ makes no difference: the probability under consideration remains μ_0, one simply replaces a phase function defined with a short τ with one defined with a long τ. For this reason, Eqs. (9.67) and (9.68) are called "transient." The transient Fluctuation Relations differ from the steady-state Fluctuation Relations also because they inherently concern ensembles of objects, or repetitions of the same experiment starting from different microscopic states but same macroscopic state. The steady-state Fluctuation Relations can be verified in a single experiment, by observing the steady-state fluctuations in time of the observables of interest. Transient relations use nonequilibrium dynamics to express properties of the initial state, commonly an equilibrium state. These, indeed, obtain nonequilibrium properties from equilibrium experiments, while the transient Fluctuation Relations obtain equilibrium properties from nonequilibrium experiments.

Given the dynamics Φ^t, different choices of f_0 are possible, hence different dissipation functions and different transient Fluctuation Relations concerning different observables. Whatever f_0 and $\delta > 0$, one obtains

$$\frac{\mu_0(\mathcal{O}|_{(-\delta,\delta)})}{\mu_0(\mathcal{O}|_{(-\delta,\delta)})} = 1, \quad \text{hence} \quad \mathbb{E}_0\left[e^{-\Omega^{(0)}_{0,\tau}}\right]_{\mathcal{O}\in(-\delta,\delta)} = 1 \tag{9.69}$$

These identities are useful, e.g., to test the accuracy of numerical simulations and to calibrate experimental equipment.

9.3.3 Steady-State Fluctuation Relation

The derivation of the transient Fluctuation Relations can be used as a starting point for the derivation of the steady-state Fluctuation Relation, which require further ingredients to hold. To begin with, the dynamics have to be capable of reaching a steady state, which, in general, is not guaranteed. Other requirements will be found as the derivation proceeds. The first step is to start averaging at a time $t > 0$, rather than at 0, and to consider the ratio

$$\frac{\mu_0(\mathscr{O}_{t,t+\tau}|_{(-A)_\delta})}{\mu_0(\mathscr{O}_{t,t+\tau}|_{(A)_\delta})} = \frac{\int_{\mathscr{O}_{t,t+\tau}|_{(-A)_\delta}} f_0(\Gamma)d\Gamma}{\int_{\mathscr{O}_{t,t+\tau}|_{(A)_\delta}} f_0(\Gamma)d\Gamma} \tag{9.70}$$

One finds that the phases concerning the numerator of Eq. (9.70) are related to those concerning the denominator: $\mathscr{O}_{t,t+\tau}|_{(-A)_\delta} = i\Phi^{2t+\tau}\mathscr{O}_{t,t+\tau}|_{(A)_\delta}$. In effect, the trajectory starting at $i\Phi^{2t+\tau}X$ traces back the one starting at X and lasting up to time $2t+\tau$. At time τ, the backward trajectory starts with the piece that the trajectory beginning at X reaches after a time $t+\tau$.

Changing coordinates in $\mathscr{M}$ so that $\Gamma = i\Phi^{2t+\tau}X$, and the Jacobian is given by $\exp\left[-\Lambda_{0,2t+\tau}(X)\right]$, we obtain

$$\int_{\mathscr{O}_{t,t+\tau}|_{(-A)_\delta}} f_0(\Gamma)\mathrm{d}\Gamma = \int_{\mathscr{O}_{t,t+\tau}|_{(A)_\delta}} f_0(i\Phi^{2t+\tau}X)e^{-\Lambda_{0,2t+\tau}(X)}\mathrm{d}X \tag{9.71}$$

$$= \int_{\mathscr{O}_{t,t+\tau}|_{(A)_\delta}} f_0(\Phi^{2t+\tau}X)e^{-\Lambda_{0,2t+\tau}(X)}\mathrm{d}X \tag{9.72}$$

$$= \int_{\mathscr{O}_{t,t+\tau}|_{(A)_\delta}} f_0(X)e^{-\Omega_{0,2t+\tau}(X)}\mathrm{d}X \tag{9.73}$$

where we have used the parity of f_0 and to Eq. (9.62). The crucial difference from the case in which averaging starts at time 0, is that the coordinate transformation that links $\mathscr{O}_{t,t+\tau}|_{(-A)_\delta}$ to $\mathscr{O}_{t,t+\tau}|_{(A)_\delta}$ now requires a time longer than the averaging interval: a lag τ is required for both A and $-A$. The result is

$$\frac{\mu_0(\mathscr{O}_{t,t+\tau}|_{(-A)_\delta})}{\mu_0(\mathscr{O}_{t,t+\tau}|_{(A)_\delta})} = \mathbb{E}_0\left[\exp\left(-\Omega^{(0)}_{0,2t+\tau}\right)\right]_{\mathscr{O}_{t,t+\tau}\in(A)_\delta} \tag{9.74}$$

where the right-hand side is the conditional average of $\exp(-\Omega^{(0)}_{0,2t+\tau})$, over phases Γ for which $\overline{\mathscr{O}}_{t,t+\tau}(\Gamma)$ is close to A. In the case that $\mathscr{O}_{t,t+\tau} = \overline{\Omega^{(0)}}_{t,t+\tau}$, one obtains

$$\frac{\mu_0(\overline{\Omega^{(0)}}_{t,t+\tau}|_{(-A)_\delta})}{\mu_0(\overline{\Omega^{(0)}}_{t,t+\tau}|_{(A)_\delta})} = \mathbb{E}_0\left[e^{-\Omega^{(0)}_{0,t}}\,e^{-\Omega^{(0)}_{t,t+\tau}}\,e^{-\Omega^{(0)}_{t+\tau,2t+\tau}}\right]_{\overline{\Omega^{(0)}}_{t,t+\tau}\in(A)_\delta} \tag{9.75}$$

$$= e^{-[A+\varepsilon(\delta,t,A,\tau)]\tau}\,\mathbb{E}_0\left[e^{-\Omega_{0,t}}\,e^{-\Omega_{t+\tau,2t+\tau}}\right]_{\overline{\Omega^{(0)}}_{t,t+\tau}\in(A)_\delta} \tag{9.76}$$

where the second equality follows because $\overline{\Omega^{(0)}}_{t,t+\tau}$ does not differ more than the accepted tolerance δ from A: the difference ε depends in principle on δ, t, A and τ, but is smaller than δ. This result is exact, for all t, τ, δ, and for any A such that both probabilities $\mu_0(\overline{\Omega^{(0)}}_{t,t+\tau}|_{(-A)_\delta})$ and $\mu_0(\overline{\Omega^{(0)}}_{t,t+\tau}|_{(A)_\delta})$ are positive. Again, the only ingredient in this derivation is the time reversibility of the dynamics and of f_0. Moreover, (9.74) and (9.76) are transient Fluctuation Relations, because they refer to f_0 for all t and τ.

The hypothesis of conservation of probability in phase space under the dynamics, like that of the mass of a fluid in real space, i.e.,:

$$\mu_t(\Phi^t E) = \mu_0(E); \quad \text{and} \quad \int_{\Phi^t E} f_t(X) dX = \int_E f_0(X) dX \tag{9.77}$$

can now be used to replace the initial distribution μ_0 of density f_0, with the distribution μ_t of density f_t, produced in a time t, as discussed at length in Chap. 6. It suffices to introduce the coordinate transformation $Y = \Phi^{-t} X$ and the corresponding Jacobian, in the left-hand side integral of Eq. (9.77). One gets

$$\left| \frac{\partial Y}{\partial X} \right| = e^{-\Lambda_{-t,0}(X)}, \quad f_t(X) = f_0(\Phi^{-t} X) e^{-\Lambda_{-t,0}(X)} = f_0(X) e^{-\Omega_{0,-t}(X)} \tag{9.78}$$

Observing that $\Phi^t \mathscr{O}_{t,t+\tau}|_{(a,b)} = \mathscr{O}|_{(a,b)}$, one also has

$$\mu_0(\mathscr{O}_{t,t+\tau}|_{(A)_\delta}) = \mu_t(\Phi^t \mathscr{O}_{t,t+\tau}|_{(A)_\delta}) = \mu_t(\mathscr{O}|_{(A)_\delta}) \tag{9.79}$$

and we may eventually rewrite Eq. (9.70) as

$$\frac{\mu_t(\mathscr{O}_{0,\tau}|_{(-A)_\delta})}{\mu_t(\mathscr{O}_{0,\tau}|_{(A)_\delta})} = \mathbb{E}_0 \left[\exp\left(-\Omega^{(0)}_{0,2t+\tau} \right) \right]_{\mathscr{O}_{t,t+\tau} \in (A)_\delta} \tag{9.80}$$

Like Eqs. (9.70), (9.80) is exact and holds for all values t, τ, δ and for appropriate values of A. Furthermore, taking $\Omega^{(0)}$ in place of $\mathscr{O}$, (9.80) implies

$$\begin{aligned} \frac{\mu_t(\overline{\Omega}^{(0)}_{0,\tau}|_{(-A)_\delta})}{\mu_t(\overline{\Omega}^{(0)}_{0,\tau}|_{(A)_\delta})} &= \mathbb{E}_0 \left[ ]e^{-\Omega^{(0)}_{0,t} - \Omega^{(0)}_{t,t+\tau} - \Omega^{(0)}_{t+\tau,2t+\tau}} \right]_{\overline{\Omega}_{t,t+\tau} \in (A)_\delta} \\ &= e^{-\tau[A+\varepsilon(\delta,t,A,\tau)]} \mathbb{E}_0 \left[e^{-\Omega^{(0)}_{0,t} - \Omega^{(0)}_{t+\tau,2t+\tau}} \right]_{\overline{\Omega}_{t,t+\tau} \in (A)_\delta} \end{aligned} \tag{9.81}$$

where the second equality derives from the fact that the condition expressed by the subscript in the conditional average makes $\Omega^{(0)}_{t,t+\tau}$ close to τA, with an error $\tau\varepsilon(\delta, t, A, \tau)$ not larger than $\tau\delta$. Now, let us assume that the distribution μ_t converges to a steady-state distribution μ_∞, i.e., that the probability $\mu_t(E)$ of any measurable set $E \subset \mathscr{M}$ tends to the steady-state probability $\mu_\infty(E)$, as $t \to \infty$. We may then identify the conditions necessary for the steady-state Fluctuation Relation to hold. To do that it is convenient to rewrite Eq. (9.81) as follows:

$$\frac{1}{\tau} \ln \frac{\mu_t(\overline{\Omega}^{(0)}_{0,\tau}|_{(A)_\delta})}{\mu_t(\overline{\Omega}^{(0)}_{0,\tau}|_{(-A)_\delta})} = A + \varepsilon(\delta, t, A, \tau) - \frac{1}{\tau} \ln \mathscr{C}_0(A, \delta, t, \tau) \tag{9.82}$$

where $\mathscr{C}_0$ is the conditional autocorrelation function for the observable

$$\mathscr{Q}_t(\Gamma) := \exp\left[-\int_0^t \Omega^{(0)}(\Phi^u \Gamma) \mathrm{d}u \right] = \exp\left[-\Omega^{(0)}_{0,t}(\Gamma) \mathrm{d}u \right] \tag{9.83}$$

that is defined by

$$\mathscr{C}_0(A,\delta,t,\tau) := \mathbb{E}_0\left[e^{-\Omega^{(0)}_{0,t}-\Omega^{(0)}_{t+\tau,2t+\tau}}\right]_{\overline{\Omega}_{t,t+\tau}\in(A)_\delta} = \mathbb{E}_0\Big[\mathscr{Q}_t\left(\mathscr{Q}_t\circ\Phi^{t+\tau}\right)\Big]_{\overline{\Omega}_{t,t+\tau}\in(A)_\delta} \tag{9.84}$$

Here, the composition of $\mathscr{Q}$ with the time evolution for a time $t+\tau$ means:

$$\begin{aligned}\left(\mathscr{Q}_t\circ\Phi^{t+\tau}\right)(\Gamma) &= \mathscr{Q}_t\left(\Phi^{t+\tau}\Gamma\right) \\ &= \exp\left[-\int_0^t \Omega^{(0)}(\Phi^u\Phi^{t+\tau}\Gamma)\mathrm{d}u\right] = \exp\left[-\int_{t+\tau}^{2t+\tau} \Omega^{(0)}(\Phi^u\Gamma)\mathrm{d}u\right]\end{aligned} \tag{9.85}$$

Notably, $\mathscr{C}_0$ is computed with respect to the initial probability distribution μ_0, that is a transient distribution and depends neither on t nor on τ.

To obtain an expression valid in the steady state, which requires the $t\to\infty$ limit, consider the quantity:

$$M_0(A,\delta,\tau) := \frac{1}{\tau}\left|\limsup_{t\to\infty}\ \ln\,\mathscr{C}_0(A,\delta,\tau,t)\right| \tag{9.86}$$

which is an upper bound for the asymptotic values of $\tau^{-1}\ln\mathscr{C}_0$, with fixed A, δ and τ. There are two possibilities: $\log\mathscr{C}_0$ is unbounded with t, or it remains finite. In the first case, one of the probabilities on the left-hand side of Eq. (9.82) vanishes in the $t\to\infty$ limit, or it does not settle, as it continues to fluctuate reaching ever closer to 0. Consequently, either A or $-A$ are not observable in the steady state. If no pair $(A,-A)$ is observable, there are no fluctuations in the steady state, and hence the steady-state Fluctuation Relation is of no interest. Conversely, if the steady-state Fluctuation Relation holds, it means that there are observable pairs $(A,-A)$ that belong to the domain of the function defined by the left-hand side of Eq. (9.82), and one may write

$$A - M_0(A,\delta,\tau) - \delta \le \frac{1}{\tau}\ln\frac{\mu_\infty(\overline{\Omega}^{(0)}_{0,\tau}|_{(A)_\delta})}{\mu_\infty(\overline{\Omega}^{(0)}_{0,\tau}|_{(-A)_\delta})} \le A + M_0(A,\delta,\tau) + \delta \tag{9.87}$$

where the term $M_0(A,\delta,\tau)$ can be considered negligible, at least in the large τ limit. The steady-state fluctuation relation (9.1) then follows from the fact that, given any $\delta>0$, the $\tau\to\infty$ limit of the central expression in Eq. (9.87) remains within $[A-\delta, A+\delta]$. For this to be the case, extra conditions, not needed in the derivation of the transient fluctuation relations, must be necessarily verified.

9.4 t-Mixing

Does $M_0(A,\delta,\tau)$ vanish under the $\tau\to\infty$ limit? Let us consider two extreme situations.

(a) Suppose that the $\Omega^{(0)}$-autocorrelation computed with respect to the initial distribution f_0 decays instantaneously. Then, the correlations between exponentials of integrals of $\Omega^{(0)}$ decay instantaneously, and the conditional average in Eq. (9.81) equals the unconditioned average:

$$\mathbb{E}_0\left[e^{-\Omega_{0,t}^{(0)}-\Omega_{t+\tau,2t+\tau}^{(0)}}\right]_{\overline{\Omega}_{t,t+\tau}\in(A)_\delta} = \mathbb{E}_0\left[e^{-\Omega_{0,t}^{(0)}-\Omega_{t+\tau,2t+\tau}^{(0)}}\right] \tag{9.88}$$

that moreover, factorizes

$$\mathbb{E}_0\left[e^{-\Omega_{0,t}^{(0)}-\Omega_{t+\tau,2t+\tau}^{(0)}}\right] = \mathbb{E}_0\left[e^{-\Omega_{0,t}^{(0)}}\right]\mathbb{E}_0\left[e^{-\Omega_{t+\tau,2t+\tau}^{(0)}}\right] \tag{9.89}$$

Combining this result with Eq. (9.69) with $\delta = \infty$, we can write

$$1 = \mathbb{E}_0\left[e^{-\Omega_{0,t}^{(0)}}\right] = \mathbb{E}_0\left[e^{-\Omega_{0,s}^{(0)}}\, e^{-\Omega_{s,t}^{(0)}}\right] = \mathbb{E}_0\left[e^{-\Omega_{0,s}^{(0)}}\right]\mathbb{E}_0\left[e^{-\Omega_{s,t}^{(0)}}\right] \tag{9.90}$$

hence

$$\mathbb{E}_0\left[e^{-\Omega_{s,t}^{(0)}}\right] = 1 \tag{9.91}$$

for all s and t. In fact, an instantaneous decay of correlations with the initial state implies an instantaneous process that does not depend on when it started. Then, $\ln M_0(A, \delta, \tau, t) = 0$ for all t and τ and the steady-state Fluctuation Relation holds without any limit to be taken. The instantaneous decay of correlations also implies immediate convergence to the steady state.

(b) Take one particle moving in empty space, under the action of a constant force $\mathbf{F}_e$ and a Gaussian isokinetic thermostat. The equations of motion are

$$\dot{\mathbf{q}} = \mathbf{p}, \qquad \dot{\mathbf{p}} = \mathbf{F}_e - \frac{\mathbf{F}_e \cdot \mathbf{p}}{\mathbf{p}\cdot\mathbf{p}}\mathbf{p}$$

If the initial momentum $\mathbf{p}$ points exactly in the direction opposite to $\mathbf{F}_e$, e.g., $\mathbf{p}(0) = -c\mathbf{F}_e$, with c any positive constant, $\mathbf{p}$ is a constant of the motion. Any other initial momentum $\mathbf{p}(0)$ rapidly turns parallel to $\mathbf{F}_e$ keeping its magnitude. This dynamics has one repelling point, $-c\mathbf{F}_e$, and one attracting point, $c\mathbf{F}_e$. The attracting point $c\mathbf{F}_e$ is also unique, its probability distribution is the Dirac δ-function $\delta(\mathbf{p} - c\mathbf{F}_e)$, and there are no fluctuations in it. Therefore, the steady-state Fluctuation Relation does not hold. In fact, not only the correlations with respect to the initial distribution do not decay in the $t \to \infty$ limit: $\mathscr{C}$ grows exponentially fast with t, because $\Omega_{0,t}^{(0)}$ grows linearly with t. On the contrary, the transient Fluctuation Relation holds even in this situation.

The above situations, one in which correlations with respect to f_0 decay instantaneously and the other in which they diverge exponentially fast, are clearly peculiar. Numerical simulations on nonequilibrium molecular dynamics (NEMD) indicate that typically M_0 is bounded for a wide range of A values (Searles et al. 2013), therefore the steady-state Ω-FR holds quite generally, with a correction of order $O(1/\tau)$, as expected on the grounds of a host of numerical as well as experimental tests.

The factorization condition (9.89) appears to be one instance of the following property, that we call *t-mixing*:

$$\lim_{t\to\infty} \left\{ \mathbb{E}_0 \left[\mathcal{O} \left(\mathcal{P} \circ \Phi^t \right) \right] - \mathbb{E}_0 \left[\mathcal{O} \right] \mathbb{E}_0 \left[\mathcal{P} \circ \Phi^t \right] \right\} = 0 \tag{9.92}$$

In particular, taking $\mathcal{O} = \Omega^{(0)}$, and recalling that $\mathbb{E}_0 \left[\Omega^{(0)} \right] = 0$, because $\Omega^{(0)}$ is odd for even f_0, the t-mixing condition (9.92) becomes

$$\lim_{t\to\infty} \mathbb{E}_0 \left[\Omega^{(0)} \left(\mathcal{P} \circ \Phi^t \right) \right] = 0 \tag{9.93}$$

Summing up, Fluctuation Relations are among the few exact relations for systems arbitrarily far from equilibrium. In particular, they are relevant for small systems, which cannot verify the conditions of local thermodynamic equilibrium. They are parameter free, hence they constitute a strong connection with the systems of interest. Transient (ensemble) Fluctuation Relations are very robust, since they require only reversibility. Moreover, they concern properties of the initial equilibrium state that are revealed by nonequilibrium experiments. In this sense, they close the circle with the Fluctuation-Dissipation relation, that obtain nonequilibrium properties from equilibrium experiments. If a system does not tend to a steady state with fluctuations, the steady-state Fluctuation Relation makes no sense. Nevertheless, the transient Fluctuation Relations, concerning the initial ensembles, hold at arbitrarily large times. Transient Fluctuation Relations differ substantially from the steady-state Fluctuation Relation, despite their similar exponential form: they are exact for all observation times τ; steady-state relations may only be approximately valid for finite τ, and require the system to be in a steady state. Furthermore, from an experimental stand point, transient Fluctuation Relations correspond to sampling a large number of identical objects, or of repetitions of the same experiment, all starting in the same initial macrostate. Steady-state Fluctuation Relations may be verified on ensembles, but also following for a sufficiently long time a single object, in its steady-state evolution.

9.5 Exact Response

Let us express the evolution of phase space averages as follows:

$$\mathbb{E}_t \left[\mathcal{P} \right] - \mathbb{E}_0 \left[\mathcal{P} \right] = \int_{\mathcal{M}} \mathcal{P}(\Gamma) \left[f_t(\Gamma) - f_0(\Gamma) \right] \mathrm{d}\Gamma = \int_0^t \frac{\mathrm{d}}{\mathrm{d}s} \mathbb{E}_s \left[\mathcal{P} \right] \, \mathrm{d}s \tag{9.94}$$

which requires not just the integrability of the phase function $\mathscr{P}$, but also the differentiability at all times of its averages. Under the hypothesis that this condition is satisfied, something common, the variation (9.94) can be computed using the identities derived below:

$$\begin{aligned}\mathscr{O}_{0,s}(\Gamma) = \int_0^s \mathscr{O}(\Phi^u\Gamma)du &= \int_\tau^{s+\tau} \mathscr{O}(\Phi^{u-\tau}\Gamma)du \\ &= \int_\tau^{s+\tau} \mathscr{O}(\Phi^{-\tau}\Phi^u\Gamma)du = \mathscr{O}_{\tau,s+\tau}(\Phi^{-\tau}\Gamma) \end{aligned} \tag{9.95}$$

and

$$\begin{aligned}\mathbb{E}_{t+s}\left[\mathscr{O}\right] = \int \mathscr{O}(\Gamma)f_{t+s}(\Gamma)d\Gamma &= \int \mathscr{O}(\Phi^s(\Phi^{-s}\Gamma))f_{t+s}(\Phi^s(\Phi^{-s}\Gamma))\left|\frac{\partial\Gamma}{\partial\Phi^{-s}\Gamma}\right|d(\Phi^{-s}\Gamma) \\ &= \int \mathscr{O}(\Phi^s(\Phi^{-s}\Gamma))f_{t+s}(\Phi^s(\Phi^{-s}\Gamma))\exp\{\Lambda_{-s,0}(\Gamma)\}\,d(\Phi^{-s}\Gamma) \\ &= \int \mathscr{O}(\Phi^s(\Phi^{-s}\Gamma))f_{t+s}(\Phi^s(\Phi^{-s}\Gamma))\exp\{\Lambda_{0,s}(\Phi^{-s}\Gamma)\}\,d(\Phi^{-s}\Gamma) \\ &= \int \mathscr{O}(\Phi^s\Gamma)f_{t+s}(\Phi^s\Gamma)\exp\{\Lambda_{0,s}(\Gamma)\}\,d\Gamma \\ &= \int \mathscr{O}(\Phi^s\Gamma)f_t(\Gamma)d\Gamma = \mathbb{E}_t\left[\mathscr{O}\circ\Phi^s\right] \end{aligned} \tag{9.96}$$

where we have used the fact that probability densities obey $f_{s+t}(\Phi^t\Gamma) = f_s(\Gamma)\exp\left(-\Lambda_{0,t}(\Gamma)\right)$, cf. Chap. 6. Now, for $0 < r < s$, note that

$$\begin{aligned}\frac{d}{ds}\mathbb{E}_s\left[\mathscr{O}\right] &= \lim_{h\to 0}\frac{1}{h}\int\Big[\mathscr{O}(\Gamma)f_{s+h}(\Gamma) - \mathscr{O}(\Gamma)f_s(\Gamma)\Big]d\Gamma \qquad (9.97)\\ &= \lim_{h\to 0}\frac{1}{h}\int \mathscr{O}(\Gamma)\Big[f_r(\Gamma)\exp\{\Omega^{f_r}_{r-s-h,0}(\Gamma)\} - f_r(\Gamma)\exp\{\Omega^{f_r}_{r-s,0}(\Gamma)\}\Big]d\Gamma \\ &= \lim_{h\to 0}\frac{1}{h}\int \mathscr{O}(\Gamma)f_r(\Gamma)\exp\{\Omega^{f_r}_{r-s,0}(\Gamma)\}\Big[\exp\{\Omega^{f_r}_{r-s-h,r-s}(\Gamma)\} - 1\Big]d\Gamma\end{aligned}$$

where

$$\begin{aligned}&\lim_{h\to 0}\frac{1}{h}\Big[\exp\{\Omega^{f_r}_{r-s-h,r-s}(\Gamma)\} - 1\Big] = \lim_{h\to 0}\frac{1}{h}\left[\exp\left\{\int_{r-s-h}^{r-s}\Omega^{f_r}(S^u\Gamma)du\right\} - 1\right] \\ &= \frac{d}{dh}\exp\left\{\int_{r-s-h}^{r-s}\Omega^{f_r}(S^u\Gamma)du\right\}\bigg|_{h=0} = \Omega^{f_r}(S^{r-s-h}\Gamma)\exp\left\{\int_{r-s-h}^{r-s}\Omega^{f_r}(S^h\Gamma)\right\}\bigg|_{h=0} \\ &= \Omega^{f_r}(S^{r-s}\Gamma)\end{aligned} \tag{9.98}$$

Consequently, one obtains

$$\frac{d}{ds}\mathbb{E}_s\left[\mathscr{O}\right] = \int \mathscr{O}(\Gamma)f_r(\Gamma)\exp\{\Omega^{f_r}_{r-s,0}(\Gamma)\}\Big[\Omega^{f_r}(S^{r-s}\Gamma)\Big]d\Gamma \tag{9.99}$$

$$= \int \mathscr{O}(\Gamma)f_s(\Gamma)\Omega^{f_r}(S^{r-s}\Gamma)d\Gamma = \mathbb{E}_s\Big[\mathscr{O}\cdot(\Omega^{f_r}\circ S^{r-s})\Big] \tag{9.100}$$

and $r = 0$ yields

$$\frac{d}{ds}\mathbb{E}_s[\mathscr{O}] = \mathbb{E}_s\left[\mathscr{O}\cdot(\Omega^{f_0}\circ S^{-s})\right] = \mathbb{E}_0\left[(\mathscr{O}\circ S^s)\cdot\Omega^{f_0}\right] \tag{9.101}$$

whereas $r = s$ yields

$$\frac{d}{ds}\mathbb{E}_s[\mathscr{O}] = \mathbb{E}_s\left[\mathscr{O}\cdot\Omega^{f_s}\right] \tag{9.102}$$

The response Eq. (9.94) can now be conveniently rewritten as

$$\mathbb{E}_t[\mathscr{P}] = \mathbb{E}_0[\mathscr{P}] + \int_0^t \mathrm{d}s\,\mathbb{E}_0\left[\Omega^{(0)}\left(\mathscr{P}\circ\Phi^s\right)\right] \tag{9.103}$$

This exact response formula concerns an ensemble of identical systems that start in the state characterized by f_0 and evolve up to time t under the perturbed dynamics expressed by the evolution operator Φ^t. If the decay of ensemble averages (9.93) proceeds at a rate faster than $O(1/t)$, the $t \to \infty$ limit can be taken, obtaining:

$$\mathbb{E}_\infty[\mathscr{P}] = \mathbb{E}_0[\mathscr{P}] + \int_0^\infty \mathrm{d}s\,\mathbb{E}_0\left[\Omega^{(0)}\left(\mathscr{P}\circ\Phi^s\right)\right] \tag{9.104}$$

which yields the asymptotic response of the initial ensemble to the driving.

The existence of the integral in Eq. (9.104) is called "*Ωt-mixing.*" Since it derives from a series of exact calculations without approximations, it is a necessary and sufficient condition for the response of an ensemble of systems. Equation (9.104) is exact, at variance with, e.g., the Green-Kubo linear response formula, which is valid "only" to first order in the perturbation.[5] In the small driving limit, however, the exact theory recovers the linear response formulae. Some useful mathematical properties of the dissipation function are illustrated in Appendix H.

9.5.1 Exact Response and GK Formalism

To compare the Green-Kubo linear response theory with the exact theory, consider a probability density f_0 on the phase space of dynamics determined by the following perturbed equations of motion:

$$\dot{\mathbf{x}} = \mathbf{G}_\varepsilon(\mathbf{x}) = \mathbf{G}_0(\mathbf{x}) + \varepsilon\,\mathbf{G}_p(\mathbf{x}) \tag{9.105}$$

where ε may or may not be small. The corresponding dissipation function writes

$$\Omega^{f_0}_{\mathbf{G}_\varepsilon}(\mathbf{x}) = \Omega^{f_0}_{\mathbf{G}_0}(\mathbf{x}) + \varepsilon\,\Omega^{f_0}_{\mathbf{G}_p}(\mathbf{x}) \tag{9.106}$$

[5] We put "only" within quotes because the linear regime is actually quite accurate in a wide set of applications.

and the evolution of f_0 is given by

$$f_t(\mathbf{x};\varepsilon) = f_0(\mathbf{x}) \exp\left[\int_{-t}^{0} d\tau\, \Omega_{\mathbf{G}_0}^{f_0}(\Phi_\varepsilon^\tau \mathbf{x})\right] \exp\left[\varepsilon \int_{-t}^{0} d\tau\, \Omega_{\mathbf{G}_p}^{f_0}(\Phi_\varepsilon^\tau \mathbf{x})\right] \tag{9.107}$$

where Φ_ε^t denotes the corresponding time evolution operator. The exact response yields

$$\mathbb{E}_{t;\varepsilon}[\mathscr{O}] = \mathbb{E}_0[\mathscr{O}] + \int_0^t \mathbb{E}_0\left[\Omega_{\mathbf{G}_0}^{f_0}(\mathscr{O} \circ \Phi_\varepsilon^\tau)\right] d\tau + \varepsilon \int_0^t \mathbb{E}_0\left[\Omega_{\mathbf{G}_p}^{f_0}(\mathscr{O} \circ \Phi_\varepsilon^\tau)\right] d\tau \tag{9.108}$$

In case $\mathbf{G}_0$ derives from a Hamiltonian H_0, as in the Green-Kubo theory, one can write

$$\Omega_{\mathbf{G}_0}^{f_0} = -\mathbf{G}_0 \cdot \nabla \ln f_0 = -\frac{1}{f_0}\{f_0, H_0\} := \frac{1}{f_0}\mathscr{L}_0 f_0 \tag{9.109}$$

and, for a Hamiltonian perturbation H_p:

$$\Omega_{\mathbf{G}_p}^{f_0} = -\mathbf{G}_p \cdot \nabla \ln f_0 = -\frac{1}{f_0}\{f_0, H_p\} := \frac{1}{f_0}\mathscr{L}_p f_0 \tag{9.110}$$

where $\{\cdot,\cdot\}$ denotes the Poisson brackets. The dissipation function can thus be written as

$$\Omega_{\mathbf{G}_\varepsilon}^{f_0} = \frac{1}{f_0}(\mathscr{L}_0 + \varepsilon \mathscr{L}_p) f_0 \tag{9.111}$$

Recalling that the expectation $\mathbb{E}_0$ amounts to integrating with $d\mu_0(\mathbf{x}) = f_0(\mathbf{x})d\mathbf{x}$, and substituting (9.111) in (9.108) yields

$$\begin{aligned}\mathbb{E}_{t;\varepsilon}[\mathscr{O}] &= \mathbb{E}_0[\mathscr{O}] \\ &+ \int_0^t d\tau \int_{\mathscr{M}} d\mathbf{x}\, \mathscr{O}(\Phi_\varepsilon^\tau \mathbf{x}) \mathscr{L}_0 f_0(\mathbf{x}) + \varepsilon \int_0^t d\tau \int_{\mathscr{M}} d\mathbf{x}\, \mathscr{O}(\Phi_\varepsilon^\tau \mathbf{x}) \mathscr{L}_p f_0(\mathbf{x})\end{aligned} \tag{9.112}$$

and in case the initial distribution $f_0(\mathbf{x})$ is the equilibrium state of the unperturbed system, so that $\mathscr{L}_0 f_0 = 0$, one obtains

$$\mathbb{E}_{t;\varepsilon}[\mathscr{O}] = \mathbb{E}_0[\mathscr{O}] + \varepsilon \int_0^t d\tau \int_{\mathscr{M}} d\mathbf{x}\, \mathscr{O}(\Phi_\varepsilon^\tau \mathbf{x})\, \mathscr{L}_p f_0(\mathbf{x}) \tag{9.113}$$

This is the exact response formula, and there is no assumption on the size of ε. An approximate formula can be obtained when ε is small, starting from Eq. (9.107), by truncating to first order the exponential expansion, and recalling that $\Omega_{\mathbf{G}_0}^{f_0} = 0$:

$$\tilde{f}_t(\mathbf{x};\varepsilon) := f_0(\mathbf{x}) + \varepsilon f_0(\mathbf{x}) \int_{-t}^{0} d\tau\, \Omega_{\mathbf{G}_p}^{f_0}(\Phi_\varepsilon^\tau \mathbf{x}) = f_0(\mathbf{x}) + \varepsilon f_0(\mathbf{x}) \int_0^t d\tau\, \Omega_{\mathbf{G}_p}^{f_0}(\Phi_\varepsilon^{-\tau} \mathbf{x}) \tag{9.114}$$

Multiplying (9.114) by $\mathscr{O}$ and integrating over the phase space, one obtains[6]:

$$\mathbb{E}^{\sim}_{t;\varepsilon}[\mathscr{O}] := \int_{\mathscr{M}} \mathscr{O}(\mathbf{x}) \tilde{f}_t(\mathbf{x};\varepsilon) d\mathbf{x} = \mathbb{E}_0[\mathscr{O}] + \varepsilon \int_0^t d\tau \, \mathbb{E}_0 \left[\mathscr{O} \left(\Omega^{f_0}_{\mathbf{G}_p} \circ \Phi^{-\tau}_{\varepsilon} \right) \right] \tag{9.115}$$

As usual, changing coordinates in the integral that defines $\mathbb{E}_0$ by $\mathbf{y} = \Phi^{-\tau}_{\varepsilon}\mathbf{x}$, the backward evolution composed with $\Omega^{f_0}_{\mathbf{G}_p}$ becomes for a Hamiltonian flow a forward evolution composed with $\mathscr{O}$ and with f_0:

$$\begin{aligned} \mathbb{E}_0 \left[\mathscr{O} \left(\Omega^{f_0}_{\mathbf{G}_p} \circ \Phi^{-\tau}_{\varepsilon} \right) \right] &= \int_{\mathscr{M}} \mathscr{O}(\mathbf{x}) \Omega^{f_0}_{\mathbf{G}_p}(\Phi^{-\tau}_{\varepsilon}\mathbf{x}) f_0(\mathbf{x}) d\mathbf{x} \\ &= \int_{\mathscr{M}} \mathscr{O}(\Phi^{\tau}_{\varepsilon}\mathbf{y}) \Omega^{f_0}_{\mathbf{G}_p}(\mathbf{y}) f_0(\Phi^{\tau}_{\varepsilon}\mathbf{y}) \left| J_{\Phi^{\tau}_{\varepsilon}}(\mathbf{y}) \right| d\mathbf{y} \end{aligned} \tag{9.116}$$

where the Jacobian determinant of the coordinate transformation, $\left| J_{\Phi^{\tau}_{\varepsilon}}(\mathbf{y}) \right|$, equals 1 for our Hamiltonian dynamics. Thus, the difference between exact and approximate response can be written as

$$\mathbb{E}_{t;\varepsilon}[\mathscr{O}] - \mathbb{E}^{\sim}_{t;\varepsilon}[\mathscr{O}] = \varepsilon \int_0^t d\tau \int_{\mathscr{M}} \mathscr{O}(\Phi^{\tau}_{\varepsilon}\mathbf{x}) \Omega^{f_0}_{\mathbf{G}_p}(\mathbf{x}) \Big[f_0(\mathbf{x}) - f_0(\Phi^{\tau}_{\varepsilon}\mathbf{x}) \Big] d\mathbf{x} \tag{9.117}$$

For every finite t, this quantity vanishes in the $\varepsilon \to 0$ limit. However, the approximate response formula (9.115) does not still express the linear response, because the intensity of the perturbation ε appears also in $\Phi^{\tau}_{\varepsilon}$. The Green-Kubo formulae, on the contrary, only use the unperturbed evolution operator Φ^{τ}_0, which is one reason why they are so successful: they average with respect to the known equilibrium ensemble, and run the unperturbed dynamics. As an example, let us follow Ref. Tuckerman (2023) taking:

$$\begin{aligned} \dot{\mathbf{q}}_i &= \mathbf{p}_i/m + \mathbf{C}_i(\Gamma) \cdot \mathbf{F}_e \,, \quad i = 1, \ldots, N \\ \dot{\mathbf{p}}_i &= \mathbf{F}_i + \mathbf{D}_i(\Gamma) \cdot \mathbf{F}_e \end{aligned} \tag{9.118}$$

derived from a Hamiltonian, $H = H_0 + H_p$. We endow the corresponding phase space with the canonical distribution, $f_0(\Gamma) = Z_0^{-1} \exp\{-\beta H_0(\Gamma)\}$. The magnitude of the external force, F_e, here plays the role of the small parameter ε, and its direction is embedded in the coefficients $\mathbf{C}_i$ and $\mathbf{D}_i$. The components of $\mathbf{G}_p$ are obtained differentiating the perturbation Hamiltonian H_p:

$$C_i F_e = \frac{\partial H_p}{\partial p_i}; \quad D_i F_e = -\frac{\partial H_p}{\partial q_i} \tag{9.119}$$

[6] Note that $\tilde{f}_t$ is only an approximation of the probability density at time t, hence it is not guaranteed to be a probability density itself.

Then, one may write

$$0 = \frac{\partial \dot{\mathbf{q}}}{\partial \mathbf{q}} + \frac{\partial \dot{\mathbf{p}}}{\partial \mathbf{p}} = \sum_{i=1}^{N} \left\{ \frac{\partial C_i}{\partial q_i} + \frac{\partial D_i}{\partial p_i} \right\} F_e, \quad \text{hence} \quad \sum_{i=1}^{N} \left\{ \frac{\partial C_i}{\partial q_i} + \frac{\partial D_i}{\partial p_i} \right\} = 0 \tag{9.120}$$

and

$$\Omega_{\mathbf{G}_p}^{f_0} = -\mathbf{G}_p \cdot \nabla_\Gamma \ln f_0 = \frac{1}{f_0} \frac{\partial f_0}{\partial H_0} \mathbf{G}_p \cdot \nabla_\Gamma H_0 = -\beta \sum_{i=1}^{N} \left[\mathbf{C}_i \frac{\partial \dot{H}_0}{\partial \mathbf{q}_i} + \mathbf{D}_i \frac{\partial \dot{H}_0}{\partial \mathbf{p}_i} \right] F_e = -\beta V J F_e \tag{9.121}$$

where J represents the dissipative flux. Thanks to the fact that $\{H_0, H_p\} = \{H, H_p\}$, the unperturbed dynamics can be used in place of the perturbed one, obtaining:

$$\begin{aligned} \mathbb{E}_{t;\varepsilon}^{\sim}[\mathscr{O}] &= \mathbb{E}_0[\mathscr{O}] - \beta F_e \int_0^t d\tau \, \mathbb{E}_0 \left[\left(\mathscr{O} \circ S_0^{t-\tau} \right) \sum_{i=1}^{N} \left(\mathbf{C}_i \frac{\partial \dot{H}_0}{\partial \mathbf{q}_i} + \mathbf{D}_i \frac{\partial \dot{H}_0}{\partial \mathbf{p}_i} \right) \right] \\ &= \mathbb{E}_0[\mathscr{O}] - \beta V F_e \int_0^t d\tau \, \mathbb{E}_0 \left[J \left(\mathscr{O} \circ S_0^{t-\tau} \right) \right] \end{aligned} \tag{9.122}$$

which is the first-order approximation of the response generated by the force of magnitude F_e. Eventually, the difference between exact and linear response is expressed by

$$\mathbb{E}_{t;\varepsilon}[\mathscr{O}] - \mathbb{E}_{t;\varepsilon}^{\sim}[\mathscr{O}] = \varepsilon \int_0^t d\tau \, \mathbb{E}_0 \left[\Omega_{\mathbf{G}_p}^{f_0} \left(\mathscr{O} \circ S_\varepsilon^\tau - \mathscr{O} \circ S_0^{t-\tau} \right) \right] \tag{9.123}$$

Note that the relations so far derived refer to ensembles of objects, not to a single object, as thermodynamics requires. Therefore, the connection between the evolution of such probabilistic entities as the ensemble averages described above, and the thermodynamic evolutions has to be justified in different ways. As discussed in detail in Sect. 6.4, this connection can be rigorously derived in certain cases, and remains reasonable in many others, although how generally it can be trusted is an open question. After all, small or far from equilibrium systems do not necessarily allow the connection between probabilities and material properties of a single object. In that case, one has to accept the probabilistic notions as the only sensible ones and keep fingers crossed, encouraged by a long list of successes.

9.6 Irreversibility and Response Theory

Response theory is strictly connected with the irreversibility in macroscopic physical phenomena. In effect, the response of a macroscopic system to a perturbation may result in an irreversible evolution toward a state different from the initial one. This is considered quite a puzzle by many because, for good reasons, the dynamics of the microscopic constituents of macroscopic objects is taken as time reversal invariant.

The puzzle is often associated with some confusion between the properties of our representation of a given phenomenon, which is a mathematical model, and the material object of interest itself. In fact, both the microscopic and the macroscopic models developed by statistical mechanics and thermodynamics embody the aspects of the phenomena of interest that are considered essential or merely interesting, and necessarily neglect other aspects. Proceeding like that, these branches of physics have greatly advanced our understanding of the physical world: informed choices allowed us to neglect large fractions of information on the enormous assemblies of atoms and molecules that constitute macroscopic objects, hence allowed the development of practical and accurate predictive theoretical frameworks. At the same time, what the microscopic description takes as essential and sufficient may not be so from the point of view of the macroscopic description of the very same object, and it is interesting to investigate what is actually needed to bridge the gap. The Fluctuation Relations constitute recent significant advances in this endeavor, since they quantify the emergence of irreversibility in a class of time reversal invariant particle systems, as the number of particles and the observation times grow to become macroscopic.

As pointed out in Sect. 1, it is quite natural that understanding requires some selection or loss of information; recall the example of the movie. On the other hand, deciding what is essential and what is superfluous is far from obvious, and only empirical assessments can corroborate or refute our assumptions. This makes delicate the investigation of aspects of the physical world that have not been considered in the past, although that is precisely what one wishes to do. Therefore, one must bear in mind that models are just that, and that they have a limited range of applicability, because they have been devised to describe certain aspects of a given phenomenon and not others.

9.6.1 Irreversible Relaxation to Equilibrium; Standard Picture

Let us summarize how relaxation to equilibrium is commonly told to arise in Hamiltonian isolated systems. The paradigmatic example refers to the expansion of a gas initially confined in one half of a container, or of the mixing of two liquids that are initially separated. The reasoning involves a combinatorial calculation that profits from the enormous number of microscopic constituents of macroscopic objects. In practice, it concludes that equilibrium states are so much more numerous than other states that a system evolving spontaneously through all kinds of states cannot avoid them, and once equilibrium is reached, it remains there except for rare and exceedingly short events. This can be understood through an idealized example of a system having a finite number of states, each of them represented by a string of N symbols, $X^{(N)} = (s_1, s_2, \ldots, s_N)$, drawn from a set $\mathscr{A} = \{\alpha_1, \alpha_2, \ldots, \alpha_L\}$. One may think of N as the number of particles, $s_i \in \mathscr{A}$ as the state of particle i, e.g., its velocity, and $X^{(N)}$ as one microscopic phase (phase space point) of the system. The empirical distribution of the symbols in a given string may be viewed as one macrostate:

$$\nu\left(X^{(N)}\right) = \{\nu_1, \nu_2, \ldots, \nu_L\}$$

where ν_j is the frequency of symbol α_j in $X^{(N)}$. Depending on how many particles have certain velocities, the total kinetic energy takes a certain value, and one may count how many different arrangements of particles yield the same kinetic energy, i.e., the same macrostate ν. Let us denote by

$$M(\nu) = \left\{X_1^{(N),\nu}, X_2^{(N),\nu}, \dots, X_n^{(N),\nu}\right\}$$

the set of strings of N symbols with same empirical distribution ν. For instance, taking $\mathscr{A} = \{a, b\}$ and $N = 3$, the set of all possible phases contains 8 strings,

$$\mathscr{M} = \{aaa, aab, aba, baa, abb, bab, bba, bbb\}$$

the possible macrostates are 4:

$$\nu(aaa) = \{1, 0\}, \qquad \nu(bbb) = \{0, 1\}$$
$$\nu(aab) = \nu(aba) = \nu(baa) = \left\{\frac{2}{3}, \frac{1}{3}\right\}, \quad \nu(abb) = \nu(bab) = \nu(bba) = \left\{\frac{1}{3}, \frac{2}{3}\right\}$$

two of which are given by a single arrangement, while the other two are each given in three different ways. One can show that the number of different macrostates is not larger than $(N+1)^L$, while the number of microstates equals L^N. For sufficiently large N, the number of strings corresponding to a given empirical distribution ν can then be approximated by

$$W(\nu) = e^{NH(\nu)}, \quad \text{with } H(\nu) = -\sum_{j=1}^{L} \nu_j \ln \nu_j$$

Then, large N implies that the macrostate $\tilde{\nu}$ corresponding to the largest number of strings, $\widetilde{W}$, contains by far the largest fraction of the total set of microscopic phases of $\mathscr{M}$, and dynamics jumping without constraints from string to string are bound to spend most of their time in $\tilde{\nu}$.[7] These conclusions merely follow from *counting* microscopic phases: no reference is made to the microscopic dynamics or to probabilistic notions, because the number of states belonging to a finite set is objective. The conclusion is that it is hard to avoid equilibrium states.

There is now a technical gap to overcome. The dynamics of particles occur in continuous time in a continuous space, rather than in discrete time and discrete space. This issue is usually overlooked, and counting states is in this case replaced by measuring volumes in phase space. If a certain macrostate, e.g., a given distribution of mass, corresponds to a larger portion of phase space than another mass distribution,

[7] Consider all arrangements of $10^3 \times 10^3$ black and white pixels. The result is a dull movie: $\mathscr{M}$ contains 2^{10^6} pictures, among which the remotely regular ones constitute a very small fraction. The overwhelming majority are gray. If pictures are shown at a rate of 25 frames a second, and the movie goes through all of them before starting again, the period is larger than 10^{301020} years! The fraction of time in which one does not see noisy arrangements is ridiculously small.

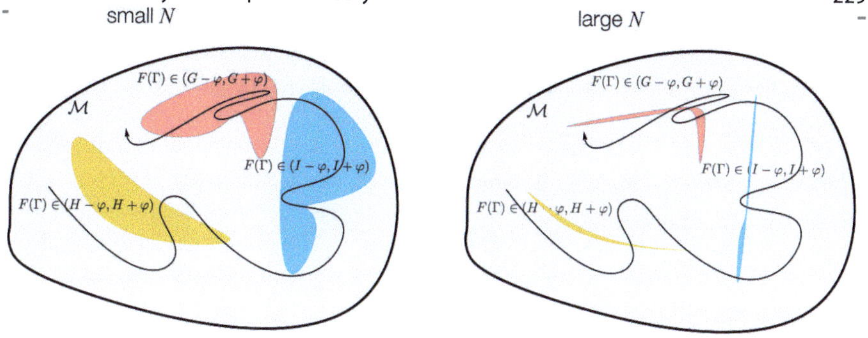

Fig. 9.1 For small N the phase space is subdivided in regions of comparable size, in which a given observable F takes different values. Fluctuations of F are large and frequent. For large N, most of the phase space is occupied by phases in which F takes its mean value. Fluctuations are rare and evanescent

the first is considered realizable in more ways than the second. This approach can be criticized, but it is supported by some mathematical arguments, and in the absence of better recipes it is accepted.

The issue of irreversibility can now be illustrated as in Fig. 9.1. The phase space $\mathcal{M}$ of a system of N particles can be subdivided in regions that correspond to different values of a certain macroscopic observable F, with a tolerance φ corresponding to the resolution of the measurement tools. In Fig. 9.1, the evolution of the microscopic phases $\Gamma(t)$ is represented by a black curve in $\mathcal{M}$ ending with an arrow indicating the direction of time. The color red represents the phases $\Gamma \in \mathcal{M}$ such that $F(\Gamma) \in (G - \varphi, G + \varphi)$, the blue region represents the $\Gamma \in \mathcal{M}$ such that $F(\Gamma) \in (I - \varphi, I + \varphi)$, and the yellow region represents the $\Gamma \in \mathcal{M}$ such that $F(\Gamma) \in (H - \varphi, H + \varphi)$. The rest of $\mathcal{M}$ (light gray) is meant to represent the phases belonging to the "equilibrium" state. Two points of view can be adopted in interpreting the observable F: it can be taken as a phase function without any further specification, e.g., the instantaneous kinetic energy of the system, $K(\Gamma)$ say, or it can be taken as the result of a measurement, hence the long time average $\overline{K}(\Gamma)$ of some other phase function K. The left panel of Fig. 9.1 portrays the situation expected for a system made of a small number of particles: a generic observable takes measurably different values which occupy comparable regions of phase space, hence must be visited with comparable frequencies. Fluctuations about the equilibrium value are clearly visible. The right panel of Fig. 9.1 shows a situation in which N is large, hence the phases that we consider as the equilibrium state are expected to occupy the vastly largest fraction of the phase space. In this case, fluctuations of F are rare and evanescent. Therefore, if the system starts outside equilibrium, it rapidly and irreversibly reaches equilibrium. For N of order $O(10^{23})$, the fluctuations are so rare that do not occur at all. Even waiting for many of the estimated lifetime of our universe, the fraction of time attributed to them would be ridiculously small, but one should at this point rather ask if there exist any object, any measurement tool, and any experiment to which the theory applies. If they do not exist, any statement on them could be as fascinating as one can imagine, but it would not be part of physics. Closer to home, Fig. 9.1 also illustrates the reason why Local Thermodynamic Equilibrium requires, to the very least, N to be large, as discussed in detail in Sect. 2.5. When N is small, the very notion of thermodynamic equilibrium is dubious, because the system does not prefer any of the regions of phase space in which the observables take different values. Furthermore, even averaging over very long times, so that fluctuations become relatively small, the mean is not bound to correspond to the value observed in the largest fraction of the phase space. Therefore, there is no compelling reason to give to one part of $\mathscr{M}$ a special status, and the thermo-hydrodynamic balances of energy, matter, etc. performed in Chap. 2, which yield the equations describing matter as a continuum, are not possible.

On dynamical grounds, these statements are hard to prove, but they have been rigorously formalized in the kinetic theory of rarefied gases, and in Khinchin's approach to the ergodic hypothesis discussed in Sect. 6.4. As demonstrated numerically by Fermi, Pasta, Tsingou and Ulam and mathematically realized by Kolmogorov, Arnold and Moser, a Hamiltonian system is not ergodic, in general (Cencini et al. 2009). The ergodic hypothesis may nevertheless be adopted because physics concerns a few special observables, and because it deals with systems of very many and properly interacting particles. In that picture, the need to distinguish a phase variable from its time average fades, because the average is indistinguishable from its, practically non-fluctuating, instantaneous value, and the balances of continuum make sense.

Regarding irreversibility, this picture faces some difficulty. As demonstrated by Fig. 9.1, a spontaneous evolution converges to the equilibrium state both forward and backward in time. This means that the ordered configuration of particles portrayed in the bottom left part of Fig. 6.3 can be the spontaneous evolution of a previous disordered state. This makes sense in time reversal invariant dynamics, because inverting the direction of time from the disordered configuration in the bottom right part of the figure is equivalent to start from such configurational disorder with inverted velocities. Neglecting how problematic the velocities inversion is, the balls will first gather to form an ordered configuration, and then will spread to become disordered again. Past and future thus appear equivalent and no arrow of time can be identified. They can be clearly distinguished, however, when a nonequilbrium state has been produced by someone or some mechanism. Indeed, as argued above, a spontaneous fluctuation resulting in a gas compressing itself in half of its container, or a fluctuation of the sand at the beach to form a footprint are so ridiculously "improbable" that one does not expect them to occur even only once in the lifetime of the universe. If those phenomena have been observed, they must have been produced by someone or something instructed to do that.

The interest then shifts to understanding how someone or something could move the gas or the sand away from their preferred disordered states, rather than toward disorder. Some other part of the universe must have provided the conditions for this to happen. Continuing like that, one is eventually forced to consult the Universe as a whole, to get an answer to the irreversibility issue, and the answer can only be hidden in its initial condition. This initial condition should have been so very far from equilibrium that it takes correspondingly long to settle to equilibrium (Penrose 2007).[8]

[8] Note: Usually this explanation of irreversibility is referred to the growth of the entropy of the universe, intended as a measure of disorder. Indeed, because entropy is a questionable concept in

Can such a special initial condition of the universe, from which all irreversibility derives, be justified? For different reasons, various authors claim that no justification is needed, cf. Refs. Barbour et al. (2014) and Callender (2012).

Whether the initial conditions of the Universe are understood or not, the irreversibility picture presented above relies on Hamiltonian dynamics as the truly and unique fundamental theoretical framework. Needless to say, the successes obtained in statistical physics adopting that point of view are countless and deep. However, different approaches have been considered, and have been the object of debate, particularly in relation to irreversible modifications of the microscopic dynamics, that could be adopted because of the T-symmetry breaking at the level of elementary particles physics. This point of view has been dismissed as referring to scales quite remote from the thermodynamic ones, and, in fact, molecular dynamics does not need irreversible modifications of the particles dynamics to simulate the relaxation of a gas to its equilibrium: classical chaos of Hamiltonian systems is enough. Similarly, having to invoke the early stages of the evolution of the Universe to explain that a bottle of water taken out of a fridge *irreversibly* relaxes to room temperature in ten minutes or that energy is dissipated when rubbing hands against each other also appears excessive. Moreover, one obtains quite a limited prediction: one concludes that the world is irreversible! but that is also obtained by the competing theories. Better would be a framework in which more specific and quantitative aspects of our daily experience, such as dissipation, can be predicted.

Because at the moment this looks out of reach, a more modest approach may prove less problematic, but still insightful. From this perspective, the models developed in nonequilibrium molecular dynamics appear quite interesting: they have been consciously developed as artificial tools to help numerical simulations of driven nonequilibrium systems, hence they introduce dissipation by hand, but the most popular of them remain time reversal invariant. They are admittedly *"synthetic"* but most successful in applications. At the same time no mathematical model, as fundamental as it may be, can be expected to be accurate over all scales of the physical world, and the phenomena of interest decide which models are best for them. In the next subsection, we thus investigate what can be learned from dissipative time reversal invariant models and the exact response theory that emerged from them.

9.7 t-Mixing and Single System Response

Consider time reversal invariant dynamics, that may or may not be driven by a dissipative field, to identify the conditions for relaxation to equilibrium or nonequilibrium steady states. As observed in Sect. 9.5, the Ωt-mixing condition, i.e., the existence of the integral

violently nonequilibrium phenomena, one merely wants to express the disorder of the Universe in terms of the numerosity of the microscopic states corresponding to its macroscopic states, as done with the Boltzmann entropy.

$$\int_0^\infty \mathrm{d}s\, \mathbb{E}_0 \left[\Omega^{(0)} \left(\mathscr{P} \circ \Phi^s \right) \right] \tag{9.124}$$

is necessary and sufficient for the relaxation of an ensemble of systems to a steady state. This avoids one limit of ergodic theory: concerning invariant states, ergodic theory offers limited information about relaxation processes. However, like standard ergodic and response theory, the response formula (9.103) is an ensemble formula; it does not tell what the behavior of a single member of the ensemble could be. Ensemble results are nevertheless commonly accepted under the assumption that they express the behavior of single objects as well. In equilibrium, this view is relatively well justified by Khinchin's approach to ergodicity, based on the fact that the observables of interest are practically constant in the phase space, hence "almost all" ensemble members have the same fate in their equilibrium states.

Nonequilibrium systems are not that well understood. In particular, "almost all" nonequilibrium steady states corresponds to "almost none" in the phase space. Indeed, such steady states are characterized by currents that in equilibrium amount to rare and fleeting fluctuations, experienced when the phase space trajectories cross very thin regions of the phase space. Therefore, nonequilibrium steady states attribute probability 1 to sets of zero phase space volume, in contrast to equilibrium states that attribute to them zero probability. This is the case whichever deterministic model one uses to describe the microscopic dynamics of nonequilibrium steady states. In particular, the models of nonequilibrium molecular dynamics obtain this effect in a very simple fashion, because they are dissipative, hence phase space volumes shrink on average, and the corresponding contraction rates take very simple forms, as illustrated in Appendix I.

Now, the t-mixing condition (9.92) applied to these models allows, if verified, the ensemble results to turn into single system results. Take an observable $\mathscr{O}$ and any integer n, let

$$E_n^{\mathscr{O},\delta} = \left\{ \Gamma : \overline{\mathscr{O}}(\Gamma) = \lim_{t\to\infty} \frac{1}{t} \int_0^t \mathscr{O}(\Phi^s \Gamma)\, \mathrm{d}s \in [n\delta, (n+1)\delta) \right\} \tag{9.125}$$

be the set of initial phases $\Gamma \in \mathscr{M}$ such that a measurement that lasts very many microscopic times yields a given value within a tolerance δ. These sets are disjoint,

$$E_n^{\mathscr{O},\delta} \cap E_{n'}^{\mathscr{O},\delta} = \emptyset, \quad \text{for } n \neq n'$$

because, for a given the initial condition, a time average cannot take two different values. Also, the $t \to \infty$ limit makes them invariant:

$$\Phi^{-t} E_n^{\mathscr{O},\delta} = E_n^{\mathscr{O},\delta}$$

If the limit in Eq. (9.125) exists for all Γ, the sets $E_n^{\mathscr{O},\delta}$ cover $\mathscr{M}$:

$$\bigcup_{n\in\mathbb{Z}} E_n^{\mathscr{O},\delta} = \mathscr{M}, \qquad \text{hence} \quad \mu_0 \left(\bigcup_{n\in\mathbb{Z}} E_n^{\mathscr{O},\delta} \right) = 1 \tag{9.126}$$

This must be an assumption: given the dynamics Φ^t, it does not follow in general that $\overline{\mathscr{O}}(\Gamma)$ exists for *all* Γ. Often one can only prove that $\overline{\mathscr{O}}(\Gamma)$ exists for almost all Γ, meaning with the exception of a set of vanishing steady-state probability, cf. Appendix B. This is not the desired result, since almost all, with respect to a steady-state distribution, could mean almost none in $\mathscr{M}$, and an experimentalist preparing the system in equilibrium, before subjecting it to dissipative forces, would find this hardly acceptable. Luckily, it is usual, in systems of physical interest, that $\overline{\mathscr{O}}(\Gamma)$ exists for almost all Γ in the sense of phase space volumes, not of steady-state probabilities. Therefore, one may assume that time averages exist with probability 1 with respect to the initial equilibrium ensemble μ_0. Then, consider the indicator function $\chi_n^{\mathscr{O},}$ of the sets $E_n^{\mathscr{O},\delta}$:

$$\chi_n^{\mathscr{O},}(\Gamma) = \begin{cases} 1 \text{ if } \Gamma \in E_n^{\mathscr{O},\delta} \\ 0 \text{ if } \Gamma \notin E_n^{\mathscr{O},\delta} \end{cases}$$

Assuming that these functions verify t-mixing, we may write

$$\mu_0\left(E_n^{\mathscr{O},\delta}\right) - \mu_0\left(E_n^{\mathscr{O},\delta}\right)^2 = \mu_0\left(E_n^{\mathscr{O},\delta} \cap \Phi^{-t} E_n^{\mathscr{O},\delta}\right) - \mu_0\left(\Phi^{-t} E_n^{\mathscr{O},\delta}\right)\mu_0\left(E_n^{\mathscr{O},\delta}\right) \overset{[}{t}\!\!\to \infty]\!\longrightarrow 0$$

It follows that

$$\mu_0\left(E_n^{\mathscr{O},\delta}\right) = 0 \text{ or } 1$$

where "= 1" holds for one and only one $n \in \mathbb{Z}$

As linear combinations of indicator functions (simple functions), are dense in the space of Lebesgue integrable functions, $L^1(\mathscr{M}, \mathscr{L})$, the t-mixing property for them suffices to turn the ensemble response to a single system response, possibly with a negligible set of exceptions. Under t-mixing, the irreversible relaxation to an equilibrium or generically steady state follows for the vast majority of the (Lebesgue almost all) initial phases in $\mathscr{M}$. This way, the objective counting to continuums of states is extended to dissipative dynamics; not just the ensemble averages relax: they do so because (almost) all single systems relax to the same state. It may be interesting to note that correlation functions are usually computed with respect to stationary distributions of microscopic states, that necessarily carry limited information about the evolution of macroscopic states. Differently, t-mixing represents the loss of memory of macrostates. This is why it looks suited to deal with evolving systems, both in the presence or absence of dissipation. The role of the number of particles N still needs to be investigated.

9.7.1 Examples: Dynamical Systems and Response Theory

The response theory described in Sect. 9.5 can be applied to dynamical systems in general. However, it arose, and it is most interesting in the case of the nonequilibrium molecular dynamics models described in Appendix I. Before illustrating some examples, let us recall the simplest of such systems, expressed by the following equations of motion:

$$
\begin{aligned}
\dot{\mathbf{q}}_i &= \mathbf{p}_i/m + \mathbf{C}_i(\Gamma)\cdot\mathbf{F}_e\,, \quad i = 1,\ldots,N \\
\dot{\mathbf{p}}_i &= \mathbf{F}_i + \mathbf{D}_i(\Gamma)\cdot\mathbf{F}_e - \alpha\mathbf{p}_i
\end{aligned}
\tag{9.127}
$$

where $\mathbf{F}_e$ is the external driving field, coupled to the particles via the "charges" $\mathbf{C}_i(\Gamma)$ and $\mathbf{D}_i(\Gamma)$. The term α is a Lagrange multiplier meant to implement some constraint so that the system reaches a steady state despite the driving field. In the literature, it is said to implement a deterministic time reversible thermostat, as illustrated in the Appendix I. If this thermostat is set to $\alpha = 0$, one obtains the adiabatic equations of motion of Eq. 9.118 that produce a continuous growth of energy. The dissipative flux, $\mathbf{J}$, is obtained from the time derivative of the internal energy

$$
H_0 = \sum_{i=1}^{N} \frac{\mathbf{p}_i\cdot\mathbf{p}_i}{2m} + \phi(\mathbf{q})
$$

under the dynamics (9.118), which represents how much energy has to be dissipated in order to maintain the steady state. The result is

$$
\dot{H}_0^{ad} = -\sum_{i=1}^{N}\left(\frac{\mathbf{p}_i}{m}\cdot\mathbf{D}_i - \mathbf{F}_i\cdot\mathbf{C}_i\right)\cdot\mathbf{F}_e \equiv -\mathbf{J}V\cdot\mathbf{F}_e \tag{9.128}
$$

The phase space is usually endowed with a probability density f_0 whose support is the whole phase space.

9.7.1.1 Simple One-Dimensional ODE

Consider the following dynamical system in the interval $[-1, 1]$:

$$
\dot{x} = x^2 - 1
$$

This equation can be solved by separation of variables:

Case 1: $x \neq \pm 1$

$$
\int \frac{dx}{x^2-1} = \int dt \tag{9.129}
$$

$$
\int -\frac{1}{2}\left[\frac{1}{1-x} + \frac{1}{1+x}\right]dx = \frac{1}{2}\ln\frac{1-x}{1+x} = t + C \tag{9.130}
$$

Denoting by Φ^t the time evolution operator, for the initial condition $x_0 \neq \pm 1$ the solution is

$$x(t; x_0) = \Phi^t x_0 = \frac{(1 + x_0) - (1 - x_0)e^{2t}}{(1 + x_0) + (1 - x_0)e^{2t}} \tag{9.131}$$

Case 2: $x = -1$: $\dot{x} = 0$; the initial condition $x_0 = -1$ is a fixed point: $\Phi^t(-1) = -1$.

Case 3: $x = 1$: $\dot{x} = 0$ so the initial condition $x_0 = 1$ is a fixed point: $\Phi^t 1 = 1$.

For time reversibility, try this operator: $ix = -x$. Check that $i\Phi^t x = \Phi^{-t} ix$ for all $x \in [0, 1]$ and all real times t. Given the solution above for case 1 we have

$$i\Phi^t x = \frac{(1 + x) - (1 + x)e^{2t}}{(1 - x) + (1 + x)e^{2t}} = -\frac{(1 + x) - (1 - x)e^{2t}}{(1 + x) + (1 - x)e^{2t}} \tag{9.132}$$

$$\Phi^{-t} ix = \Phi^{-t}(-x) = \frac{(1 - x) - (1 + x)e^{-2t}}{(1 - x) + (1 + x)e^{-2t}} = -\frac{(1 + x) - (1 - x)e^{2t}}{(1 + x) + (1 - x)e^{2t}} \tag{9.133}$$

Reversibility condition verified: system is time reversal invariant. For the generalized Liouville's equation, we need the phase space variation rate $\Lambda = \mathrm{div}G$. Here $G = x^2 - 1$ is one dimensional, so

$$\Lambda = \frac{d}{dx}(x^2 - 1) = 2x \tag{9.134}$$

The generalized Liouville's equation takes the form:

$$\frac{\mathrm{d}f}{\mathrm{d}t} = -2xf \tag{9.135}$$

Because Λ does not vanish, the dynamics are not conservative. To know whether they are dissipative, we have to calculate the time average of Λ starting from any initial condition. At the fixed points, the average equals the value, because there is no variation of Λ, therefore $\overline{\Lambda}(-1) = -2$, negative means dissipative; $\overline{\Lambda}(1) = 2$, non-dissipative. For all other points

$$\overline{\Lambda}(x) = \lim_{t \to \infty} \frac{1}{t} \int_0^t 2\frac{(1 + x) - (1 - x)e^{2t'}}{(1 + x) + (1 - x)e^{2t'}} dt' \tag{9.136}$$

the integrand converges to -2 whatever x one takes so the long time limit will be $\overline{\Lambda}(x) = -2$, except for $x = 1$. The dynamics are dissipative. For the evolution of f_0 we first compute

$$\int_0^t \Lambda(\Phi^s x) ds = \int_0^t \frac{A - Be^{2s}}{A + Be^{2s}} ds = t - \ln\frac{A + Be^{2t}}{2} \tag{9.137}$$

where $A = 1 + x$ and $B = 1 - x$. We then use

$$f^{(t)}(x) = e^{-\int_{-t}^{0} \Lambda(\Phi^s x)ds} f^0(\Phi^{-t}x) = e^{-t} \frac{2}{A + Be^{-2t}} \frac{1}{2} \tag{9.138}$$

For the Ω formalism, let us simply observe that in this case

$$\int_{-t}^{0} \Omega(\Phi^s x)ds = -\int_{-t}^{0} \Lambda(\Phi^s x)ds \tag{9.139}$$

because f_0 is constant in [0, 1]. So, the two give the same answer as necessary, but in a very simple way.

A normalized quadratic distribution in $[-1, 1]$ is $f_0(x) = 3x^2/2$. Then Λ does not change, while Ω changes so that

$$\int_{-t}^{0} \Omega(\Phi^s x)ds = 2 \ln \frac{(1+x) - (1-x)e^{-2t}}{(1+x) + (1-x)e^{-2t}} - 2 \ln x - \int_{-t}^{0} \Lambda(\Phi^s x)ds \tag{9.140}$$

Apply now the same formulae:

$$f^{(t)}(x) = \exp\left\{-\Lambda_{-t,0}(x)\right\} f^{(0)}(\Phi^{-t}x) \tag{9.141}$$

$$= \exp\left\{\Omega^{(0)}_{-t,0}(x)\right\} f^{(0)}(x); \tag{9.142}$$

where

$$f^{(0)}(\Phi^{-t}x) = \frac{3}{2} \left[\frac{(1+x) - (1-x)e^{-2t}}{(1+x) + (1-x)e^{-2t}}\right]^2 \tag{9.143}$$

Again, the solutions are the same, but this time in two different ways; one is called Eulerian (with Ω), the other is called Lagrangian. The response is given by

$$\mathbb{E}_t [\mathcal{O}] = \mathbb{E}_0 [\mathcal{O}] + \int_0^t \mathrm{d}s \; \mathbb{E}_0 \left[\left(\mathcal{O} \circ \Phi^s\right) \Omega^{(0)}\right] \tag{9.144}$$

Consider the simple case $f_0(x) = 1/2$. Then

$$\mathbb{E}_t [\mathcal{O}] = \int_{-1}^{1} \mathcal{O}(x) f^{(t)}(x)dx = \frac{1}{2} \int_{-1}^{1} \mathcal{O}(x)dx + \int_0^t ds \frac{1}{2} \int_{-1}^{1} dx \mathcal{O}(\Phi^s x)\Omega(x) \tag{9.145}$$

For instance, let $\mathcal{O}(x) = x$, one has

$$\mathbb{E}_t [\mathcal{O}] = \frac{1}{2} \int_{-1}^{1} xdx + \int_0^t ds \frac{1}{2} \int_{-1}^{1} dx \mathcal{O}(\Phi^s x)\Omega(x) \tag{9.146}$$

Because $\mathcal{O}(\Phi^t x) \to -1$ for $t \to \infty$ except for $x = 1$, and because $\Omega(x) = -2x$, one can easily get an asymptotic formula for large t. The time average is

$$\overline{x} = \lim_{t \to \infty} \frac{1}{t} \int_0^t \frac{(1 + x_0) - (1 - x_0)e^{2s}}{(1 + x_0) + (1 - x_0)e^{2s}} ds = -1 \tag{9.147}$$

Finally, if the original dynamical system is perturbed to turn

$$\dot{x} = x^2$$

$x_0 = 0$ is a fixed point and the solution with initial condition x_0 exists for all t: $\Phi^t x_0 = x_0 = 0$. For the other initial conditions $x_0 \neq 0$, the solutions does not exist for all t, since it takes the form

$$\Phi^t x_0 = \frac{x_0}{1 - x_0 t}, \quad x_0 \neq 0 \tag{9.148}$$

9.7.1.2 Less Simple One-Dimensional ODE

Consider the dynamical system

$$\dot{x} = x - x^3 \tag{9.149}$$

that has two asymptotically stable fixed points $x_- = -1$ and $x_+ = 1$, and one unstable fixed point, $x_0 = 0$. Take the invariant set $\mathcal{M} = [-1, 1]$ as the phase space of the system. The flow is given by

$$\Phi^t x = \frac{x \exp(t)}{\sqrt{1 + x^2 \left[\exp(2t) - 1\right]}}, \quad x \in \mathcal{M} \tag{9.150}$$

Assuming $f_0(x) = 1/2$, uniform on $\mathcal{M}$, the dissipation function is expressed by $\Omega^{f_0}(x) = -\Lambda(x) = 3x^2 - 1$, hence it changes sign in $\mathcal{M}$. The evolved distribution is given by

$$f_t(x) = \frac{1}{2}\chi_{[-1,1]}(x) \frac{\exp(-t)}{\left[1 - x^2 + x^2 \exp(-2t)\right]^{3/2}} \tag{9.151}$$

Then, the density develops two singularities at $x = \pm 1$, as those points are attractors. If, on the other hand, $f_0(x) = \frac{3}{4}(1 - x^2)$, we have

$$\Omega^{f_0}(x) = 5x^2 - 1 \tag{9.152}$$

that also changes sign, and the density at time t becomes

$$f_t(x) = \frac{3}{4} \frac{(1 - x^2) \exp(-t)}{\left[1 - x^2 + x^2 \exp(-2t)\right]^{5/2}} \tag{9.153}$$

Both Eqs. (9.151) and (9.153) show that $d\mu_t = f_t(x)dx$ converges weakly to $\frac{1}{2}(\delta_{-1} + \delta_1)$. This is correct, because the flow vanishes at the boundaries of the phase space: $\mathcal{M}$ is invariant; hence the points inside $\mathcal{M}$ have no preimage outside $\mathcal{M}$. In this case, the non-differentiability of the uniform density in $\mathbb{R}$ has no bearing on the theory. For the observable $\mathcal{O}(x) = x$ the response starting from the parabolic density f_0 is given by

$$\mathbb{E}_t(\mathcal{O}) = \mathbb{E}_0(\mathcal{O}) + \int_0^t \mathrm{d}s \int_{-1}^{1} \mathrm{d}x \, \frac{3}{4}(1 - x^2)(5x^2 - 1) \frac{x \exp(t)}{\sqrt{1 + x^2 \left[\exp(2t) - 1\right]}} = 0 \tag{9.154}$$

It does not depend on time, because the integrand is odd in x, and it is computed over an interval symmetric with respect to the point 0.

9.7.1.3 ε-Perturbation

Consider the one-dimensional dynamical system:

$$\dot{x} = x - \varepsilon x^3, \quad 0 \le \varepsilon \le 1 \tag{9.155}$$

The point $x_0 = 0$ is now unstable, while $x_- = -1/\sqrt{\varepsilon}$ and $x_+ = 1/\sqrt{\varepsilon}$ are stable, and $\mathcal{M}_\varepsilon = [-1/\sqrt{\varepsilon}, 1/\sqrt{\varepsilon}] \supset [-1, 1]$ is invariant. The flow is expressed by

$$\Phi_\varepsilon^t x = \frac{xe^t}{\sqrt{1 + \varepsilon x^2(e^{2t} - 1)}} = \Phi_0^t x - \frac{\varepsilon}{2} x^3 e^{t(e^{2t}-1)} + O\left(\varepsilon^2 x^5 e^{t(e^{2t}-1)^2}\right), \quad \text{for } x \in \mathcal{M}_\varepsilon$$

where $\Phi_0^t x = xe^t$. The ε expansion at $\varepsilon = 0$

$$\Phi_\varepsilon^t x = \Phi_0^t x - \frac{\varepsilon}{2} x^3 e^{t(e^{2t}-1)} + \ldots \tag{9.156}$$

has an error that grows without limit as t increases, which shows the unsuitability of the linear response. In this case, always assuming $f_0(x) = 1/2$ in $[-1, 1]$, one has

$$\begin{aligned} f_t(x; \varepsilon) &= \frac{1}{2}\chi_{[-1,1]}(x) e^{-\int_{-t}^0 d\tau} e^{3\varepsilon \int_{-t}^0 \frac{x^2 e^{2\tau}}{1+\varepsilon x^2 (e^{2\tau}-1)} d\tau} \\ &= \frac{1}{2}\chi_{[-1,1]}(x)\, e^{-t(1-\varepsilon x^2 + \varepsilon x^2 e^{-2t})^{-\frac{3}{2}}} \end{aligned} \tag{9.157}$$

being $\Omega_{V_0}^{f_0}(x) = -1$ and $\Omega_{V_p}^{f_0}(x) = 3x^2$, since $V_0(x) = x$ and $V_p(x) = -x^3$ Observe that $f_t(\pm 1/\sqrt{\varepsilon}; \varepsilon) \to \infty$ while $f_t(x; \varepsilon) \to 0$ for $-1/\sqrt{\varepsilon} < x < 1/\sqrt{\varepsilon}$ as $t \to \infty$. Thus the measure

$$d\mu_t^\varepsilon(x) = f_t(x; \varepsilon)\, dx$$

converges weakly to

$$\left[\frac{1}{2}\delta_{-1/\sqrt{\varepsilon}} + \frac{1}{2}\delta_{1/\sqrt{\varepsilon}}\right]$$

9.7.1.4 Second Kind of ε-Perturbation

Let us consider the dynamical system $\dot{x} = 0$ and its perturbation

$$\dot{x} = \varepsilon(x - x^3), \quad -1 \le \varepsilon \le 1 \tag{9.158}$$

and assume the invariant interval $\mathscr{M} = [-1, 1]$ as the phase space. For $\varepsilon \neq 0$ the fixed points are $x_0 = 0$, $x_- = -1$ and $x_+ = 1$. For $\varepsilon > 0$, the point 0 is unstable, while -1 and 1 are stable. The opposite holds for $\varepsilon < 0$ For $\varepsilon = 0$, all the points are fixed. The flow is given by

$$\Phi_\varepsilon^t x = \frac{x e^{\varepsilon t}}{\sqrt{1 + \varepsilon x^2 (e^{2\varepsilon t} - 1)}} \tag{9.159}$$

Taking the initial uniform distribution on $[-1, 1]$, we obtain $\Omega_V^{f_0} = -\varepsilon(1 - 3x^2)$, and the evolved density is expressed by

$$f_t(x; \varepsilon) = \frac{1}{2} \chi_{[-1,1]}(x) \exp\left[-\varepsilon t \left(1 - \varepsilon x^2 + \varepsilon x^2 \exp -2\varepsilon t\right)^{-\frac{3}{2}}\right] \tag{9.160}$$

For $\varepsilon > 0$ the distribution $d\mu_t^\varepsilon(x) = f_t(x; \varepsilon)\, dx$ converges to $\frac{1}{2}\delta_{-1} + \frac{1}{2}\delta_1$, while for $\varepsilon < 0$ the limit is δ_0. The response is

$$\mathbb{E}_t [\mathscr{O}] = \mathbb{E}_t [\mathscr{O}] - \frac{\varepsilon}{2} \int_0^t d\tau \int_{-1}^1 \mathscr{O}(\Phi_\varepsilon^\tau x)(1 - 3\varepsilon x^2) dx \tag{9.161}$$

The linear response is not a good approximation if ε is not small.

9.7.1.5 One Particle with a One-Particle Environment

Consider a test particle of mass M interacting harmonically with a field particle labeled 1, having mass m, in a one-dimensional space. The equations of motion are given by

$$\dot{x} = v\,; \quad M\dot{v} = -\gamma v - x_1 \tag{9.162}$$

$$\dot{x}_1 = v_1\,; \quad m_1 \dot{v}_1 = -k_1 x_1 + x_1, \quad \text{hence} \quad \dot{v}_1 = -\eta^2 x_1 \tag{9.163}$$

where k_1 is chosen so that $\eta^2 = (k_1 - 1)/m_1 > 0$, and a "shade" of the action-reaction principle is preserved. These equations can be solved with initial conditions $x(0) = 0$, $v(0) = 0$, $x_1(0) = x_{10}$, $v_1(0) = 0$, even if initial position and velocity of the field particle are unknown. First of all, note that the field particle is not affected by the test particle, so its equations of motion can be independently solved. The result is

$$x_1(t; x_{10}, v_{10}) = x_{10} \cos \eta t; \quad v_1(t; x_{10}, v_{10}) = -x_{10}\eta \sin \eta t \tag{9.164}$$

where x_{10} is the unknown initial position. Substituting in Eq. (9.162), and solving, one obtains

$$v(t; 0, 0, x_{10}, 0) = \frac{x_{10}}{M^2\eta^2 + \gamma^2}\left[\gamma\, e^{-\gamma t/M} - \gamma \cos \eta t - M\eta \sin \eta t\right] := x_{10}\, g(t) \quad (9.165)$$

$$x(t; 0, 0, x_{10}, 0) = \frac{x_{10}}{M^2\eta^2 + \gamma^2}\left[-Me^{-\gamma t/M} - \frac{\gamma}{\eta} \sin \eta t + M \cos \eta t\right] \quad (9.166)$$

where the time-dependent function g has been introduced for simplicity.

Now, take x_{10} uniformly distributed in $[-1, 1]$, and compute the steady-state velocity autocorrelation function of the test particle and its power spectrum. The time-dependent average velocity vanishes, because

$$\mathbb{E}\left[v(t)\right] = \mathbb{E}\left[x_{10}\right] g(t) = 0$$

The time autocorrelation of the velocity is given by

$$\Phi_V(t, t') = \mathbb{E}\left[x_{10}^2\right] g(t)g(t') = \frac{1}{3}\, g(t)g(t') \quad (9.167)$$

In the steady state, when the exponential term has vanished, only the time difference $\tau = t - t'$ matters, and

$$g(t)g(t') = g(0)g(\tau) = \gamma^2 \cos \eta\tau + M\eta\gamma \sin \eta\tau$$

so the spectrum takes the form:

$$\begin{aligned} I_V(\omega) &= \frac{1}{6\pi}\int_{-\infty}^{\infty} \Phi_V(\tau)e^{-i\omega\tau}\, \mathrm{d}\tau \\ &= \frac{\gamma}{12\pi}\int_{-\infty}^{\infty}\left[(\gamma + M\eta)\, e^{i\tau(\eta-\omega)} + (\gamma - M\eta)\, e^{-i\tau(\eta+\omega)}\right] \mathrm{d}\tau \quad (9.168) \\ &= \frac{\gamma}{6}\Big[(\gamma + M\eta)\, \delta(\omega - \eta) + (\gamma - M\eta)\, \delta(\omega + \eta)\Big] \end{aligned}$$

and has two frequencies of different intensity.

If the test particle interacts with N independent field particles of mass m_i attached to harmonic springs of elastic constant k_i, $i = 1, \ldots, N$, and the equations of motion are given by

$$\begin{aligned} \dot{x} = v; \quad & M\dot{v} = -\gamma v - \sum_{i=1}^{N} x_i \\ \dot{x}_i = v_i; \quad & m_i \dot{v}_i = -k_i x_i + x_i \end{aligned}$$

various scenarios arise, depending on the relative masses and on N. In general, the motion becomes more and more complex, gaining frequencies, until it appears practically chaotic, although it continues to be integrable.

9.7.1.6 Gaussian Isoenergetic System

For the isoenergetic dynamics illustrated in Appendix I.0.2, in which the internal energy is conserved, i.e., H_0 is fixed and $\dot{H}_0 = 0$, the equilibrium ensemble is microcanonical. Therefore, for an equilibrium Hamiltonian N particle system in d dimensions, one has

$$\alpha(\Gamma) = -\frac{\mathbf{J}(\Gamma)V \cdot \mathbf{F}_e}{2K(\Gamma)} \tag{9.169}$$

while

$$\Lambda(\Gamma) = \sum_{i=1}^{N} \left(\frac{\partial \mathbf{C}_i(\Gamma)}{\partial \mathbf{q}_i} + \frac{\partial \mathbf{D}_i(\Gamma)}{\partial \mathbf{p}_i} \right) \cdot \mathbf{F}_e - dN\alpha(\Gamma) + O\left(\frac{1}{N}\right) = -dN\alpha(\Gamma) + O\left(\frac{1}{N}\right) \tag{9.170}$$

as in the original paper (Evans et al. 1993). Then, one may write

$$\Omega^{(0)}_{t,t+\tau}(\Gamma) = \ln \frac{f^{(0)}(\Phi^t \Gamma)}{f^{(0)}(\Phi^{t+\tau} \Gamma)} - \Lambda_{t,t+\tau}(\Gamma) = -\Lambda_{t,t+\tau}(\Gamma)$$

9.7.1.7 Gaussian Isokinetic System

In this case the kinetic energy K is fixed, and one may formally write $K = (dN - d - 1)k_B T = (dN - d - 1)/\beta$, if also the center of mass is fixed, to prevent contributions from drift, so that

$$\alpha(\Gamma) = -\frac{(\dot{H}_0(\Gamma) + \mathbf{J}(\Gamma)V \cdot \mathbf{F}_e)\beta}{dN - d - 1} \tag{9.171}$$

In turn, Λ is still expressed by (9.170), with a different $O(1/N)$ term. In this case, f_0 writes

$$f_0(\Gamma) = \frac{1}{Q} e^{-\beta H_0(\Gamma)} \delta(K(\Gamma) - K_0)$$

where Q is the normalization constant. The result for the dissipation function is

$$\overline{\Omega}_{0,\tau}(\Gamma) = \frac{\beta}{\tau} \int_0^{\tau} \dot{H}_0(\Phi^s \Gamma) ds - \overline{\Lambda}_{0,\tau}(\Gamma) = -\overline{(\mathbf{J} \cdot \mathbf{F}_e)}_{0,\tau} V\beta + O\left(\frac{1}{N}\right) \tag{9.172}$$

Combining this with Eqs. (9.171) and (9.170), one obtains

$$\sigma(\Gamma) = -\mathbf{J}(\Gamma) \cdot \mathbf{F}_e V\beta + O\left(\frac{1}{N}\right) \tag{9.173}$$

where σ is the entropy production, if local thermodynamic equilibrium is verified.

9.7.1.8 Nosé-Hoover Thermostat

In this case, described in Appendix I.0.3, there is one extra equation to be added to the equations of motion (9.127):

$$\dot{\zeta} = \frac{1}{\tau}\Big[2K(\Gamma) - dNk_B T\Big] \tag{9.174}$$

where τ sets the relaxation time scale of the thermostat and T is the imposed average temperature. Then, for unperturbed Hamiltonian dynamics, one may write

$$\Lambda = \nabla \cdot \dot{\Gamma} + \frac{\partial \dot{\zeta}}{\partial \zeta} = -\, dN\zeta \tag{9.175}$$

In this case, f_0 is given by

$$f_0(\Gamma) = \frac{1}{M} e^{-\beta(H_0(\Gamma) + \frac{1}{2} Q\zeta^2(\Gamma))}$$

where M is the normalization. It follows that

$$\frac{f_0(\Gamma)}{f_0(\Phi^\tau \Gamma)} = e^{\beta \int_0^\tau (\dot{H}_0(\Phi^s \Gamma) + Q\zeta(\Phi^s \Gamma)\dot{\zeta}(\Phi^s \Gamma))ds}$$

Moreover, we obtain

$$\Omega(\Gamma) = -\mathbf{J}(\Gamma) \cdot \mathbf{F}_e V \beta \tag{9.176}$$

which tells that the driven nonequilibrium evolution of systems with equilibrium Hamiltonian dynamics, is characterized by the dissipation function, which equals the entropy production, in case local thermodynamic equilibrium holds.

9.7.1.9 One Isoenergetic Particle

For a particle in a two-dimensional space, let $H_0 = (p_x^2 + p_y^2)/2$ be the unperturbed Hamiltonian. Take the canonical ensemble $f_0 = Z_0^{-1} \exp(-\beta H_0)$ as the initial distribution. Introduce the perturbation $H_p = -Eq_x$, to obtain the Hamiltonian $H = H_0 + H_p$. The corresponding equations of motion are

$$\begin{cases} \dot{q}_x = p_x, \\ \dot{q}_y = p_y, \\ \dot{p}_x = E, \\ \dot{p}_y = 0, \end{cases} \tag{9.177}$$

with the dissipation function

$$\Omega_E^{f_0} = \Omega_0^{f_0} + \Omega_p^{f_0} = \beta E p_x$$

in which the momentum in x-direction plays the role of internal dissipative flux, $J = -p_x$, and $F_e = E$ the corresponding thermodynamic force. This example is an instance of the adiabatic system (9.118), in which $C_1 = C_2 = 0$ and $D_1 = 1, D_2 = 0$. Denote the phases by $\Gamma = (q_x, q_y, p_x, p_y)$, and observe that the solution of (9.177) with initial condition Γ gives $p_x(\Gamma, t) = p_x + Et$ and $p_y(\Gamma, t) = p_y$. Then, we compute the time evolution of the probability distribution function f_t starting from the integrated dissipation function:

$$\Omega^{f_0}_{-t,0}(\Gamma) = \beta E \int_{-t}^{0} p_x(\Gamma, s)\mathrm{d}s = \beta E p_x t - \beta E^2 \frac{t^2}{2}$$

obtaining

$$f_t(\Gamma) = f_0(\Gamma) e^{\Omega^{f_0}_{-t,0}(\Gamma)} = f_0(\Gamma) \exp\left[\beta E p_x t - \beta E^2 \frac{t^2}{2}\right]$$

We expand the exponential in the limit of a small external field E, and we neglect the terms of order higher than linear in E, to obtain the linear response for the evolving density:

$$\widetilde{f}_t(\Gamma) = f_0(\Gamma)(1 + \beta E p_x t)$$

The difference between exact and linear response can be clarified by looking at Liouville's equation for f_t and proceeding a la Green-Kubo. Let us start introducing the Liouville operator of the perturbed system:

$$\mathscr{L} = \mathscr{L}_0 + \Delta\mathscr{L}$$

where $\mathscr{L}_0 = p_x \partial_{q_x} + p_y \partial_{q_y}$ corresponds to the unperturbed system and $\Delta\mathscr{L} \equiv \mathscr{L}_p = E\partial_{p_x}$ corresponds to the perturbation. Splitting the evolving ensemble distribution function as

$$f_t(\Gamma) = f_0(\Gamma) + \Delta f_t(\Gamma)$$

Liouville's equation takes the form:

$$\partial_t(f_0(\Gamma) + \Delta f_t(\Gamma)) + (\mathscr{L}_0 + \Delta\mathscr{L})(f_0(\Gamma) + \Delta f_t(\Gamma)) = 0$$

Recalling that the equilibrium distribution f_0 is invariant under the unperturbed dynamics—so that $(\partial_t + \mathscr{L}_0) f_0(\Gamma) = 0$—we obtain the evolution equation for the deviation $\Delta f_t(\Gamma)$:

$$(\partial_t + \mathscr{L}_0)\Delta f_t(\Gamma) = -\Delta\mathscr{L} f_0(\Gamma) - \Delta\mathscr{L}\Delta f_t(\Gamma)$$

Then, assuming that $\Delta\mathscr{L}\Delta f(\Gamma)$ is negligible w.r.t. the other terms, we arrive at

$$(\partial_t + \mathscr{L}_0)\Delta f_t(\Gamma) = -\Delta\mathscr{L} f_0(\Gamma) \tag{9.178}$$

Since $\Delta\mathscr{L} f_0(\Gamma) = E\partial_{p_x}\left[Z_0^{-1}\exp(-\beta H_0)\right] = -\beta E p_x f_0(\Gamma)$, Eq. (9.178) can be rewritten as

$$(\partial_t + p_x\partial_{q_x} + p_y\partial_{q_y})\Delta f_t(\Gamma) = \beta E p_x f_0(\Gamma) \tag{9.179}$$

where $\beta f_0 = -\partial f_0/\partial H_0$, $E = F_e$ and $p_x = -J$ The solution to the evolution equation for the variation of the ensemble distribution function is given by

$$\Delta f_t(\Gamma) = -\int_0^t d\tau \exp[-(t-\tau)\mathscr{L}_0]\Delta\mathscr{L}(\tau) f_0(\Gamma) \tag{9.180}$$

$$= -\beta E\int_0^t d\tau \exp[-(t-\tau)\mathscr{L}_0](-p_x) f_0(\Gamma) \tag{9.181}$$

$$= -\beta F_e\int_0^t d\tau \exp[-(t-\tau)\mathscr{L}_0]J(\Gamma) f_0(\Gamma) \tag{9.182}$$

where we have used the fact that the external perturbation is time independent. Then, the phase space average of an observable $\mathscr{O}$, with respect to $f_t(\Gamma) = f_0(\Gamma) + \Delta f_t(\Gamma)$, writes

$$\begin{aligned} \mathbb{E}_t^{\sim}[\mathscr{O}] &= \mathbb{E}_0[\mathscr{O}] - \beta F_e\int_0^t d\tau\int_{\mathscr{M}} d\Gamma f_0(\Gamma)J(\Gamma)\exp[-(t-\tau)\mathscr{L}_0]\mathscr{O}(\Gamma) \\ &= \mathbb{E}_0[\mathscr{O}] + \beta E\int_0^t d\tau\int_{\mathscr{M}} d\Gamma f_0(\Gamma)p_x(\mathscr{O}\circ\Phi_0^{t-\tau}(\Gamma)) \\ &= \mathbb{E}_0[\mathscr{O}] + \int_0^t d\tau\, \mathbb{E}_0\left[(\mathscr{O}\circ\Phi_0^{t-\tau})\Omega_E^{f_0}\right] \end{aligned} \tag{9.183}$$

noting that Φ_0^t is the flow of the unperturbed dynamics. As an example, take $\mathscr{O} = p_x$, and compute its time evolving linear response. Since $\mathbb{E}_0[p_x] = 0$ and $\mathbb{E}_0\left[p_x^2\right] = \beta^{-1}$, we have

$$\mathbb{E}_t^{\sim}[p_x] = \frac{\beta E}{Z_0}\int_0^t ds\int dp_y e^{-\frac{\beta}{2}p_y^2}\int dp_x e^{-\frac{\beta}{2}p_x^2}p_x^2 = \beta\mathbb{E}_0\left[p_x^2\right]Et = \beta\frac{1}{\beta}Et = Et \tag{9.184}$$

Now, denoting by Φ_E^t the flow of the perturbed system, and recalling that $p_x\circ\Phi_E^t = p_x + Et$ and $\Omega_E^{f_0} = \beta E p_x$, the exact response formula (6.56) yields

$$\mathbb{E}_t[p_x] = \int_0^t d\tau\int_{\mathscr{M}} d\Gamma f_0(\Gamma)(p_x\circ\Phi_E^{\tau}(\Gamma))\Omega_E^{f_0}(\Gamma) \tag{9.185}$$

$$= \beta E\int_0^t d\tau\int_{\mathscr{M}} d\Gamma f_0(\Gamma)(p_x^2 + Ep_x\tau) = \beta\mathbb{E}_0\left[p_x^2\right]Et = Et \tag{9.186}$$

For such an observable, linear and exact responses do coincide. In general, we have

$$\mathbb{E}_t[\mathscr{O}] - \mathbb{E}_t^{\sim}[\mathscr{O}] = \int_0^t ds\, \mathbb{E}_0\left[\Omega_E^{f_0}\left[(\mathscr{O}\circ S_E^{\tau}) - (\mathscr{O}\circ S_0^{t-\tau})\right]\right] \tag{9.187}$$

9.7.1.10 Perturbation of a Harmonic Oscillator

Consider the Hamiltonian of one harmonic oscillator $H_0 = (p^2 + \omega^2 q^2)/2$, and let a perturbation take the form $H_p = -Eq$. The equations of motion of $H = H_0 + H_p$ are

$$\begin{cases} \dot{q} = p, \\ \dot{p} = E - \omega^2 q \end{cases} \tag{9.188}$$

with flow

$$\Phi^t_E \begin{pmatrix} q \\ p \end{pmatrix} = \begin{pmatrix} \left(q - \frac{E}{\omega^2}\right) \cos\omega t + \frac{p}{\omega} \sin\omega t + \frac{E}{\omega^2} \\ -\omega \left(q - \frac{E}{\omega^2}\right) \sin\omega t + p \cos\omega t \end{pmatrix} \tag{9.189}$$

The dissipation function with initial distribution $f_0 = \exp\{-\beta H_0\}/Z_0$ is given by

$$\Omega^{f_0}_E = -G \cdot \nabla \ln f_0 = (p, E - \omega^2 q) \cdot \beta \begin{pmatrix} \omega^2 q \\ p \end{pmatrix} = \beta E p \tag{9.190}$$

since the vector field is Hamiltonian and Λ vanishes. Moreover, we see that $\Omega^{f_0}_E$ coincides with the dissipation function $\Omega^{f_0}_p$ of the vector field of H_p, because f_0 is the equilibrium distribution of H_0. Thus the exact and the linear response formulas are respectively given by

$$\mathbb{E}_t\left[\mathcal{O}\right] = \mathbb{E}_0\left[\mathcal{O}\right] + \int_0^t \mathrm{d}s\, \mathbb{E}_0\left[\beta E p \left(\mathcal{O} \circ \Phi^s_E\right)\right] \tag{9.191}$$

and

$$\mathbb{E}^{\sim}_t\left[\mathcal{O}\right] = \mathbb{E}_0\left[\mathcal{O}\right] + \int_0^t \mathrm{d}s\, \mathbb{E}_0\left[\beta E p \left(\mathcal{O} \circ \Phi^{t-s}_0\right)\right] \tag{9.192}$$

For instance, take $\mathcal{O} = q$. In this case $\mathcal{O} \circ \Phi^s_E$ and $\mathcal{O} \circ \Phi^s_0$ are given by (9.189). For $E \neq 0$ we have

$$\begin{aligned} &\mathbb{E}_0\left[\beta E p \left(\mathcal{O} \circ S^s_E\right)\right] \qquad (9.193) \\ &= \int\!\!\int \mathrm{d}q\, \mathrm{d}p \frac{\exp -\beta(\frac{p^2}{2} + \frac{\omega^2}{2} q^2)}{Z_0} \beta E p \left[\left(q - \frac{E}{\omega^2}\right) \cos\omega t + \frac{p}{\omega} \sin\omega t + \frac{E}{\omega^2}\right] \\ &= \frac{\beta E}{\omega} \sin\omega t \int_{-\infty}^{+\infty} \mathrm{d}q \int_{-\infty}^{+\infty} \mathrm{d}p \frac{\exp -\beta(\frac{p^2}{2} + \frac{\omega^2}{2} q^2)}{Z_0} p^2 = \frac{E}{\omega} \sin\omega t \end{aligned}$$

since

$$\mathbb{E}_0\left[p\right] = \int\!\!\int \mathrm{d}q\, \mathrm{d}p\, f_0(q, p) p = 0 \tag{9.194}$$

$$\mathbb{E}_0\left[q\right] = \int\!\!\int \mathrm{d}q\, \mathrm{d}p\, f_0(q, p) q = 0 \tag{9.195}$$

$$\mathbb{E}_0\left[p^2\right] = \int\!\!\int \mathrm{d}q\, \mathrm{d}p\, f_0(q, p) p^2 = \beta^{-1} \tag{9.196}$$

The response formula (9.191) gives

$$\mathbb{E}_t\left[p\right] = \mathbb{E}_0\left[p\right] + \int_0^t \mathrm{d}s\,\frac{E}{\omega}\,\sin\omega\,s = \frac{E}{\omega^2}(1-\cos\omega t) \tag{9.197}$$

Analogously, the unperturbed dynamics, given by $E = 0$ in (9.189), yields

$$\begin{aligned}
&\mathbb{E}_0\left[\beta E p\left(\mathscr{O}\circ S_0^{t-s}\right)\right] \\
&= \int\!\!\int \mathrm{d}q\,\mathrm{d}p\frac{\exp-\beta(\frac{p^2}{2}+\frac{\omega^2}{2}q^2)}{Z_0}\beta E p\left[q\cos\omega\left(t-s\right)+\frac{p}{\omega}\sin\omega\left(t-s\right)\right] \\
&= \frac{E}{\omega}\sin\omega\left(t-s\right)
\end{aligned} \tag{9.198}$$

and from (9.192)

$$\mathbb{E}_t^{\sim}\left[p\right] = \mathbb{E}_0\left[p\right] + \int_0^t \mathrm{d}s\,\frac{E}{\omega}\,\sin\omega\left(t-s\right) = \frac{E}{\omega^2}(1-\cos\omega t) \tag{9.199}$$

This shows that the perturbed and unperturbed responses of the observable $\mathscr{O} = p$ coincide. If we consider the observable $\mathscr{O} = q\,p$, the situation is different. In the first place,

$$\begin{aligned}
\mathscr{O}\circ\Phi_E^t = &-\omega\left(q-\frac{E}{\omega^2}\right)^2\frac{1}{2}\sin 2\omega t + p\left(q-\frac{E}{\omega^2}\right)\cos^2\omega t - p\left(q-\frac{E}{\omega^2}\right)\sin^2\omega t \\
&+p^2\frac{1}{2\omega}\sin 2\omega t - \frac{E}{\omega}\left(q-\frac{E}{\omega^2}\right)\sin\omega t + p\frac{E}{\omega^2}\cos\omega t
\end{aligned} \tag{9.200}$$

that must be multiplied by p before taking the ensemble averages in (9.191) and (9.192). This gives a linear combination of monomials $p,\ pq,\ pq^2,\ p^2q,\ p^3$, whose average with respect to f_0 vanishes, and the term

$$p^2\frac{E}{\omega^2}\left[-\cos^2\omega t + \sin^2\omega t + \cos\omega t\right]$$

Thus, recalling that $\mathbb{E}_0\left[p^2\right] = \beta^{-1}$, we have

$$\begin{aligned}
\mathbb{E}_t\left[qp\right] &= \beta E\int_0^t \mathrm{d}s\frac{E}{\omega^2}\mathbb{E}_0\left[p^2\right]\left[-\cos^2\omega s + \sin^2\omega s + \cos\omega s\right] \\
&= \frac{E^2}{\omega^3}\left[\sin\omega t + \frac{1}{2}\sin 2\omega t\right]
\end{aligned} \tag{9.201}$$

On the other hand, from (9.200) with $E = 0$, we find that $p(\mathscr{O}\circ\Phi_0^t)$ is a linear combination of $q^2p,\ qp^2,\ p^3$. This implies that $\mathbb{E}_0\left[\beta E p\left(\mathscr{O}\circ\Phi_0^{t-s}\right)\right] = 0$ and, from (9.192), that $\mathbb{E}_t^{\sim}\left[qp\right] = 0$, showing that the perturbed and unperturbed responses substantially differ in the case of $\mathscr{O} = q\,p$.

9.8 Current Research

Fluctuation relations and response formulae for nonequilibrium systems are at the forefront of present-day theoretical research, given their involvement in the development of the science and technology of small or strongly nonequilibrium systems. Once more, the variety of current lines of research in this field is impressive; therefore, we mention only part of the literature with which we are more familiar; the readers can easily find more on the web.

As the categories of large and small are relative notions, small means any system made of a moderate number of interacting components, hence quite large systems, such as the Earth climate (Ghil and Lucarini 2020; Yin et al. 2024) or the cosmic microwave background (Baiesi et al. 2020). As molecular dynamics constitutes the field within which most of the theory illustrated above has been developed, very many applications pertain to that field. In particular, the exact response theory finds application in situations in which the ratio signal to noise is poor (Maffioli et al. 2022), as in complex tribology experiments. The interpretation of molecular biology experiments is also a very active field (Monge et al. 2018). Finally, we note that Machine Learning and Artificial Intelligence absorb information from all branches of statistical mechanics, and recently, a number of researchers are focusing on the techniques described in this chapter (Rotskoff and Vanden-Eijnden 2019; Carbone et al. 2024).

10 Finale

The scientist does not study nature because it is useful; he studies it because he delights in it, and he delights in it because it is beautiful.

Science and method, Henri Poincaré

All things, at least those we know, contain number; for it is evident that nothing whatever can either be thought or known, without number.

Philolaus, Fragment 2
Life of Pythagoras, Iamblichus

Throughout this book, we have tried to keep technical and mathematical tools of use in the study of nonequilibrium phenomena together with the intellectual reflections that accompanied their development. We deem important to keep in mind such foundational aspects of statistical mechanics and nonequilibrium physics, as further and even unpredictable progress is assured. Moving in uncharted territory using the same techniques that have proved successful in the past is, in fact, as inevitable as fraught with danger. We also tried to highlight the golden thread linking the past to the present, with an eye on the future, and constituting the unifying frame of a subject that is still evolving, hence appears often quite fragmented. For this reason, we have ordered the chapters of the book in an approximately chronological order, with some notable exception meant to keep the sequence logically consistent. Therefore, thermo-hydrodynamics comes before statistical mechanics, and among the topics of statistical mechanics the Brownian motion has been placed before the kinetic theory of gases, because using that theory it tackles the issue of fluctuations, on which all modern developments hinge, and which gave the ultimate green light to probabilities in physics. Initially, fluctuations that are small compared to the observed signals have been considered, and treated by and large within a Gaussian formalism. The most

C. Mejía-Monasterio and L. Rondoni, *Nonequilibrium Statistical Mechanics I*,
Theoretical and Mathematical Physics,
https://doi.org/10.1007/978-3-031-98249-1_10

important entailing result is the theory of linear response of perturbed equilibrium systems. Successively, "big" fluctuations came to the fore, particularly relevant as "small" systems attracted growing interest, taking advantage of the development of the large deviations formalism. That led to fluctuation relations and an extension of linear response to stationary states of nonequilibrium systems, first, and to a general exact, non-perturbative, response theory. Analogous frameworks have been developed within the realm of dynamical systems, motivated by the uncontainable progress of molecular dynamics, as well as in the stochastic realm, where explicit calculations can be more easily performed, although more stringent conditions must be satisfied. The result is a wide open field, as old as statistical mechanics and still looking like just at the beginning. Good luck to those who will like to contribute, our book simply being a modest gesture of encouragement.

Simple Facts About Legendre Transforms

A

Current controversies about mind and brain are strikingly similar to debates about atoms, molecules, chemical structures, and reactions, and related matters, which were very much alive well into the twentieth century. Similar, and in ways that I think are instructive.

On Nature and Language, Noam Chomsky

The common definition of Legendre transforms given in mathematics textbooks, not concerned about specific physical applications, follows a different convention from that of Sect. 5.1. The advantage is a stronger symmetry in the definition of the transform and its inverse. Here we adopt this general convention, as done in Sect. 5.3. Consider a real function defined on an interval, $f : I \to \mathbb{R}$. For a given slope $q \in \mathbb{R}$ and for each point $x \in I$, consider the (vertical) increment from the ordinate of the function to the ordinate of the straight line of slope $y = qx$, both computed at x, namely

$$F(q, x) = qx - f(x)$$

If at some point x the function is above the line, the value of this difference is negative; otherwise, it is positive. Consider the supremum of F at fixed q, i.e., as a function of the slope:

$$f_*(q) = \sup_{x \in I} (qx - f(x))$$

The function f_* is defined for all slopes $q \in \mathbb{R}$ and takes values in $\mathbb{R} \cup +\infty$. It is convex because it is the sup of a family of affine functions, and it is called the convex conjugate of f.

C. Mejía-Monasterio and L. Rondoni, *Nonequilibrium Statistical Mechanics I*, Theoretical and Mathematical Physics,
https://doi.org/10.1007/978-3-031-98249-1

Assume moreover that I is open and f is differentiable in I. If the sup in the definition of f_* is attained at some point $\overline{x}$, then the derivative

$$\frac{d}{dx}(qx - f(x)) = q - f'(x)$$

is zero at $\overline{x}$, and $q = f'(\overline{x})$. We would like to solve this equation to get $\overline{x}$ as a function of q. This requires the extra assumption of the monotonicity of f' and leads to the following definition:

Definition *Let f be differentiable on the open interval I and strictly convex,* i.e., *the derivative f' is strictly increasing. Let $J = f'(I)$ be the set of all possible slopes of the tangents to f. The Legendre transform of f is defined as the following function:*

$$J \ni q \mapsto q f'^{-1}(q) - f[f'^{-1}(q)] := f_*(q)$$

where f'^{-1} is the inverse of the derivative of f.

It follows that

Lemma *The Legendre transform of f is the restriction to J of the convex conjugate of f to the open interval I.*

To prove that, we check that the value of the Legendre transform is the value of the function $x \mapsto qx - f(x)$ at the (unique) point where its derivative is 0. From the previous computations, we have that

$$f_*(q) = q f'^{-1}(q) - f[f'^{-1}(q)] \geq qx - f(x), \quad x \in I, q \in J$$

which means f_* is the sup of F in I. Conversely, we can recover f from f_*. In fact, this comes from the relation (Young inequality, see below):

$$qx \leq f(x) + f_*(q)$$

with equality if $q = f'(x)$ is symmetric. Hence, f defined on I is the convex conjugate of f_* defined on J.

Some examples help to understand the above definitions.

Example 1 If $f(x) = \frac{1}{4}x^4$, $I = \mathbb{R}$, then $f'(x) = x^3$, with $J = \mathbb{R}$ and inverse $f'^{-1}(y) = y^{1/3}$. It follows that the Legendre transform is

$$y \mapsto y y^{1/3} - \frac{1}{4}\left(y^{1/3}\right)^4 = \frac{3}{4}y^{4/3}$$

Note the conjugate exponents $\frac{1}{4} + \frac{1}{4/3} = 1$ and the inequality $xy \leq \frac{1}{4}x^4 + \frac{3}{4}y^{4/3}$.

Example 2 If $f(x) = \exp x - 1 - x$, $I = \mathbb{R}$, $f'(x) = \exp x - 1$, then $J =]-1, +\infty[$ and $f'^{-1}(y) = \log(1+y)$. The value of the Legendre transform for each $y > -1$ is

$$\begin{aligned} f_*(y) &= y\log(1+y) - f(\log(1+y)) = y\log(1+y) - (1+y) + 1 + \log(1+y) \\ &= (1+y)\log(1+y) - y \end{aligned} \tag{A.1}$$

The Young inequality is

$$yx \leq \exp x - 1 - x + (1+y)\log(1+y) - y$$

Example 3 If $f(p) = p\log p$, $p > 0$, then $f'(p) = \log p + 1$ whose image is $J = \mathbb{R}$. The inverse of the derivative is $f'^{-1}(u) = \exp(u-1)$. The Legendre transform is

$$f_*(u) = u\exp(u-1) - \exp(u-1)(u-1) = \exp(u-1)$$

The Young inequality is

$$up \leq p\log p + \exp(u-1)$$

If p_x is a probability with support on the finite set $\mathbb{X}$, and u is a random variable on $\mathbb{X}$, then the entropy is

$$H(p) = -\sum_{x\in\mathbb{X}} p_x \log p_x \leq -\sum_{x\in\mathbb{X}} u_x p_x + \sum_{x\in\mathbb{X}} \exp(u_x - 1)$$

Example 4 (*Mechanics*)
Consider the Lagrangian function of one harmonic oscillator on the real line:

$$L(v,q) = \frac{1}{2}v^2 - \frac{k}{2}q^2, \quad v \in \mathbb{R}$$

For each given q, the domain is $v \in I =]-\infty, +\infty[$. The partial derivative mapping is

$$v \mapsto \frac{\partial}{\partial v}L(v,q) = v$$

i.e., the identity function on $\mathbb{R}$ with range $J =]-\infty, +\infty[$ and the inverse of this partial derivative mapping is then again the identity. If we denote by p the "slope" variable for every given position q:

$$p = \frac{\partial}{\partial v}L(v,q)$$

and denote by $H(\cdot, q)$ the Legendre transform:

$$H(p, q) = p^2 - \left(\frac{1}{2}p^2 - \frac{k}{2}q^2\right) = \frac{1}{2}p^2 + \frac{k}{2}q^2$$

we obtain the Hamiltonian. Then, $L(v, q)$ and $H(p, q)$ are related by the inequality

$$pv \leq L(v, q) + H(p, q)$$

with equality if $v = p$.

The Legendre transform enjoys the following properties.

Theorem *Let $\mathscr{L}$ denote the Legendre transform operator. Then*

$$\mathscr{L} f_* = \mathscr{L}(\mathscr{L} f) = \mathscr{L}^2 f = f$$

i.e., *the Legendre transform is an involution.*

Corollary *Take $f_* = \mathscr{L} f$, and consider the family of straight lines of equation $y = px - f_*(p)$. The envelop of such lines is a curve of equation $y = f(x)$, where $f = \mathscr{L} f_*$.*

Young's inequality: *Let $f(x)$ and $f_*(p)$ be Legendre transforms of one another. Then*

$$px \leq f(x) + f_*(p)$$

Birkhoff's Ergodic Theorem

B

All happy families are alike; each unhappy family is unhappy in its own way.

Anna Karenina, Lev Tolstoy

A fundamental question in dynamics is to understand when the phase space averages are equal to time averages. In real experiments, the values of the physical observables correspond to the time averages, but these are too difficult to compute; therefore, one hopes that there exists a probability measure μ that yields the time average by calculation of an integral over the phase space, called the phase space average. The ergodic theorem of Birkhoff states the conditions under which the time average exists and is related to the phase space average.

Let us start by defining the time and phase space average. Let the tuple $(\mathcal{M}, \Phi, \mu)$ be a dynamical system and a function $f : \mathcal{M} \longrightarrow I\!R$. For every $x \in \mathcal{M}$, we define the time average as

$$\overline{f}(x) = \lim_{t\to\infty} \frac{1}{t} \sum_{s=0}^{t-1} f\left(\Phi^s x\right) \tag{B.1}$$

if time is discrete, $t \in \mathbf{Z}$, and as

$$\overline{f}(x) = \lim_{t\to\infty} \frac{1}{t} \int_0^t f\left(\Phi^s x\right) \mathrm{d}s \tag{B.2}$$

if time is continuous, $t \in I\!R$. Additionally, for every $f \in L^1(\mathcal{M}, \mu)$, the phase average is defined by

$$\langle f \rangle_\mu = \int_{\mathcal{M}} f \mathrm{d}\mu \tag{B.3}$$

C. Mejía-Monasterio and L. Rondoni, *Nonequilibrium Statistical Mechanics I*, Theoretical and Mathematical Physics,
https://doi.org/10.1007/978-3-031-98249-1

The existence of these quantities is not a trivial issue. For example, think of a process in which 0 and 1 alternate, each time equating the total length of the previous strings of values:

$$\underset{1\,element}{0}\quad \underset{2\,elements}{11}\quad \underset{3\,elements}{000}\quad \underset{6\,elements}{111111}\quad \underset{12\,elements}{000000000000}\ldots$$

The corresponding time average does not exist: at finite times, it takes values that go from $1/3$ to $2/3$ and back, never settling down:

$$0\,,\ \frac{2}{3}\,,\ \frac{1}{3}\,,\ \frac{2}{3}\,,\ \frac{1}{3}\,,\ldots$$

Much as this could seem an artificial example, it is in fact allowed by many chaotic dynamics. Therefore, what follows is relevant.

Birkhoff's ergodic theorem: *Let* $(\mathcal{M}, \Phi, \mu)$ *be a dynamical system. Then:*

(a) $f \in L^1(\mathcal{M}, \mu)$ *implies that* $\overline{f}(x)$ *exists for* μ*-almost every* $x \in \mathcal{M}$*;*
(b) $f \in L^p(\mathcal{M}, \mu)$*, with* $1 \leq p < \infty$*, implies that* $\overline{f} \in L^p(\mathcal{M}, \mu)$ *and that*

$$\lim_{n\to\infty} \left\| \overline{f} - \frac{1}{n}\sum_{j=0}^{n-1} f \circ \Phi^j \right\|_p = 0 \tag{B.4}$$

with $\overline{f}(\Phi x) = \overline{f}(x)$*, for every* $x \in \mathcal{M}$*;*
(c) $f \in L^p(\mathcal{M}, \mu)$*, with* $1 \leq p < \infty$*, implies that*

$$\int_{\mathcal{M}} \overline{f}\mathrm{d}\mu = \int_{\mathcal{M}} f\mathrm{d}\mu \tag{B.5}$$

Proof This proof is based on the validity of the following

Lemma (Maximal Ergodic Theorem): *Given* $f \in L^1$*, consider*

$$F_t(x) := \sum_{j=0}^{t-1} f\left(\Phi^j x\right), \quad \text{with} \quad x \in \mathcal{M} \quad \text{and} \quad t \in \mathbb{N}, \tag{B.6}$$

and introduce $E(f) := \left\{x \in \mathcal{M} : \sup_{t\geq 1} F_t(x) > 0\right\}$*. Then,* $E(f)$ *is measurable and*

$$\int_{E(f)} f\,\mathrm{d}\mu \geq 0 \tag{B.7}$$

Note that $F_t(x) > 0$ is the same as $F_t(x)/t > 0$, where $F_t(x)/t$ is the time average of f along the trajectory up to time t, or the approximation up to time t of $\overline{f}(x)$.

Therefore, the Lemma states that the integral of f over $E(f)$ is non-negative, where $E(f)$ is the set in which the finite-time average F_t/t is positive at least for one t. Indeed, dividing by t does not change the sign of the expression.

Also, by definition:

$$\sup_{t\geq 1} F_t(x) = \sup_{t\geq 1}\left[f(x),\, f(x) + f(\Phi x),\, f(x) + f(\Phi x) + f\left(\Phi^2 x\right), ...\right] \tag{B.8}$$

This Lemma can be used to prove part (a) of Birkhoff's theorem. To do that, take $a < b \in \mathbb{R}$, and introduce

$$E_{a,b} = \left\{x \in \mathcal{M} : \liminf_{t\to\infty} \frac{1}{t} F_t(x) < a < b < \limsup_{t\to\infty} \frac{1}{t} F_t(x)\right\} \tag{B.9}$$

where the lim inf and the lim sup of a sequence are defined by

$$\liminf x_n = \sup_{k\geq 1} \inf_{n\geq k} x_n\ , \qquad \limsup x_n = \inf_{k\geq 1} \sup_{n\geq k} x_n$$

The set $E_{a,b}$ is measurable, since lim inf and lim sup of measurable functions are measurable. Moreover, the set is invariant under the action of Φ, because of the limits in t: $\Phi E_{a,b} = E_{a,b}$. For a point $x \in E_{a,b}$, the time average does not exist, because the lim inf and the lim sup do not coincide, and the set of all points for which the time average does not exist is then given by

$$E = \bigcup_{a<b} E_{a,b} \tag{B.10}$$

Actually, to separate lim inf from lim sup in all cases x in which they differ, it suffices to take $a, b \in \mathbb{Q}$, because even so a and b can be made as close as one wants. Therefore, it suffices to prove that each $E_{a,b}$ with rational a and b has vanishing measure.

As we are going to show, the Lemma allows us to write

$$b\, \mu\left(E_{a,b}\right) \leq \int_{E_{a,b}} f\, \mathrm{d}\mu \leq a\, \mu\left(E_{a,b}\right) \tag{B.11}$$

which obviously implies $\mu\left(E_{a,b}\right) = 0$. Indeed, let us apply the Lemma to the function

$$g(x) = \begin{cases} f(x) - b & \text{for } x \in E_{a,b} \\ -1 & \text{for } x \notin E_{a,b} \end{cases} \tag{B.12}$$

The set of points such that $G_t = \sum_{j=0}^{t-1} g \circ \Phi^j$ is non-negative is precisely $E_{a,b}$. Indeed, $G_t(x) = F_t(x) - bt$ for $x \in E_{a,b}$, and $G_t(x) = -t$ for $x \notin E_{a,b}$, because $E_{a,b}$ is invariant, so either $\Phi^j x \in E_{a,b}$ for all j or $\Phi^j x \notin E_{a,b}$ for all j. Then:

i. if $x \in E_{a,b}$, then by definition of $E_{a,b}$ there is one t such that $b < F_t(x)/t = G_t/t + b$, hence $G_t(x) > 0$;
ii. if $x \notin E_{a,b}$, then $G_t(x) = -t < 0$.

Then, $E(g) = E_{a,b}$ and the Lemma then implies

$$0 \leq \int_{E_{a,b}} g \, \mathrm{d}\mu = \int_{E_{a,b}} f \, \mathrm{d}\mu - b\, \mu\left(E_{a,b}\right) \,, \quad i.e., \quad b\, \mu\left(E_{a,b}\right) \leq \int_{E_{a,b}} f \, \mathrm{d}\mu \tag{B.13}$$

Analogously, we may consider

$$g(x) = \begin{cases} a - f(x) & \text{for } x \in E_{a,b} \\ -1 & \text{for } x \notin E_{a,b} \end{cases} \tag{B.14}$$

Once again, the set of points such that $G_t = \sum_{j=0}^{t-1} g \circ \Phi^j = ta - F_t$ is non-negative for at least one t is precisely $E_{a,b}$, and

$$0 \leq \int_{E_{a,b}} g \, \mathrm{d}\mu = a\, \mu\left(E_{a,b}\right) - \int_{E_{a,b}} f \, \mathrm{d}\mu \,, \quad i.e., \quad a\, \mu\left(E_{a,b}\right) \geq \int_{E_{a,b}} f \, \mathrm{d}\mu \tag{B.15}$$

Then, $\mu\left(E_{a,b}\right) = 0$ for all $a < b$ and $\mu(E) = 0$ as well. In other words, the time average exists for μ-almost all $x \in \mathscr{M}$.

We can now use the fact that (a) holds. To prove (b), take $f \in L^p$, with $p \geq 1$ and observe that

$$\left|\overline{f}(x)\right| = \left| \lim_{n\to\infty} \frac{1}{n} \sum_{j=0}^{n-1} f\left(\Phi^j x\right) \right| = \lim_{n\to\infty} \frac{1}{n} \left| \sum_{j=0}^{n-1} f\left(\Phi^j x\right) \right| \leq \lim_{n\to\infty} \frac{1}{n} \sum_{j=0}^{n-1} \left| f\left(\Phi^j x\right) \right|$$

for μ-almost every $x \in \mathscr{M}$. The second equality also implies that for every $\varepsilon > 0$, $\exists\, n_\varepsilon \in I\!N$ such that

$$-\varepsilon < \left|\overline{f}(x)\right| - \frac{1}{n} \left| \sum_{j=0}^{n-1} f\left(\Phi^j x\right) \right| < \varepsilon \,, \quad \text{if} \quad n > n_\varepsilon \tag{B.16}$$

which means that $F_n := \left(\sum_{j=0}^{n-1} f \circ \Phi^j\right)/n$ is dominated by an L^1 function, for large n. Indeed:

$$|F_n(x)| = \frac{1}{n} \left| \sum_{j=0}^{n-1} f\left(\Phi^j x\right) \right| < \left|\overline{f}(x)\right| + \varepsilon \,, \quad \text{with} \int_{\mathscr{M}} \left(\left|\overline{f}(x)\right| + \varepsilon\right) \mathrm{d}\mu = \left\| \, \left|\overline{f}\right| + \varepsilon \, \right\|_1 < \infty \tag{B.17}$$

because $(\overline{f}+\varepsilon) \in L^1$, as we are going to show. First of all, observe that $|f| \in L^1$ if $f \in L^1$, which implies that the time average of $|f|$

$$\overline{|f|}(x) = \lim_{n\to\infty} \frac{1}{n} \sum_{j=0}^{n-1} \left| f\left(\Phi^j x\right)\right| \equiv \lim_{n\to\infty} y_n(x) \,, \qquad y_n = \frac{1}{n} \sum_{j=0}^{n-1} \left| f \circ \Phi^j \right| \tag{B.18}$$

exists for μ almost every $x \in \mathscr{M}$. Then,

$$|\overline{f}(x)|^p = \left(\left| \lim_{n\to\infty} \frac{1}{n} \sum_{j=0}^{n-1} f\left(\Phi^j x\right) \right| \right)^p \le \lim_{n\to\infty} \left(\frac{1}{n} \sum_{j=0}^{n-1} \left| f\left(\Phi^j x\right)\right| \right)^p = \lim_{n\to\infty} y_n^p(x) \,, \tag{B.19}$$

where $\lim_{n\to\infty} y_n^p(x)$ exists for μ almost all x, because $\lim_{n\to\infty} y_n(x)$ does. Moreover, because $\left|\overline{f}\right|^p$ is non-negative, it is integrable if $\lim_{n\to\infty} y_n^p$ is and, by Fatou's lemma, this limit, which equals the lim inf for almost all x, is integrable if the lim inf of the sequence is.[1]

Is the lim inf y_n^p integrable?

Recalling that $\| \, |.| \, \|_p = \|.\|_p$, we can write

$$\begin{aligned} \int_{\mathscr{M}} \left(\frac{1}{n} \sum_{j=0}^{n-1} \left| f \circ \Phi^j \right| \right)^p \mathrm{d}\mu &= \left\| \frac{1}{n} \sum_{j=0}^{n-1} \left| f \circ \Phi^j \right| \right\|_p^p \\ &\le \left(\frac{1}{n} \sum_{j=0}^{n-1} \left\| \, |f \circ \Phi^j| \, \right\|_p \right)^p = \left(\frac{1}{n} \sum_{j=0}^{n-1} \left\| f \circ \Phi^j \right\|_p \right)^p \end{aligned} \tag{B.20}$$

Observe that $\|f\|_p = \|f \circ \Phi\|_p$. Indeed, consider first a simple function: $f(x) = \sum_k c_k \chi_k(x)$, where the c_k's are constants and χ_k is the indicator function of the set E_k, with $\mathscr{M} = \cup_k E_k$ and $E_k \cap E_j = \emptyset$ if $k \neq j$. Then, we have

$$|f(x)|^p = \left| \sum_k c_k \chi_k(x) \right|^p = \sum_k |c_k|^p \, \chi_k(x) \tag{B.21}$$

because for every $x \in \mathscr{M}$, only one of the χ_k equals 1, while all the others equal 0. At the same time,

$$|f(\Phi x)|^p = \sum_k |c_k|^p \, \chi_k(\Phi x) = \sum_k |c_k|^p \, \chi_{\Phi^{-1} E_k}(x) \tag{B.22}$$

[1] **Fatou's Lemma:** *Let $g_n \in L^+$, namely positive functions (not necessarily integrable), for every n, and $g = \liminf g_n$. Then $\int g \mathrm{d}\mu \le \liminf \int g_n \mathrm{d}\mu$.*

because $\Phi x \in E_k$ iff $x \in \Phi^{-1} E_k$. Integrating Eq. (B.21), we get

$$\|f\|_p^p = \int_{\mathcal{M}} |f|^p \, \mathrm{d}\mu = \sum_k |c_k|^p \, \mu\,(E_k) \tag{B.23}$$

while integrating Eq. (B.22), and recalling that μ is invariant for Φ, we get

$$\|f \circ \Phi\|_p^p = \sum_k |c_k|^p \, \mu \left(\Phi^{-1} E_k\right) = \sum_k |c_k|^p \, \mu\,(E_k) = \|f\|_p^p \,, \quad i.e., \quad \|f \circ \Phi\|_p = \|f\|_p \tag{B.24}$$

Because simple functions are dense in L^p, the same holds for all $f \in L^p$. Therefore, the issue of the integrability of $\left|\overline{f}\right|^p$, which has been related to Eq. (B.21), can be continued observing that

$$\int_{\mathcal{M}} \left(\frac{1}{n} \sum_{j=0}^{n-1} \left| f \circ \Phi^j \right| \right)^p \mathrm{d}\mu \le \left(\frac{1}{n} \sum_{j=0}^{n-1} \left\| f \circ \Phi^j \right\|_p \right)^p = \left(\frac{1}{n} \sum_{j=0}^{n-1} \|f\|_p \right)^p = \|f\|_p^p \tag{B.25}$$

which holds for every $n \in I\!N$. Then,

$$\begin{aligned}
\|\overline{f}\|_p^p &= \int_{\mathcal{M}} \left|\overline{f}\right|^p \mathrm{d}\mu \le \int_{\mathcal{M}} \lim_{n\to\infty} y_n^p \, \mathrm{d}\mu = \int_{\mathcal{M}} \liminf_{n\to\infty} y_n^p \, \mathrm{d}\mu \le \liminf_{n\to\infty} \int_{\mathcal{M}} y_n^p \, \mathrm{d}\mu \\
&= \liminf_{n\to\infty} \int_{\mathcal{M}} \left(\frac{1}{n} \sum_{j=0}^{n-1} \left| f \circ \Phi^j \right| \right)^p \mathrm{d}\mu = \liminf_{n\to\infty} \left(\left\| \frac{1}{n} \sum_{j=0}^{n-1} \left| f \circ \Phi^j \right| \right\|_p \right)^p \\
&\le \liminf_{n\to\infty} \left(\frac{1}{n} \sum_{j=0}^{n-1} \left\| f \circ \Phi^j \right\|_p \right)^p = \liminf_{n\to\infty} \left(\frac{1}{n} \sum_{j=0}^{n-1} \|f\|_p \right)^p = \|f\|_p^p
\end{aligned} \tag{B.26}$$

Therefore, the p-norm of the time average is not larger than the p-norm of the function, i.e., $\overline{f} \in L^p$.

Remark *This proof heavily relies on the invariance of the measure μ; it is an* "if and only if".

To prove (B.4) under the assumption that (a) holds, let us begin with $f \in L^\infty \subset L^p$, if $p < \infty$, where

$$\|f\|_\infty = \inf \left\{ a \ge 0 : \mu\left(\{x : |f(x)| > a\}\right) = 0 \right\} \tag{B.27}$$

The fact that (a) holds means that

$$\left| \overline{f} - \frac{1}{n} \sum_{j=0}^{n-1} f \circ \Phi^j \right| \to 0 \,, \text{ hence } \left| \overline{f} - \frac{1}{n} \sum_{j=0}^{n-1} f \circ \Phi^j \right|^p \to 0 \text{ for } \mu\text{-almost every } x \in \mathcal{M} \tag{B.28}$$

but this does not imply convergence in norm. Now, the properties of absolute value and of the ∞-norm imply that

$$\left|F_n(x)\right| = \left|\frac{1}{n}\sum_{j=0}^{n-1} f\left(\Phi^j x\right)\right| \leq \frac{1}{n}\sum_{j=0}^{n-1}\left|f\left(\Phi^j x\right)\right| \leq \frac{1}{n}\sum_{j=0}^{n-1}\|f\|_\infty = \|f\|_\infty \tag{B.29}$$

for μ-almost every x. In other words, the sequence $\{F_n\}$ is dominated by the constant $\|f\|_\infty$, which is in L^1. Moreover,

$$\begin{aligned}\left|\overline{f}(x) - \frac{1}{n}\sum_{j=0}^{n-1} f\left(\Phi^j x\right)\right| &\leq \left|\overline{f}(x)\right| + \left|\frac{1}{n}\sum_{j=0}^{n-1} f\left(\Phi^j x\right)\right| \\ &\leq \left|\overline{f}(x)\right| + \frac{1}{n}\sum_{j=0}^{n-1}\left|f\left(\Phi^j x\right)\right| = \left|\ \left|\overline{f}(x)\right| + \frac{1}{n}\sum_{j=0}^{n-1}\left|f\left(\Phi^j x\right)\right|\ \right|\end{aligned} \tag{B.30}$$

and rising to the p-th power,

$$\left|\overline{f}(x) - \frac{1}{n}\sum_{j=0}^{n-1} f\left(\Phi^j x\right)\right|^p \leq \left|\ \left|\overline{f}(x)\right| + \frac{1}{n}\sum_{j=0}^{n-1}\left|f\left(\Phi^j x\right)\right|\ \right|^p \tag{B.31}$$

Because $f \in L^\infty$, for μ-almost every x, we can write

$$\left|\overline{f}(x)\right| = \left|\lim_{n\to\infty}\frac{1}{n}\sum_{j=0}^{n-1} f\left(\Phi^j x\right)\right| \leq \lim_{n\to\infty}\frac{1}{n}\sum_{j=0}^{n-1}\left|f\left(\Phi^j x\right)\right| \leq \lim_{n\to\infty}\frac{1}{n}\sum_{j=0}^{n-1}\|f\|_\infty = \|f\|_\infty \tag{B.32}$$

which, together with Eqs. (B.29) and (B.31), leads to

$$\left|\overline{f}(x) - \frac{1}{n}\sum_{j=0}^{n-1} f\left(\Phi^j x\right)\right|^p \leq \left|\ \|f\|_\infty + \|f\|_\infty\ \right|^p = 2^p\,\|f\|_\infty^p < \infty \tag{B.33}$$

Therefore, for $f \in L^\infty$, Eq. (B.4) holds. Indeed, the dominated convergence theorem yields

$$\begin{aligned}\lim_{n\to\infty}\left\|\overline{f} - \frac{1}{n}\sum_{j=0}^{n-1} f\circ\Phi^j\right\|_p^p &= \lim_{n\to\infty}\int_{\mathcal{M}}\left|\overline{f} - \frac{1}{n}\sum_{j=0}^{n-1} f\circ\Phi^j\right|^p \mathrm{d}\mu \\ &= \int_{\mathcal{M}}\lim_{n\to\infty}\left|\overline{f} - \frac{1}{n}\sum_{j=0}^{n-1} f\circ\Phi^j\right|^p \mathrm{d}\mu = \int_{\mathcal{M}} 0\,\mathrm{d}\mu = 0\end{aligned} \tag{B.34}$$

Take now $f \in L^p$, and one $\varepsilon > 0$. Because L^∞ is dense in L^p we may pick $f_0 \in L^\infty$, such that $\| f_0 - f \|_p \leq \varepsilon/3$. Because $L^p \subset L^1$, part (a) of the theorem applies to f_0, so we can choose $N \in \mathbb{N}$ such that

$$\left\| \overline{f}_0 - \frac{1}{n} \sum_{j=0}^{n-1} f_0 \circ \Phi^j \right\|_p \leq \frac{\varepsilon}{3}, \quad \text{if} \quad n > N \tag{B.35}$$

Then we can write

$$\begin{aligned} \left\| \overline{f} - \frac{1}{n} \sum_{j=0}^{n-1} f \circ \Phi^j \right\|_p &= \left\| \overline{f} - \overline{f}_0 + \overline{f}_0 - \frac{1}{n} \sum_{j=0}^{n-1} f_0 \circ \Phi^j + \frac{1}{n} \sum_{j=0}^{n-1} f_0 \circ \Phi^j - \frac{1}{n} \sum_{j=0}^{n-1} f \circ \Phi^j \right\|_p \\ &\leq \| \overline{f} - \overline{f}_0 \|_p + \left\| \overline{f}_0 - \frac{1}{n} \sum_{j=0}^{n-1} f_0 \circ \Phi^j \right\|_p + \left\| \frac{1}{n} \sum_{j=0}^{n-1} (f - f_0) \circ \Phi^j \right\|_p \end{aligned}$$

Because the time average is not larger than the norm of the function, cf. below Eq. (B.25), we can write

$$\| \overline{f} - \overline{f}_0 \|_p = \| \overline{f - f_0} \|_p \leq \| f - f_0 \|_p \leq \frac{\varepsilon}{3} \tag{B.36}$$

Also,

$$\left\| \frac{1}{n} \sum_{j=0}^{n-1} (f - f_0) \circ \Phi^j \right\|_p \leq \frac{1}{n} \sum_{j=0}^{n-1} \left\| (f - f_0) \circ \Phi^j \right\|_p = \frac{1}{n} \sum_{j=0}^{n-1} \| f - f_0 \|_p = \| f - f_0 \|_p \leq \frac{\varepsilon}{3} \tag{B.37}$$

Therefore, for $n > N$ we have

$$\left\| \overline{f} - \frac{1}{n} \sum_{j=0}^{n-1} f \circ \Phi^j \right\|_p \leq \varepsilon \tag{B.38}$$

which completes the proof of (B.4), assuming that (a) holds.

It remains to prove that time averages are constants of the motion. Indeed,

$$\begin{aligned} \overline{f}(\Phi x) &= \lim_{n\to\infty} \frac{1}{n} \sum_{j=0}^{n-1} f\left(\Phi^j \Phi x\right) = \lim_{n\to\infty} \left(\frac{1}{n} \sum_{j=0}^{n} f\left(\Phi^j x\right) - \frac{f(x)}{n} \right) \\ &= \lim_{n\to\infty} \left(\frac{n+1}{n} \frac{1}{n+1} \sum_{j=0}^{n} f\left(\Phi^j x\right) - \frac{f(x)}{n} \right) = \lim_{n\to\infty} \frac{1}{n+1} \sum_{j=0}^{n} f\left(\Phi^j x\right) = \overline{f}(x) \end{aligned} \tag{B.39}$$

because $n + 1/n \to 1$ and $f(x)/n \to 0$ for almost every $x \in \mathcal{M}$ (otherwise $f \notin L^p$).

This completes the proof of (b), provided (a) holds.

To prove (c), let us take $f \in L^p$. Then, by (a)

$$F_n := \frac{1}{n}\sum_{j=0}^{n-1} f \circ \Phi^j \tag{B.40}$$

converges μ-almost everywhere to $\overline{f}$, and by (b) it also converges in L^1 to $\overline{f}$

$$\lim_{n\to\infty} \left\| \overline{f} - \frac{1}{n}\sum_{j=0}^{n-1} f \circ \Phi^j \right\|_1 = 0\,, \quad \text{or} \quad \frac{1}{n}\sum_{j=0}^{n} f \circ \Phi^j \underset{L^1}{\longrightarrow} \overline{f} \tag{B.41}$$

For large N, the sequence F_n is then dominated by $|\overline{f} + \varepsilon|$, which is an L^1 function, because $\overline{f} \in L^p \subset L^1$ and every constant lies in L^1. Therefore, we may apply the dominated convergence theorem to state that

$$\begin{aligned} \int_{\mathcal{M}} \overline{f} \mathrm{d}\mu &= \int_{\mathcal{M}} \lim_{n\to\infty} F_n \,\mathrm{d}\mu = \lim_{n\to\infty} \int_{\mathcal{M}} F_n \,\mathrm{d}\mu \\ &= \lim_{n\to\infty} \frac{1}{n}\sum_{j=0}^{n-1} \int_{\mathcal{M}} f \circ \Phi^j \,\mathrm{d}\mu = \lim_{n\to\infty} \frac{1}{n}\sum_{j=0}^{n-1} \int_{\mathcal{M}} f \,\mathrm{d}\mu = \int_{\mathcal{M}} f \,\mathrm{d}\mu \end{aligned} \tag{B.42}$$

where we relied in the same calculation of Eq. (B.22) for simple functions, and we extended it to L^1 functions. This completes the proof of the Birkhoff-Khinchin theorem, provided the Maximal Lemma holds.

The Maximal Lemma has the following corollary:

Corollary *Let $A \subset E(f)$ be measurable and invariant ($\Phi^{-1}A = A$). Then*

$$\int_A f \,\mathrm{d}\mu \geq 0 \tag{B.43}$$

Proof Let χ_A be the indicator function of A. Then $E\,(f\chi_A) = A$. The reason is that $\Phi x \in A$ if $x \in A$, because A is invariant, and $f(x)\chi_A(x) = f(x)$ for $x \in A$. Therefore, the fact such an $x \in A \subset E(f)$ implies $x \in E(f\chi_A)$. Moreover, a point y outside A does not belong to $E\,(f\chi_A)$, because $f\chi_A$ vanishes in such a point, and its evolution remains outside A, so that

$$\sum_{j=0}^{n-1} f(\Phi^j y)\chi_A(\Phi^j y) = 0\,, \quad \text{for all} \quad n \in \mathbb{N} \tag{B.44}$$

and never > 0. The Maximal Lemma now leads to

$$\int_{E(f\chi_A)} f\chi_A \,\mathrm{d}\mu \geq 0\,, \quad \text{where} \quad \int_{E(f\chi_A)} f\chi_A \,\mathrm{d}\mu = \int_A f\chi_A \,\mathrm{d}\mu = \int_A f \,\mathrm{d}\mu \tag{B.45}$$

which proves the Corollary.

It only remains to prove the Maximal Lemma. To this purpose, introduce

$$f_n(x) = \max\left\{ f(x),\, f(x) + f(\Phi x),\, ...,\, \sum_{j=0}^{n-1} f\left(\Phi^j x\right) \right\} \tag{B.46}$$

Obviously,

$$E(f) = \bigcup_{n=1}^{\infty} \{x : f_n(x) > 0\} \tag{B.47}$$

This union is monotonic because by definition one has

$$f_{n+1}(x) = \max\left\{ f(x),\, ...,\, \sum_{j=0}^{n-1} f\left(\Phi^j x\right),\, \sum_{j=0}^{n} f\left(\Phi^j x\right) \right\} \geq \max\left\{ f(x),\, ...,\, \sum_{j=0}^{n-1} f\left(\Phi^j x\right) \right\} = f_n(x)$$

hence $\{x : f_{n+1}(x) > 0\} \supset \{x : f_n(x) > 0\}$. Then it suffices to prove that

$$\int_{\{x : f_n(x) > 0\}} f \, \mathrm{d}\mu \geq 0 \tag{B.48}$$

for one n. To do so, observe that

$$f_n(x) \geq \sum_{j=0}^{m} f\left(\Phi^j x\right), \quad \text{if } 0 \leq m \leq n-1$$

This implies that

$$f_n(\Phi x) + f(x) = f(x) + \max\left\{ f(\Phi x),\, ...,\, \sum_{j=1}^{n} f\left(\Phi^j x\right) \right\} \geq f(x) + \sum_{j=1}^{m+1} f\left(\Phi^j x\right) = \sum_{j=0}^{m+1} f\left(\Phi^j x\right), \tag{B.49}$$

for $0 \leq m \leq n-1$, because the sum $\sum_{j=1}^{m+1}$ is just one of the sums among which the maximum is to be found.

Note: for every x there is an $\hat{m}, 0 \leq \hat{m} \leq n-1$, such that $f_n(x) = \sum_{j=0}^{\hat{m}} f\left(\Phi^j x\right)$, since $f_n(x)$ is one of the sums over which we maximize. Analogously, for every x there is $\tilde{m}$, $0 \leq \tilde{m} \leq n$, such that $f_{n+1}(x) = \sum_{j=0}^{\tilde{m}} f\left(\Phi^j x\right)$. However, in general, $f_{n+1}(x)$ is not expressed by one of the sums that may appear as the last term in Eq. (B.49), because $m \geq 0$, hence $\sum_{j=0}^{m+1}$ contains two terms at least, and it cannot express $f(x)$ alone. But $f(x)$ is one of the elements that must be considered in the maximization defining $f_{n+1}(x)$.

We thus need to consider two cases: $f_n(\Phi x) \geq 0$ and $f_n(\Phi x) < 0$.

- $f_n(\Phi x) \geq 0$. In this case we have $f_n(\Phi x) + f(x) \geq f(x)$, thus:

$$\begin{aligned} f_n(\Phi x) + f(x) &= \max\{f(x), f_n(\Phi x) + f(x)\} \\ &= \max\left\{f(x), f(x) + \max\left\{f(\Phi x), ..., \sum_{j=1}^{n} f\left(\Phi^j x\right)\right\}\right\} = f_{n+1}(x) \end{aligned} \tag{B.50}$$

which proves that $f_n(\Phi x) \geq 0$ implies

$$f_n(\Phi x) + f(x) \geq f_n(x)\,, \quad \text{and} \quad f(x) \geq f_n(x) - f_n \circ \Phi(x) \tag{B.51}$$

Then, decomposing the integral in Eq. (B.48) as

$$\int_{\{f_n>0\}} f\, \mathrm{d}\mu = \int_{\{f_n>0\}\cap\{f_n\circ\Phi<0\}} f\, \mathrm{d}\mu + \int_{\{f_n>0\}\cap\{f_n\circ\Phi\geq 0\}} f\, \mathrm{d}\mu \tag{B.52}$$

we can write

$$\begin{aligned} \int_{\{f_n>0\}} f\, \mathrm{d}\mu \geq & \int_{\{f_n>0\}\cap\{f_n\circ\Phi<0\}} f\, \mathrm{d}\mu \\ & + \int_{\{f_n>0\}\cap\{f_n\circ\Phi\geq 0\}} f_n\, \mathrm{d}\mu - \int_{\{f_n>0\}\cap\{f_n\circ\Phi\geq 0\}} f_n \circ \Phi\, \mathrm{d}\mu \end{aligned} \tag{B.53}$$

- $f_n(\Phi x) < 0$. In this case, the fact that $f_n(x) \geq 0$ implies $f_n(x) = f(x)$ and $f(x) > 0$. Indeed, $f_n(\Phi x) < 0$ means that

$$\begin{aligned} 0 > f_n(\Phi x) &= \max\left\{f(\Phi x), f(\Phi x) + f\left(\Phi^2 x\right), ..., \sum_{j=1}^{m} f\left(\Phi^j x\right), ..., \sum_{j=1}^{n} f\left(\Phi^j x\right)\right\} \\ &\geq \sum_{j=1}^{m} f\left(\Phi^j x\right) \end{aligned} \tag{B.54}$$

for $1 \leq m \leq n-1$. At the same time, $f_n(x) \geq 0$ implies that there exists $\hat{m}$, $0 \leq \hat{m} \leq n-1$, such that

$$\sum_{j=0}^{\hat{m}} f\left(\Phi^j x\right) = f(x) + \sum_{j=1}^{\hat{m}} f\left(\Phi^j x\right) \geq 0 \tag{B.55}$$

Then, taking $m = \hat{m}$ in Eq. (B.54), it follows that $f(x) > 0$, hence that

$$f_n(x) = \max\left\{f(x), ..., f(x) + \sum_{j=1}^{m} f\left(\Phi^j x\right), ..., f(x) + \sum_{j=1}^{n-1} f\left(\Phi^j x\right)\right\} = f(x) \tag{B.56}$$

because all sums added to $f(x)$ are negative.

Collecting the results for the two cases $f_n(\Phi x) \geq 0$ and $f_n(\Phi x) < 0$, we may now write

$$\int_{\{f_n>0\}\cap\{f_n\circ\Phi<0\}} f \, \mathrm{d}\mu + \int_{\{f_n>0\}\cap\{f_n\circ\Phi\geq 0\}} f_n \, \mathrm{d}\mu = \tag{B.57}$$
$$\int_{\{f_n>0\}\cap\{f_n\circ\Phi<0\}} f_n \, \mathrm{d}\mu + \int_{\{f_n>0\}\cap\{f_n\circ\Phi\geq 0\}} f_n \, \mathrm{d}\mu = \int_{\{f_n>0\}} f_n \, \mathrm{d}\mu$$

which implies

$$\begin{aligned} \int_{\{f_n>0\}} f \, \mathrm{d}\mu &\geq \int_{\{f_n>0\}} f_n \, \mathrm{d}\mu - \int_{\{f_n>0\}\cap\{f_n\circ\Phi\geq 0\}} f_n \circ \Phi \, \mathrm{d}\mu \\ &= \int_{\Phi^{-1}\{f_n>0\}} f_n \circ \Phi \, \mathrm{d}\mu - \int_{\{f_n>0\}\cap\{f_n\circ\Phi\geq 0\}} f_n \circ \Phi \, \mathrm{d}\mu \\ &= \int_{\{f_n\circ\Phi>0\}} f_n \circ \Phi \, \mathrm{d}\mu - \int_{\{f_n>0\}\cap\{f_n\circ\Phi\geq 0\}} f_n \circ \Phi \, \mathrm{d}\mu \geq 0 \end{aligned} \tag{B.58}$$

because

$$\Phi^{-1}\{x : f_n(x) > 0\} = \{x : f_n(\Phi x) > 0\} \supset \{x : f_n(x) > 0\} \cap \{x : f_n(\Phi x) \geq 0\}$$

This concludes the proof of the Maximal Lemma.

Probability Theory

C

Science is made up of mistakes, but they are mistakes which it is useful to make, because they lead little by little to the truth.

Journey to the Center of the Earth, Jules Verne

C.1 Probability Space

One of the central concepts in probability theory is measurability. Roughly speaking, measurability is essential to associate to each event in the probability space a value for the likelihood to observe it.

Let Ω be a non-empty set, $\mathscr{F}$ a σ-algebra of Ω[2] and $P : \mathscr{F} \to \mathbb{R}$ a measure on Ω. If P satisfies the following conditions

1. $P(F) > 0 \quad \forall F \in \mathscr{F}$
2. $P(\Omega) = 1$
3. For any set $\{F_i\}_{i=1}^{\infty}$ of elements of $\mathscr{F}$, $P(\bigcup_{i=1}^{\infty} F_i) \leq \sum_{i=1}^{\infty} P(F_i)$. The equality holds when $\{F_i\}_{i=1}^{\infty}$ are disjoint.

[2] A σ-algebra is a collection of subsets of Ω that satisfy the following conditions:

1. $\emptyset, \Omega \in \mathscr{F}$
2. If $F \in \mathscr{F}$ then its complement $F^c \in \mathscr{F}$
3. If $\{F_i\}_{i=1}^{\infty} \in \mathscr{F}$ then $\bigcap_{i=1}^{\infty} F_i \in \mathscr{F}$ and $\bigcup_{i=1}^{\infty} F_i \in \mathscr{F}$.

C. Mejía-Monasterio and L. Rondoni, *Nonequilibrium Statistical Mechanics I*, Theoretical and Mathematical Physics,
https://doi.org/10.1007/978-3-031-98249-1

then P is a probability measure and the triad $(\Omega, \mathscr{F}, P)$ a probability space. The properties above constitute the three Kolmogorov's axioms of probability stating positivity, normalization and countable additivity (Kolmogorov 1956). The last property implies the ordering property stating that if $F_i \subset F_j$ then $P(F_i) \leq P(F_j)$.

The set Ω is the sample "space" of a given system, and can be thought of as a set containing all possible fates, outcomes, or states of it. Instead the set σ-algebra[3] $\mathscr{F}$ represents information. Any $F \in \mathscr{F}$ is an event, namely a possible observation of the system, and the probability that the event F is observed is given by $P(F)$. The probability measure assigns a precise value (a measure) to every event in $\mathscr{F}$.

C.2 Random Variables

In probability theory, the notion of measurability is fundamental to extract information concerning the state of the system. This information is obtained through random variables. Here we restrict ourselves to real-valued random variables and will refer to them simply as random variables.

Let $(\Omega, \mathscr{F}, P)$ a probability space. A *measurable function* $\xi : \Omega \to \mathbb{R}$ is called a $\mathbb{R}$-valued random variable if for each $B \in \mathscr{B}$, $\xi^{-1}(B) \in \mathscr{F}$, namely if the value a random variable takes corresponds to an event. $\mathscr{B}$ is the Borel σ-algebra defined as the smallest σ-algebra containing all open subsets of $\mathbb{R}$. The content of $\mathscr{B}$ may be thought as the collection of all subsets of $\mathbb{R}$ well behaved with respect to the Lebesgue measure, and a random variable as a measurable function from the probability space into $\mathbb{R}$.

The indicator function

Let $(\Omega, \mathscr{F}, P)$ a probability space and $F \in \mathscr{F}$. The indicator function of the event F defined by

$$\chi_F(\omega) = \begin{cases} 1 \text{ if } \omega \in F \\ 0 \text{ if } \omega \notin F \end{cases}$$

is a random variable.

The simple function

Let $(\Omega, \mathscr{F}, P)$ a probability space and $\{F_i\}_{i=1}^m \in \mathscr{F}$ a finite sequence of mutually disjoint sets $F_i \bigcap F_j = \emptyset$, and let $\{x_i\}_{i=1}^m \in \mathbb{R}$. If $\{F_i\}$ is a partition of the sample space $\bigcup_{i=1}^m F_i = \Omega$ then the simple function defined as

$$\xi = \sum_{i=1}^{m} x_i \chi_{F_i}(\omega)$$

is a random variable. If $m = \infty$, then the ξ is called elementary.

[3] In probability the terminology σ stands for countable.

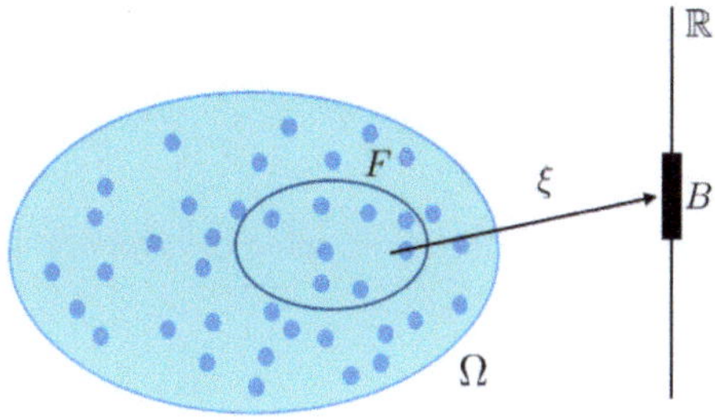

Fig. C.1 Action of a real-valued random variable ξ with probability space $(\Omega, \mathscr{F}, P)$. The random variable maps each event $F \in \mathscr{F}$ to a set $B \in \mathscr{B}$, with measure $\mu_\xi(B)$

C.3 Distribution Function

Let $\xi : \Omega \to \mathbb{R}$ a random variable describing the outcome of an experiment. Then the probability that the measurement has a value $x \in B \in \mathscr{B}$ is given by $P(\xi^{-1}(B))$. This interpretation of a physical measurement makes sense only if ξ is measurable. Therefore, for any random variable ξ we can define a function on the sets $B \in \mathscr{B}$ as

$$\mu_\xi(B) = P(\xi^{-1}(B)) \tag{C.1}$$

This function is a measurable function, and the triad $(\mathbb{R}, \mathscr{B}, \mu_x i)$ is therefore a probability space. The measure μ_ξ is called the distribution of the random variable ξ. A pictorial representation of the action of the random variable on the probability space is shown in Fig. C.1.

In terms of the distribution μ_ξ we can distinguish discrete from continuous random variables. If for a countable set of $\mathbb{R}$, μ_ξ is measure 1 then ξ is called a discrete random variable. Conversely, if μ_ξ is measure zero 0 for every countable set of $\mathbb{R}$ then ξ is called a continuous random variable.

The distribution μ_ξ is often given in terms of the distribution function of a random variable. The distribution function of a random variable $\xi : \Omega \to \mathbb{R}$ is

$$\tilde{P}_\xi(x) = P(\xi < x) = \mu_\xi(-\infty, x]$$

satisfying the following conditions

1. $\lim_{x\to-\infty} \tilde{P}_\xi(x) = 0$
2. $\lim_{x\to\infty} \tilde{P}_\xi(x) = 1$
3. $\tilde{P}_\xi(x)$ is a non-decreasing function of x
4. $\tilde{P}_\xi(x)$ is right continuous, namely $\lim_{\varepsilon\to 0} \tilde{P}_\xi(x + \varepsilon)$ exists and is equal to $\tilde{P}_\xi(x)$

Conversely, any function $\tilde{P}$ satisfying the conditions above is the distribution function of a random variable $\mu_\xi(-\infty, x] = \tilde{P}_\xi(x)$.

In physics, $\tilde{P}_\xi(x)$ is referred to as the cumulative distribution function of ξ. If ξ is a continuous random variable then we can define its probability density function that not always exists.

Let ξ a continuous random variable. If a function $f_\xi : \mathbb{R} \to [0, \infty)$ exists such that

$$P(\xi \in B) = \int_B f_\xi(x)\mathrm{d}x \tag{C.2}$$

then the function f_ξ is called the probability density of the random variable ξ. If the density exists then

$$\tilde{P}_\xi(x) = \int_{-\infty}^{x} f(s)\mathrm{d}s \tag{C.3}$$

It follows that

$$\int_{\mathbb{R}} f(x)\mathrm{d}x = 1\,. \tag{C.4}$$

Gaussian distribution

A continuous random variable ξ is said to be Gaussian distributed with mean μ and variance σ^2 if it has a probability density

$$f(x) = \frac{1}{\sqrt{2\pi\sigma^2}}\mathrm{e}^{-\frac{(x-\mu)^2}{2\sigma^2}}$$

defined over $x \in \mathbb{R}$.

C.4 Expectation Value

The mean, or expectation value of a random variable gives the average value the variable attains during a measurement.

Let $(\Omega, \mathscr{F}, P)$ be a probability space and $\xi : \Omega \to \mathbb{R}$ a continuous random variable. The expectation value of ξ is defined as

$$\mathbb{E}(\xi) = \int_\Omega \xi \mathrm{d}P \tag{C.5}$$

where $\int$ the Lebesgue integral. In terms of its distribution function $\mu\xi$, the expectation value is evaluated after a Riemann-Stieltjes integral

$$\mathbb{E}(\xi) = \int_{\mathbb{R}} x\mathrm{d}\mu_\xi$$

Furthermore, if ξ has a probability distribution density f then the measure is simply $\mu_\xi = f(x)\mathrm{d}x$ and the expectation value

$$\mathbb{E}(\xi) = \int_{\infty}^{\infty} xf(x)\mathrm{d}x \tag{C.6}$$

Also, if $g\colon \mathbb{R} \to \mathbb{R}$ is a function of ξ then

$$\mathbb{E}\left(g(\xi)\right) = \int_{\infty}^{\infty} g(x) f(x) \mathrm{d}x$$

The definition of expectation value parallels the definition of the Lebesgue integral of measurable functions: infinite expected values are allowed, and if $\mathbb{E}\left(\xi\right) < \infty$ then we say ξ is integrable. Let $(\Omega, \mathscr{F}, P)$ a probability space, ξ, ν random variables and $a, b \in \mathbb{R}$ constants. Then

1. (equivalence) If $\xi = \nu$ *a.s.* then $\mathbb{E}\left(\xi\right) = \mathbb{E}\left(\nu\right)$.
2. (dominance) If $\xi \leq \nu$ *a.s.* then $\mathbb{E}\left(\xi\right) \leq \mathbb{E}\left(\nu\right)$.
3. (linearity) $\mathbb{E}\left(a\xi + b\nu\right) = a\mathbb{E}\left(\xi\right) + b\mathbb{E}\left(\nu\right)$.
4. $|\mathbb{E}\left(\xi\right)| \leq \mathbb{E}\left(|\xi|\right)$.

If ξ is a discrete simple random variable then its expectation value is given by

$$\mathbb{E}\left(\xi\right) = \sum_{i=1}^{m} x_i P(F_i) \tag{C.7}$$

Extension of the expectation value to elementary random variables requires a special care for the convergence of the sum when $m = \infty$ and the x_i can attain negative values (Breiman 1968). Definition of non-negative random variables solves convergence issues and the definition (C.7) satisfies the properties above.

Jensen's inequality

Let $(\Omega, \mathscr{F}, P)$ a probability space, ξ a random variable and $g(\xi)$ a convex function. If $\mathbb{E}\left(\xi\right) < \infty$ it follows that

$$\mathbb{E}\left(g(\xi)\right) \geq g(\mathbb{E}\left(\xi\right)) \tag{C.8}$$

Chebyshev's inequality

Let $(\Omega, \mathscr{F}, P)$ a probability space and ξ a random variable. For any constant $a > 0$ it follows that

$$P(\xi| > a) \leq \frac{\mathbb{E}\left(\xi\right)}{a} \tag{C.9}$$

This inequality is often called Markov's inequality.

C.5 Moments of the Distribution

The moments of a distribution, or moments of a random variable constitute the statistics of the random variable. Let $(\Omega, \mathscr{F}, P)$ a probability space and ξ a random variable. We call

$$m_\xi^n \equiv \mathbb{E}\left(\xi^n\right) = \int_{\mathbb{R}} x^n \mathrm{d}\mu_\xi$$

the moment of order n of ξ that we denote as m_ξ^n. The moments of lower order such as the mean

$$\mu = m_\xi^1 = \mathbb{E}(\xi)$$

and the variance

$$\sigma^2 = m_\xi^2 = \mathbb{E}\left(\xi^2\right) - \mathbb{E}(\xi)^2$$

have a special relevance in applications.

Additional common definitions of moments are the central moments $\mathbb{E}((\xi - \mathbb{E}(\xi))^n)$, the absolute moments $\mathbb{E}(|\xi|^n)$, and the absolute central moments $\mathbb{E}(|\xi - \mathbb{E}(\xi)|^n)$.

C.6 The Characteristic Function

Let $(\Omega, \mathscr{F}, P)$ a probability space, $\xi : \Omega \to \mathbb{R}$ a random variable and $\lambda \in \mathbb{R}$. The characteristic function if ξ is defined as

$$\phi_\xi(\lambda) = \mathbb{E}\left(\mathrm{e}^{i\lambda\xi}\right) \tag{C.10}$$

If ξ is a continuous random variable and has a probability density f_ξ, then Eq. C.10 states that the characteristic function is the inverse Fourier transform of the probability density f_ξ

$$\phi_\xi(\lambda) = \int_{-\infty}^{\infty} \mathrm{e}^{i\lambda x} f(x)\mathrm{d}x \tag{C.11}$$

The importance of $\phi_\xi(\lambda)$ relies on the fact that the moments of the random variable are determined by the derivatives of the characteristic function as

$$m_{\xi,k} \equiv \mathbb{E}\left(\xi^k\right) = i^{-k} \left.\frac{\mathrm{d}^n \phi_\xi(\lambda)}{\mathrm{d}\lambda^n}\right|_{\lambda=0}$$

Moreover, if $\{\xi_i\}_{i=1}^n$ are independent random variables then

$$\phi_{\xi_1+\xi_2+\cdots+\xi_n}(\lambda) = \phi_{\xi_1}(\lambda) \cdot \phi_{\xi_2}(\lambda) \cdot \ldots \cdot \phi_{\xi_n}(\lambda) \quad \forall\, \lambda$$

is a very useful property.

Finally, for ξ, ν random variables, if $\phi_\xi(\lambda) = \phi_\nu(\lambda)$, for all λ then $\tilde{P}_\xi = \tilde{P}_\nu$. We say that the characteristic function $\phi_\xi(\lambda)$ determines the distribution function of ξ.

On certain occasions, it is more convenient to substitute in (C.10) the Fourier transform by its Laplace counterpart $M_\xi(\lambda) = \mathbb{E}\left(e^{\lambda\xi}\right)$, called the moment-generating function. The characteristic and moment-generating functions are equivalent in certain sense and the use of one or the other is determined by the convenience of working on Fourier's or Laplace's space and their associated convergence properties. However, it should be noted that the moment-generating function does not always exist, e.g., for dynamics for which some moments diverge.

From a physical perspective, the moment-generating function plays a central role in statistical mechanics. In the canonical ensemble, the energy of the system E is a fluctuating quantity and as such, it is a continuous real-valued random variable. Straightforwardly we recognize the central quantity of statistical mechanics, namely the partition function $Z(\beta) = \mathbb{E}\left(\exp(-\beta E)\right)$ as the moment-generating function of the energy. This connection extends to other ensembles to which other fluctuating thermodynamic quantities are added.

C.7 Cumulants of the Distribution

Consider the characteristic function of a random variable ξ. Expanding out the exponential on the left-hand side of Eq. C.11 and taking logarithms we obtain

$$\ln\left(\phi_\xi(\lambda)\right) = \sum_{n=1}^{\infty} \frac{(i\lambda)^n}{n!}\kappa_{\xi,n} \tag{C.12}$$

where $\kappa_{\xi,n}$ is called the n-th cumulant of the distribution $\tilde{P}_\xi$ or simply the n-th cumulant of the random variable ξ

The function $\ln\left(\phi_\xi(\lambda)\right)$ is called the cumulant-generating function since the cumulants are obtained as its derivatives

$$\kappa_{\xi,n} = i^{-n} \left.\frac{d^n \ln\left(\phi_\xi(\lambda)\right)}{d\lambda^n}\right|_{\lambda=0}. \tag{C.13}$$

The cumulants, also called cumulative moments, have a special relevance when a sequence of random variables is considered. Then the cumulant $\kappa_{\xi,n}$ is a function of moments $m_{\xi,k}$ of order up to $k = n$ plus covariances of orders lower than n of the different random variables.

Except for the lower moments (κ_1 is the mean, κ_2 the variance, κ_3 the third central moment), there is not an easy way to express a relation between cumulants and moments. They are particularly useful when a comparison between a probability density and a Gaussian distribution is needed, since the later is characterized by having all zero cumulants of order higher than 2. The Gaussian distribution is the only distribution having this property.

As for the moment-generating function, in statistical mechanics cumulants appear naturally in the canonical ensemble. There, the thermodynamic potential is the Helmholtz free energy $F(\beta) = -\beta^{-1} \ln Z(\beta)$. Physical quantities such as the internal energy, specific heat, *etc.* are cumulants of the energy, as they are obtained through derivatives of $F(\beta)$.

C.8 The Law of Large Numbers and Central Limit Theorem

Sums of independent random variables with arbitrary probability distribution shares common properties (expected values), when the sums considered are made of a large amount of terms. Extracting this common information is the content of the law of large numbers and the central limit theorem, and the fundamental basis of statistical theories.

We have not discussed yet independence. Let $(\Omega, \mathscr{F}, P)$ probability space. A finite sequence of random variables $\{\xi_i\}_{i=1}^n$, so that $\xi_i : \Omega \to \mathbb{R}$ for all i, are said to be independent if for any sets $B_1, B_2, \dots, B_n \in \mathscr{B}$,

$$P\left(\xi_1 \in B_1, \xi_2 \in B_2, \dots, \xi_n \in B_n\right) = \prod_{i=1}^{n} P(\xi_i \in B_i)$$

Strong law of large numbers

Let $(\Omega, \mathscr{F}, P)$ a probability space and $\{\xi_i \mid \xi_i : \Omega \to \mathbb{R}\}_{i=1}^n$ a finite sequence of independent, identically distributed random variables, usually denoted as *i.i.d.* random variables. Then

$$P\left(\lim_{n\to\infty} \frac{\xi_1 + \xi_2 + \ldots + \xi_n}{n} = \mathbb{E}\left(\xi\right)\right) = 1 \qquad \text{(C.14)}$$

where $\mathbb{E}\left(\xi\right) = \mathbb{E}\left(\xi_i\right)\ \forall i$ is the expectation value of the random variables.

The law of large numbers states that the arithmetic averages of a finite number of *i.i.d.* random variables converges in distribution to the mean of their distribution.

Central limit theorem

Analogously to the law of large numbers, the Central Limit Theorem (CLT) deals whit sums of *i.i.d.* random variables and estimates the fluctuations that are expected in the limit of very large number of variables.

Let $(\Omega, \mathscr{F}, P)$ a probability space and $\{\xi_i \mid \xi_i : \Omega \to \mathbb{R}\}_{i=1}^n$ a finite sequence of *i.i.d.* random variables with mean $\mathbb{E}\left(\xi_i\right) = \mu$ and variance $\mathbb{E}\left(\xi_i^2\right)\mathbb{E}\left(\xi_i\right)^2 = \sigma^2 > 0$ for all $i = 1, 2, \dots, n$, and $a, b \in \mathbb{R}$ constants. Set $S_n = \xi_1 + \xi_2 + \cdots + \xi_n$. Then

$$\lim_{n\to\infty} P\left(a \le \frac{S_n - n\mu}{\sqrt{n}\sigma} \le b\right) = \frac{1}{\sqrt{2\pi}} \int_a^b \mathrm{e}^{-\frac{x^2}{2}} \mathrm{d}x \qquad \text{(C.15)}$$

Consider an experiment in which the outcome of a measurement is ξ_i. The experiment is repeated n times, and each realization is independent of the rest. Then the central limit theorem specifies how large the deviation from the average measurement is when the number of realizations tends to infinity. The details of the distribution function of ξ are irrelevant. Therefore, the CLT concerns universal properties of all experiments that solely depend on coarse properties of the distribution.

Change of Variables in Stochastic Processes

D

è proprio all'interno di questa generale insensatezza che si costituisce il senso
(precisely within this general nonsense, sense is formed)

Documanità, Maurizio Ferraris

Stochastic processes are quite different from deterministic ones for many reasons. First of all their variables are intrinsically probabilistic in nature, and that has a number of consequences even in the apparently most obvious calculus rules. For example, consider differentials and change of variables, or observables that are functions of other observables for the following process (Gardiner 1985):

$$\mathrm{d}x(t) = a(x(t), t)\mathrm{d}t + b(x(t), t)\mathrm{d}W(t)$$

where $a(x(t), t)$ is the drift term and $b(x(t), t)$ the noise term. This stochastic equation has a formal solution given by

$$x(t) = x(t_0) + \int_{t_0}^{t} dt' a(x(t'), t')dt + \int_{t_0}^{t} dt' b(x(t'), t')dW(t') \tag{D.1}$$

In the case of the "grain of pollen" of the Brownian motion in one dimension, the corresponding differential equation for the velocity v becomes

$$\mathrm{d}v(t) = -\gamma v\mathrm{d}t + \mathrm{d}W \tag{D.2}$$

$$\text{so that } \; a = -\gamma v; \;\; b = 1 \tag{D.3}$$

Let us now examine a new variable function of the process, say $f(x)$ e.g., $f(v) = v^2$. Naively, since $\mathrm{d}v^2 = 2v dv$ one could directly substitute this in Eq. D.2 to obtain

C. Mejía-Monasterio and L. Rondoni, *Nonequilibrium Statistical Mechanics I*, Theoretical and Mathematical Physics,
https://doi.org/10.1007/978-3-031-98249-1

that v^2 follows the equation $dv^2 = 2v(-\gamma v dt + dW)$. To realize why this is not correct, let us expand df to second order in dW:

$$\begin{aligned} \mathrm{d}f(x(t)) &= f(x(t)+\mathrm{d}x(t)) - f(x(t)) = \\ &= f'(x(t))\mathrm{d}x(t) + \frac{1}{2}f''(x(t))(\mathrm{d}x(t))^2 + \cdots = \\ &= f'(x(t))[a(x(t),t)\mathrm{d}t + b(x(t),t)\mathrm{d}W(t)] + \frac{1}{2}f''(x(t))b^2(x(t),t)\mathrm{d}W^2(t) + \cdots \end{aligned}$$

The second-order terms in $\mathrm{d}W$ do not need to be negligible, while at the same time, $\mathrm{d}t^\alpha$, for $\alpha > 1$ is negligible. Indeed, one can prove that $dW^2 = dt$ (Gardiner 1985). Moreover, using the independence of x and W, averages involving terms with differential $\mathrm{d}x\mathrm{d}W$ yield 0. It also follows that one can take $\mathrm{d}t\mathrm{d}W$ as $\mathrm{d}t^{3/2}$ which is negligible compared to $\mathrm{d}t$.

We can now write the equation for the variable $v^2(t)$, in the sense of Itô[4]:

$$\mathrm{d}v^2(t) = 2v(t)\mathrm{d}v + \mathrm{d}W^2(t) = \tag{D.4}$$

$$= 2v(t)(-\gamma v\mathrm{d}t + \mathrm{d}W) + \mathrm{d}W^2(t) = \tag{D.5}$$

$$= (1 - 2\gamma v(t)^2)\,\mathrm{d}t + \sqrt{v(t)^2}\,\mathrm{d}W(t) \tag{D.6}$$

[4] Stratonovich is considered natural for real (non-white) noise, with finite correlation time τ, but it makes calculations almost impossible. However, for a precise limit $\tau \to 0$ there is equivalence.

Boltzmann Equation and H-Theorem: Detailed Derivation

E

According to this theory a dissolved molecule is differentiated from a suspended body solely by its dimension

Investigations on the theory of the Brownian Movement, Albert Einstein

In this Appendix, a detailed derivation of the Boltzmann equation and of the H-Theorem is given, following the approach of Ref. Cercignani et al. (1994). Consider a system of N identical rigid spheres of diameter $\sigma > 0$ and mass $m = 1$, subjected to no external forces, and interacting only via elastic collisions, when the distance between their centers equals σ. Neglecting rotational contributions to the dynamics of the system, because they are unaffected by the collisions, the phase space of the system is $6N$-dimensional and can be written as $\mathscr{M} = V^N \times \mathbb{R}^{3N}$, where $V \subseteq \mathbb{R}^3$ is the container within which the particles are confined. One point $\Gamma = (x_1, ..., x_N, \xi_1, ..., \xi_N) \in \mathscr{M}$, with $x_i \in V$ and $\xi_i \in \mathbb{R}^3$, $i = 1, ..., N$, has $3N$ components for the N positions x_i of the particles, and $3N$ components for their N velocities ξ_i. One such point describes one microscopic state, or phase of the system. Given the initial condition $\Gamma_0 \in \mathscr{M}$, let us denote by $\Gamma = \Phi^t \Gamma_0 \in \mathscr{M}$ its position at time t, where $\Phi^t : \mathscr{M} \to \mathscr{M}$ is the flow operator. Let $P(\Gamma, t)$ be a probability density on $\mathscr{M}$, that evolves with the dynamics, as described in Chap. 6. Between two collisions, the particles move along straight lines with constant speed, hence Liouville's equation takes the form:

$$\frac{\partial P}{\partial t} + \sum_{i=1}^{N} \xi_i \cdot \frac{\partial P}{\partial x_i} = 0 \tag{E.1}$$

where there is no contribution from the gradient of P with respect to the velocities, because the accelerations vanish, due to the absence of forces. When a collision takes

C. Mejía-Monasterio and L. Rondoni, *Nonequilibrium Statistical Mechanics I*, Theoretical and Mathematical Physics,
https://doi.org/10.1007/978-3-031-98249-1

place, the positions do not change, while the velocities change discontinuously, therefore Eq. (E.1) stops, and immediately restarts with post-collision velocities. Collisions occur on a lower dimensional subset of the phase space, identified by the condition that the distance among the centers of the colliding particles equals σ, and they occur at discrete times. Thus, no mathematical difficulty is associated with them: it suffices to consider the following. Because collisions are elastic, mass, momentum, and energy are conserved. Therefore, the variation of velocities they induce obey the following equations:

$$\begin{aligned} \xi_1 + \xi_2 &= \xi_1^{'} + \xi_2^{'} \\ |\xi_1|^2 + |\xi_2|^2 &= |\xi_1^{'}|^2 + |\xi_2^{'}|^2 \end{aligned} \tag{E.2}$$

where ξ_1 and ξ_2 denote the velocities of particles 1 and 2 after the impact, and ξ_1' and ξ_2' those before the impact. Then, letting

$$n_{12} = \frac{x_1 - x_2}{|x_1 - x_2|} \tag{E.3}$$

be the unit vector along the line joining the centers of the colliding spheres, the conservation laws imply:

$$\begin{aligned} \xi_1^{'} &= \xi_1 - n\,(n \cdot (\xi_1 - \xi_2)) \\ \xi_2^{'} &= \xi_2 + n\,(n \cdot (\xi_1 - \xi_2)) \end{aligned} \tag{E.4}$$

As particles i and j cannot overlap, P vanishes at the phase points with $|x_i - x_j| < \sigma$, $\forall i \neq j$, and it is convenient to introduce the subset $U \subset \mathcal{M}$ of the phases with no particles overlap, which obey $|x_i - x_j| \geq \sigma$ for all pairs of particles. The boundary conditions

$$P(\Gamma, t) = P(\Gamma^{'}, t), \quad \text{for } \Gamma \in \partial U \tag{E.5}$$

must then be introduced, with Γ' indicating the microscopic phase immediately before collision, and Γ that immediately after. Explicitly, this means:

$$\begin{aligned} &P(x_1, \xi_1, ..., x_i, \xi_i, ..., x_j, \xi_j, ..., x_N, \xi_N, t) = \\ &P(x_1, \xi_1, ..., x_i, \xi_i - n_{ij}\left(n_{ij} \cdot (\xi_i - \xi_j)\right), ..., x_j, \xi_j + n_{ij}\left(n_{ij} \cdot (\xi_i - \xi_j)\right), ..., x_N, \xi_N, t) \end{aligned} \tag{E.6}$$

for $|x_i - x_j| = \sigma, \forall i \neq j$, where $n_{ij} = (x_i - x_j)/|x_i - x_j|$.

E.0.1 Boltzmann Equation

The quantity $P(\Gamma; t)$ represents the probability density for finding the N spheres around the point $\Gamma \in \mathcal{M}$ at time t. Integrating it over the positions and velocities of all particles from 2 to N produces the marginal distribution

$$P^{(1)}(x_1, \xi_1, t) = \int P(x_1, ..., x_N, \xi_1, ..., \xi_N, t) \prod_{i=1}^{N} dx_i d\xi_i \tag{E.7}$$

that represents the probability density for finding particle 1 around the point (x_1, ξ_1) at time t. The evolution of $P^{(1)}$ is obtained explicitly taking into consideration the effect of collisions, which requires knowledge of the probability density $P^{(2)}(x_1, \xi_1, x_2, \xi_2, t)$ of finding particle 1 in position x_1 with velocity ξ_1 and particle 2 in position x_2 with velocity ξ_2 at time t, where (x_1, ξ_1, x_2, ξ_2) corresponds to a collision point. Given a small volume Δx_1 in V around x_1, and a small volume $\Delta\xi_1$ in $\mathbb{R}^3$ around ξ_1, the probability of finding particle 1 in it is given by $P^{(1)}(x_1, \xi_1, t)\Delta x_1 \Delta\xi_1$. It varies because points flow in and out of Δx_1 and because at collisions ξ_1 can be pushed inside $\Delta\xi_1$ from outside, or kicked out of $\Delta\xi_1$ from inside. Therefore, in the small volume limit, conservation of probability leads to the following equation:

$$\frac{\partial P^{(1)}}{\partial t} + \xi_1 \cdot \frac{\partial P^{(1)}}{\partial x_1} = G - L$$

where $L\Delta x_1 \Delta\xi_1$ and $G\Delta x_1 \Delta\xi_1$ respectively are the expected numbers of particles that leave and that enter the volume $\Delta\xi_1$, in the unit time, because of collisions. When particles 1 and 2 collide, the event takes place on a surface area element of particle 1 centered around the normal n_{12}, of size $\sigma^2 dn_{12}$.

Consider the cylinder with this surface element as its basis, and height given by $|\xi_2 - \xi_1|dt$, directed along $\xi_2 - \xi_1$. Its volume is $\sigma^2|\xi_2 - \xi_1|dtdn_{12}$. The probability of collisions is then given by the probability that particles lie inside this cylinder, at time t. Particle 1 can collide with $(N-1)$ identical particles, not only with particle 2, so the balance equation for particle 1 includes collisions with any of them. Also, collisions of three and more particles at once can be neglected, because they correspond to a lower dimensional subset of the binary collisions subset of U, hence their probability vanishes, compared to that of binary collisions. Therefore, we can write $L = (N-1)l$ and $G = (N-1)g$ where l indicates the contribution of a single particle, which is equal to that of any other particle, including particle 2. Then, we integrate over the whole surface, which means over all vectors n_{12}, and we integrate over all possible velocities of sphere 2. For sake of simplicity, we use the symbol n in place of n_{12}, and we consider the following quantity:

$$l dx_1 d\xi_1 dt = dx_1 d\xi_1 dt \int_{\mathbb{R}^3} \int_{\mathbb{S}^2_-} P^{(2)}(x_1, \xi_1, x_1 + \sigma n, \xi_2, t)|(\xi_2 - \xi_1) \cdot n|\sigma^2 dn d\xi_2 \tag{E.8}$$

where $\mathbb{S}^2_-$ is the hemisphere corresponding to $(\xi_2 - \xi_1) \cdot n < 0$, which means that particles are moving toward each other, before the collision, and $P^{(2)}$ is the probability density of having two particles in x_1 and $x_1 + \sigma n$, with velocity ξ_1 and ξ_2 respectively. The result is

$$L = (N-1)\sigma^2 \int_{\mathbb{R}^3} \int_{\mathbb{S}^2_-} P^{(2)}(x_1, \xi_1, x_2, \xi_2, t)|(\xi_2 - \xi_1) \cdot n| dn d\xi_2 \tag{E.9}$$

Analogously, we obtain

$$G = (N-1)\sigma^2 \int_{\mathbb{R}^3} \int_{\mathbb{S}^2_+} P^{(2)}(x_1, \xi_1, x_2, \xi_2, t)|(\xi_2 - \xi_1) \cdot n| dn d\xi_2$$

for the gain term G, where $\mathbb{S}^2_+$ is the hemisphere corresponding to $(\xi_2 - \xi_1) \cdot n > 0$, which means that the particles are moving away from each other after the collision.

Now, a rarefied gas is characterized by large N and small σ, so that particles collide, but the volume they occupy is negligible. Mathematically, this corresponds to the Boltzmann-Grad limit: $N \to \infty, \sigma \to 0$ with positive total cross section $N\sigma = const$, that implies vanishing occupied volume in the $N \to \infty$ limit: $N\sigma^{3/2} \to 0$. Then, particles can be considered independent before they collide, the molecular chaos hypothesis

$$P^{(2)}(x_1, \xi_1, x_2, \xi_2, t) = P^{(1)}(x_1, \xi_1, t)P^{(1)}(x_2, \xi_2, t) \tag{E.10}$$

appears justified, and one may write

$$L = N\sigma^2 \int_{\mathbb{R}^3} \int_{\mathbb{S}^2_-} P^{(1)}(x_1, \xi_1, t)P^{(1)}(x_2, \xi_2, t)|(\xi_2 - \xi_1) \cdot n| dn d\xi_2 \tag{E.11}$$

$$G = N\sigma^2 \int_{\mathbb{R}^3} \int_{\mathbb{S}^2_+} P^{(1)}(x_1, \xi_1^{'}, t)P^{(1)}(x_2, \xi_2^{'}, t)|(\xi_2 - \xi_1) \cdot n| dn d\xi_2 \tag{E.12}$$

For the gain term G, Eq. (E.12) follows interpreting $\xi_1^{'}, \xi_2^{'}$ as velocities entering the hemisphere $(\xi_2^{'} - \xi_1^{'}) \cdot n < 0$, considering that $(\xi_2 - \xi_1) \cdot n = -(\xi_2^{'} - \xi_1^{'}) \cdot n$, and that the transformation $(\xi_1, \xi_2) \to (\xi_1^{'}, \xi_2^{'})$ preserves volumes: $d\xi_1 d\xi_2 = d\xi_1^{'} d\xi_2^{'}$. The Boltzmann equation follows:

$$\frac{\partial P^{(1)}}{\partial t} + \xi_1 \cdot \frac{\partial P^{(1)}}{\partial x_1} = N\sigma^2 \times \tag{E.13}$$

$$\int_{\mathbb{R}^3} \int_{\mathbb{S}^2_-} \left(P^{(1)}(x_1, \xi_1^{'}, t)P^{(1)}(x_2, \xi_2^{'}, t) - P^{(1)}(x_1, \xi_1, t)P^{(1)}(x_2, \xi_2, t) \right) |(\xi_2 - \xi_1) \cdot n| dn d\xi_2$$

E.0.2 Collision Integral and Collision Invariants

Denoting by $f = f(x, \xi, t)$, $f_* = f(x_*, \xi_*, t)$ the densities before a collision and by $f^{'} = f(x^{'}, \xi^{'}, t)$, $f_*^{'} = f(x_*^{'}, \xi_*^{'}, t)$ those after the collision, the Boltzmann equation writes

$$\frac{\partial f}{\partial t} + \xi \cdot \nabla_x f = N\sigma^2 Q(f, f) = \alpha \int_{\mathbb{R}^3} \int_{\mathbb{S}^2_-} \left(f^{'} f_*^{'} - f f_* \right) |n \cdot (\xi - \xi_*)| dn d\xi_* \tag{E.14}$$

with $\alpha = N\sigma^2$, and the collision operator Q defined by

$$Q(f, g) = \frac{1}{2} \int_{\mathbb{R}^3} \int_{\mathbb{S}^2_+} \left(f^{'} g_*^{'} + g^{'} f_*^{'} + f g_* - g f_* \right) |n \cdot (\xi - \xi_*)| dn d\xi_* \tag{E.15}$$

Take an observable $\phi = \phi(\xi)$ such that the integral

$$\int_{\mathbb{R}^3} Q(f,g)\phi(\xi)d\xi = \frac{1}{2}\int_{\mathbb{R}^3}\int_{\mathbb{R}^3}\int_{\mathbb{S}^2_+} \left(\left(f^{'}g^{'}_* + g^{'}f^{'}_* + fg_* - gf_*\right)\phi(\xi)\right)|n\cdot(\xi-\xi_*)|dnd\xi_*d\xi \tag{E.16}$$

exists and the order of integration does not matter. Swapping the starred and unstarred variables, the symmetry of the integrand yields

$$\int_{\mathbb{R}^3} Q(f,g)\phi(\xi)d\xi = \frac{1}{2}\int_{\mathbb{R}^3}\int_{\mathbb{R}^3}\int_{\mathbb{S}^2_+} \left[\left(f^{'}g^{'}_* + g^{'}f^{'}_* + fg_* - gf_*\right)\phi(\xi_*)\right]|n\cdot(\xi-\xi_*)|\,dnd\xi d\xi_* \tag{E.17}$$

and also

$$\begin{aligned}\int_{\mathbb{R}^3} Q(f,g)\phi(\xi)d\xi &= \\ &\frac{1}{2}\int_{\mathbb{R}^3}\int_{\mathbb{R}^3}\int_{\mathbb{S}^2_+} \left[\left(fg_* + gf_* + f^{'}g^{'}_* - g^{'}f^{'}_*\right)\phi(\xi^{'})\right]|n\cdot(\xi^{'}-\xi^{'}_*)|dnd\xi^{'}d\xi^{'}_* \\ &= \frac{1}{2}\int_{\mathbb{R}^3}\int_{\mathbb{R}^3}\int_{\mathbb{S}^2_+} \left[\left(fg_* + gf_* + f^{'}g^{'}_* - g^{'}f^{'}_*\right)\phi(\xi^{'})\right]|n\cdot(\xi-\xi_*)|dnd\xi d\xi_*\end{aligned} \tag{E.18}$$

where the last equality follows from the fact that $(\xi-\xi_*)\cdot n = -(\xi^{'}-\xi^{'}_*)\cdot n$, and that the volume is conserved by the transformation $(\xi,\xi_*)\to(\xi^{'},\xi^{'}_*)$, so that $d\xi d\xi_* = d\xi^{'}d\xi^{'}_*$. Because this integral is again symmetric under the exchange of starred with non-starred variables, we may write

$$\int_{\mathbb{R}^3} Q(f,g)\phi(\xi)d\xi = \frac{1}{2}\int_{\mathbb{R}^3}\int_{\mathbb{R}^3}\int_{\mathbb{S}^2_+} \left[\left(fg_* + gf_* + f^{'}g^{'}_* - g^{'}f^{'}_*\right)\phi(\xi^{'}_*)\right]|n\cdot(\xi-\xi_*)|dnd\xi d\xi_* \tag{E.19}$$

Then, summing the four equations and dividing by 4, we obtain

$$\int_{\mathbb{R}^3} Q(f,g)\phi(\xi)d\xi = \frac{1}{8}\int_{\mathbb{R}^3}\int_{\mathbb{R}^3}\int_{\mathbb{S}^2_+} \left(fg_* + gf_* + f^{'}g^{'}_* - g^{'}f^{'}_*\right)\left(\phi + \phi_* - \phi^{'} - \phi^{'}_*\right)|n\cdot(\xi-\xi_*)|dnd\xi d\xi_* \tag{E.20}$$

which vanishes, independently of the functions f and g, if

$$\phi + \phi_* = \phi^{'} + \phi^{'}_* \tag{E.21}$$

The functions ϕ satisfying this relation are called collision invariants, and are conserved in time by the evolution determined by the Boltzmann equation.

Depending on the boundary conditions, in a three-dimensional space, one may have five conserved quantities: the number of particles, the three components of the total momentum and the total energy. This is the case, for example, when the gas is free to move in the infinite space, and the flow of its properties out of a sphere of radius R vanishes for $R \to \infty$. It is also true in periodic boundary conditions. For a gas within a compact domain with reflecting walls, momentum is not conserved, so there are only two collision invariants. Boltzmann assumed ϕ to be differentiable twice and concluded that the most general solution of the relation is given by

$$\phi(\xi) = A + B \cdot \xi + C|\xi|^2$$

A fundamental result is the following. Let $P^{(s)}(x_1, \xi_1, ..., x_s, \xi_s)$ be the initial condition for the s-particles distribution, i.e., the marginal of the phase space initial probability density, in which the coordinates and velocities of particles $s + 1, ..., N$ have been integrated out, and consider

$$f^{(s)}(x_1, \xi_1, ..., x_s, \xi_s, t) = \prod_{i=1}^{s} f(x_i, \xi_i, t)$$

be the product of s copies of the solution f of the Boltzmann equation. In the Boltzmann-Grad limit, and under appropriate conditions, loosely speaking, Lanford proved that (Lanford 1981)

$$\lim_{N\to\infty} P^{(2)}(x, \xi_1, x + n\sigma, \xi_2, t) = f(x, \xi_1, t) f(x, \xi_2, t) \tag{E.22}$$

$$\lim_{N\to\infty} P^{(1)}(x, \xi, t) = f(x, \xi, t) \tag{E.23}$$

for $t \in [0, t_0]$, where $t_0 = \tau/5$ is one fifth of the mean free time τ. Later, the value of t_0 was extended to half a mean free time. This means that for at least a fraction of time, the evolution of the projected solution of Liouville's equation, $P^{(1)}$, follows, in the Boltzmann-Grad limit, the solution of the Boltzmann equation. As half of the N particles collide in the time interval $[0, t_0]$, this result is quite satisfactory to prove that the Boltzmann molecular chaos hypothesis does not contradict classical mechanics, provided the system is sufficiently large, in particular when it is macroscopic.

Conceptually, the 1-particle distribution, which is a projection of the phase space ensemble, represents the quantity of ensemble members whose particle 1 lies close to x_1 with velocity close to ξ_1, whatever positions and velocities the remaining particles have. Identifying this probability with the fraction of particles in a volume element, one gets an equation for the evolution of the density of mass. As discussed in Chap. 6, this is a delicate step to take, but it was worth trying, and proved to be successful.

E.1 Entropy and H-Theorem

Let us begin with the Boltzmann inequality:

$$\int_{\mathbb{R}^3} \log(f) Q(f, f) d\xi \leq 0 \tag{E.24}$$

where equality holds if and only if $\log(f)$ is a collision invariant. This means that

$$f = \exp\left(a + b \cdot v + c|\xi|^2\right), \quad \text{with} \quad a, c \in \mathbb{R}, \quad b \in \mathbb{R}^3$$

One can indeed write (Cercignani et al. 1994):

$$\int_{\mathbb{R}^3} \log(f) Q(f, f) d\xi = 1/4 \int_{\mathbb{R}^3} \int_{\mathbb{R}^3} \int_{\mathbb{S}^2_+} \log(f f_* / f^{'} f^{'}_*)(f^{'} f^{'}_* - f f_*)|(\xi - \xi_*) \cdot n| d\xi_* dn \tag{E.25}$$

and use the following inequality:

$$(x - y) \log(y/x) \leq 0, \quad \forall x, y \in \mathbb{R}^+ \tag{E.26}$$

easily derived from the fact that the terms $(x - y)$ and $\log(y/x)$ have opposite signs if $x \neq y$, and that one can set $(x - y) \log(y/x) = 0$ for $x = y$ It follows that Eq. (E.24) holds if and only if f is a Maxwellian. Indeed, given f such that

$$f^{'} f^{'}_* = f f_* \tag{E.27}$$

one obtains

$$\log(f^{'}) + \log(f^{'}_*) = \log(f) + \log(f_*) \tag{E.28}$$

which is verified if and only if (Cercignani et al. 1994):

$$\log(f) = a + b \cdot v + c|\xi|^2 \tag{E.29}$$

and setting $c = -\beta$, $b = 2\beta v$ with v constant vector, we obtain the Maxwell distribution:

$$f = A \exp(-\beta|\xi - v|^2)$$

with A the normalization constant, and the vector v as mean velocity of the gas particles:

$$\mathbb{E}[\xi] = \frac{\int_{\mathbb{R}^3} \xi f d\xi}{\int_{\mathbb{R}^3} f d\xi} = v \tag{E.30}$$

E.1.1 H-Theorem

Defining the functional $\mathscr{H}$ as

$$\mathscr{H} = \int_{\mathbb{R}^3} f \log(f) d\xi \tag{E.31}$$

and introducing its integral over the volume V as

$$H(f) = \int_V \mathscr{H} dx = \int_{V \times \mathbb{R}^3} f \log(f)\, d\xi dx \tag{E.32}$$

the H-theorem states

$$\frac{d}{dt} H(f) = \frac{d}{dt} \int_V \mathscr{H} dx \leq 0 \tag{E.33}$$

Indeed, we have

$$\begin{aligned} \frac{d}{dt} H(f) &= \frac{d}{dt} \int_{V \times \mathbb{R}^3} f \log(f)\, d\xi dx = \int_{V \times \mathbb{R}^3} \frac{\partial}{\partial t} f \log(f)\, d\xi dx \\ &= \int_{V \times \mathbb{R}^3} \Big[1 + \log(f)\Big] \frac{\partial f}{\partial t} d\xi dx \end{aligned} \tag{E.34}$$

When f is solution of the Boltzmann equation, the partial derivative with respect to time is given by that equation, and we can write

$$\frac{d}{dt} H(f) = \int_{V \times \mathbb{R}^3} \Big[1 + \log(f)\Big] (-\xi \cdot \nabla_x f + Q(f, f))\, d\xi dx \tag{E.35}$$

Then, imposing, for example, periodic boundary conditions, one gets

$$\int_V -\xi \cdot \nabla_x f dx = 0 \tag{E.36}$$

and

$$\int_{V \times \mathbb{R}^3} \xi \cdot \nabla_x f \log(f) dx d\xi = \int_{V \times \mathbb{R}^3} \xi \cdot \nabla_x \left(f \log(f) - f\right) dx d\xi = 0 \tag{E.37}$$

Furthermore, the integral of Q vanishes

$$\int_{V \times \mathbb{R}^3} Q(f, f) = 0 \tag{E.38}$$

since $\phi^0 = 1$ obviously is a collision invariant. Therefore, we can write

$$\frac{d}{dt} H(f) = \int_{V \times \mathbb{R}^3} \log(f) Q(f, f) d\xi dx \leq 0 \tag{E.39}$$

where the equality holds if and only if f is a Maxwellian.

The H-theorem shows that the Boltzmann equation has an irreversible behavior, which looks in contrast with the time reversal invariant laws of classical mechanics governing the motion of particles. In reality, various approximations and hypotheses, that are extraneous to the dynamics of an N particle system, have been introduced to pass from the evolution of probabilities in phase space to the Boltzmann equation. In particular, the N particle system has been approximated by a system consisting of infinitely many particles, through the Boltzmann-Grad limit, and this has invited the molecular chaos hypothesis that substantially changed the nature of the dynamics, breaking its original time reversibility. The point is that approximations and molecular chaos perfectly represent the physics of the systems of interest: rarefied gases in first instance, but many more.

Ensembles Theory

F

As soon as you begin to understand the immensity of Super K, you will realize that just being finite isn't much of a limitation [...] Infinity is a red herring.

Things a computer scientist rarely talks about,

Donald E. Knuth

In this Appendix we summarize a few facts about the theory of ensembles, which are relevant for the subject of this book. The subject is quite technical, hence we refer the interested readers to the relevant chapters of Refs. Gallavotti (1999) and Gallavotti (2014), which we summarize in part. As we will see, although ensembles refer to particle systems with dynamics characterized by their energy, their derivation makes no use of the dynamics. The original purpose of ensembles is indeed to model equilibrium thermodynamics, and for that it suffices to avoid contradictions with the dynamical properties of the systems of interest. Derivations from the dynamics are object of ergodic theory, but as we have seen in this book, they meet insurmountable difficulties, except in special cases. Nevertheless, the success of statistical mechanics is deeply rooted in the applicability of ensembles. In brief, the incredibly large number of degrees of freedom of the microscopic constituents of macroscopic objects makes the statistics overwhelm the details of the microscopic dynamics (Chibbaro et al. 2014). This can be appreciated under various lights. Below, within the realm of classical mechanics, we focus on the equivalence of ensembles in the thermodynamic limit. This way, we demonstrate that different dynamics result in the same macroscopic properties, confirming that, under certain conditions, the fine details of the dynamics can be safely and profitably neglected. This approach is ultimately justified by experience, by the fact that ensembles yield the correct thermodynamic properties.

C. Mejía-Monasterio and L. Rondoni, *Nonequilibrium Statistical Mechanics I*, Theoretical and Mathematical Physics,
https://doi.org/10.1007/978-3-031-98249-1

Given their success, ensembles are also used more and more to investigate the properties of small or far from equilibrium systems. In these cases, one should be cautious, since the bounds of thermodynamics may have been exceeded, and the properties of the dynamics may be crucial.

F.1 Statistical Ensembles and Thermodynamics

Consider a system made of N identical atoms, for simplicity, and denote their coordinates and momenta by $\Gamma = \{(q_i, p_i)\}_{i=1}^N \in \mathcal{M}$, q_i and p_i being $d-$dimensional vectors, and $\mathcal{M}$ the phase space. If each coordinate and momentum can be measured with at best an accuracy δq and δp, a given microstate is one point hidden in a cell Δ of volume $h^{dN} = \delta q^{dN} \delta p^{dN}$.[5] Then, a theory describing such measurements has Δ as its basic element, and the measurable properties of the microstate Γ are properties of the cell Δ to which Γ belongs. In particular, this parallels the facts described in Sects. 2.1 and 2.5, which explain that the quantities obeying the thermodynamic differential equations make sense for measurements performed within space and time scales much larger than the microscopic ones. Consequently, the energy of a Hamiltonian system whose particles interact via a central potential φ must be written as

$$E(\Gamma) = K(p) + \phi(q) = \sum_{i=1}^{N} \frac{p_i^2}{2m} + \frac{1}{2} \sum_{ij=1}^{N} \varphi(q_i - q_j) := E(\Delta) \qquad \text{(F.1)}$$

where it is understood that it makes no sense to distinguish the energy of different points within the same cell Δ, consistently with the thermodynamic level of description. Of course, one may be legitimately interested in a microscopic description which probes finer scales than that of Δ. In that case, a different theory applies. Remaining in our present framework, one realizes that a stability condition must be satisfied: the energy must be bounded from below:

$$E(\Gamma) \geq U^0 := \min_{\Gamma} E(\Gamma) > BN\,, \quad B \in \mathbb{R} \qquad \text{(F.2)}$$

and we assume that to be the case. Consider now an invariant probability distribution μ that attributes a probability $\mu(\Delta)$ to each cell $\Delta \subset \mathcal{M}$ that identifies a microstate. The collection $\mathcal{E}$ of the invariant probabilities on $\mathcal{M}$, each of which attributes given values to all the cells Δ whose union makes $\mathcal{M}$, represents a family of macroscopic equilibrium states. Gibbs called *ensemble* one element of $\mathcal{E}$; Boltzmann called it *monode*. Given one such probability $\mu \in \mathcal{E}$, i.e., given one equilibrium thermodynamic state, an observable quantity $\mathcal{O}$ is defined as the average of a function of the cells, $f_{\mathcal{O}}$:

$$\mathcal{O}(\mu) := \mathbb{E}_\mu [\mathcal{O}] = \sum_{\Delta \subset \mathcal{M}} \mu(\Delta) f_{\mathcal{O}}(\Delta) \qquad \text{(F.3)}$$

[5] Despite the handy symbol, h has no quantum content, because macroscopic measurements involve scales much larger than the microscopic ones.

The main quantities of thermodynamic interest are then expressed by

$$
\begin{aligned}
\text{internal energy} &\rightarrow U(\mu) = \sum_{\Delta} \mu(\Delta) E(\Delta) \\
\text{volume} &\rightarrow V(\mu) = f_V(\Delta) \\
\text{kinetic energy} &\rightarrow K(\mu) = \sum_{\Delta} \mu(\Delta) f_K(\Delta) \\
\text{pressure} &\rightarrow P(\mu) = \sum_{\Delta} \mu(\Delta) f_P(\Delta)
\end{aligned} \tag{F.4}
$$

where f_V is the volume available to the system (which is the same for all $\Delta \subset \mathscr{M}$), f_K is the sum of the kinetic energies of the particles in Δ, and f_P is the variation of momentum of the particles per unit time and unit area, due to collisions with the walls of the container. The above definitions are considered appropriate if they lead to the known thermodynamic relations. In particular, varying the ensemble μ within the family $\mathscr{E}$, the following must hold:

$$
\frac{dU + PdV}{T} = \text{exact differential} = dS \tag{F.5}
$$

where T is proportional to the kinetic energy: $T = 2K(\mu)/3Nk$, with k empirically determined, and $S = S(\mu)$ can be interpreted as the function of state known as *entropy*. Ensembles for which this condition is satisfied were called *orthodes* by Boltzmann. The family of *canonical* probabilities is made of probabilities taking the exponential form:

$$
\mu(\Delta) = \frac{e^{-\beta E(\Delta)}}{Z(\beta, V)} \quad \text{with} \quad Z(\beta, V) = \sum_{\Delta} e^{-\beta E(\Delta)} \tag{F.6}
$$

where the normalization $Z(\beta, V)$ is called *canonical partition function*. The family of *microcanonical* probabilities is made of uniform probabilities taking the form:

$$
\mu(\Delta) = \begin{cases} \dfrac{1}{\mathscr{N}(U, V)} & \text{if} \quad E(\Delta) \in [U - \delta E, U] \\ 0 & \text{else} \end{cases} \tag{F.7}
$$

where $\delta E \ll U$, and

$$
\mathscr{N}(U, V) = \sum_{\Delta : E(\Delta) \in [U - \delta E, U]} 1 \tag{F.8}
$$

the number of cells with energy in $[U - \delta E, U]$, is called *microcanonical partition function*.

Note: the above are not only discrete, but also finite sums, because the energy of physical systems, including physical baths, is finite; they are not integrals. For sufficiently small cells, and only under proper conditions, which fundamentally limit the choice of the functions to be averaged, they can be approximated by integrals. Unlike common opinions, the discrete version of the above formulae is the exact representation, and the continuous version is but an approximation, which makes calculations easier to perform. In the case, the approximation is accurate, which requires observables to be slowly varying, the canonical partition function can be written as

$$Z(\beta, V) = \int \mathrm{d}q\mathrm{d}p \frac{1}{N!h^{3N}} e^{-\beta[K(p)+\Phi(q)]} \tag{F.9}$$

where $h = \delta q_i \delta p_i$, the single particle cell, is the same for all i if particles are identical, and gives the full phase space cell when raised to $3N$ for particles with 3 degrees of freedom. In turn, the factor $N!$ accounts for the permutations allowed when particles are indistinguishable. In this framework, a configuration Δ is identified by the numbers of particles, n_i, in the cubes c_i of size h^3, in which a (six-dimensional) point represents a particle.[6] Replacing the sums by the integrals, one makes two errors:

1. an analytic error due to the fact that one point (q, p), e.g., the center of Δ, is required to represent the cell Δ, and an analytic error if the continuum expression extends to infinity in positions or momenta;
2. a combinatorial error due to the fact that in the sums, hence in the integrals when they are good approximations of the sums, the state described by $n_1, n_2, ...$ is counted $N!/(n_1!n_2!...)$ times, instead of $N!$ times, because permutations within the cells are neglected by the definition of minimal measurement cell, or maximum measurement accuracy.

In the limit of small h, the analytical error becomes negligible for averages of functions that have a small variation in all cells, i.e., everywhere in $\mathscr{M}$ or, at least, in its parts *"of interest."* The small h limit also solves the combinatorial problem if most of the cells are then occupied by 0 or by a single particle. This analysis is rather technical and goes beyond the scope of the present brief account. Therefore we refer the readers to Ref. Gallavotti (1999) in particular. What matters is being consistent with experience, something that can be checked *a posteriori*, comparing theoretical and experimental results. The conclusion for a perfect gas, having in mind smooth phase functions such as those considered in Eq. (F.4), is the following: the temperature must be sufficiently high:

$$T > \frac{h^2}{mk_B} \left(\frac{N}{V} \right)^{2/3} \tag{F.10}$$

and each component of δq must be smaller than the mean molecular distance, while each component of δp must be smaller than the average momentum of a particle.

[6] The practical and conceptual difference between the six-dimensional (real) cubes $\delta q \delta p$ and the $6N$-dimensional (abstract phase space) cells Δ is remarkable.

This does not contradict the fact that measurements concern much larger scales, since it only refers to the mathematical needs for the continuous description to be safely used as an approximation of the discrete one. The physical and the mathematical needs are reconciled if the phase functions of interest negligibly vary over the cells of experimental interest, which is to say that gradients of such quantities must be correspondingly small, as required also in Sects. 2.1 and 2.5. Restricting to the cells of equilibrium states, this is obviously realized, and the combinatorial error may then be absorbed as an irrelevant constant. As a concrete example, take $N = 2.7\,10^{19}$ atoms of hydrogen, having mass $m = 3.34\,10^{-24}$ g, in a volume of $1\,\text{cm}^3$. Then, taking $h = 6.62\,10^{-27}$ erg/K leads to the conclusion that it suffices to have $T > 1$ K. Although that bound varies wildly when changing materials and conditions, this is a consoling result.[7] Let us now assume that all conditions are met. In the canonical ensemble we may thus write

$$\begin{aligned} K(\mu) &= \int \mathrm{d}p\mathrm{d}q \frac{e^{-\beta[K(p)+\Phi(q)]}}{N!h^{3N}Z(\beta, V}) \sum_{i=1}^{N} \frac{p_i^2}{2m} \\ v &= \frac{V}{N} \\ U(\mu) &= -\frac{\partial}{\partial \text{fi}} \log Z(\beta, V) \\ P(\mu) &= \sum_{Q} N \int \mathrm{d}q_2 ... \mathrm{d}q_N \mathrm{d}p_1 ... \mathrm{d}p_N \frac{e^{-\beta[K(p)+\Phi(q)]}}{Z(\beta, V)N!h^{3n}} 2mv^2 \frac{a}{A} \end{aligned} \tag{F.11}$$

where the sum runs over the small cubes Q touching the boundary of V in a surface of area a, $A = \sum_Q a$ is the area of the (cubic) container, and q_1 is the center of Q. This then yields

$$P = \frac{1}{\beta} \frac{\partial}{\partial \text{V}} \log Z(\beta, V) \tag{F.12}$$

which means that, introducing

$$F = -\frac{1}{\beta} \log Z(\beta, V) \quad \text{and} \quad S = \frac{U - F}{T} \tag{F.13}$$

one obtains

$$T = \frac{2\,K(\mu)}{3k_B N} = \frac{1}{k_B \beta} \quad \Longrightarrow \quad \frac{\mathrm{d}T}{T} = -\frac{\mathrm{d}\beta}{\beta} \tag{F.14}$$

[7] The value chosen for h intentionally coincides with Planck's constant, as an undoubtedly small quantity. However, here it only represents a limit to our observations which have no limit, in principle, in the classical picture. When such a picture does not apply, an altogether different formalism is required.

That allows the derivation of the standard thermodynamic relations, for instance:

$$\begin{aligned} \mathrm{d}F &= \left[\frac{1}{\beta^2}\log Z(\beta, V) - \frac{1}{\beta}\frac{\partial \log}{\partial \mathrm{fi}} Z(\beta, V)\right]\mathrm{d}\beta - \frac{1}{\beta}\frac{\partial}{\partial \mathrm{V}}\log Z(\beta, V)\mathrm{d}V \\ &= -\frac{1}{\beta}\log Z(\beta, V)\frac{\mathrm{d}T}{T} + \frac{\partial \log}{\partial \mathrm{fi}} Z(\beta, V)\frac{\mathrm{d}T}{T} - P\mathrm{d}V \\ &= \frac{F-U}{T}\mathrm{d}T - P\mathrm{d}V = -S\mathrm{d}T - P\mathrm{d}V \end{aligned} \tag{F.15}$$

The desired expression follows:

$$T\mathrm{d}S = \mathrm{d}U + P\mathrm{d}V \tag{F.16}$$

if F is interpreted as the free energy of the system. Formally, the above requires no conditions on the value of N, which could then also be small. This is special to the canonical ensemble, and is not shared by the other ensembles, that explicitly require the thermodynamic limit. However, even here the possibility for N to be small is only formal for two reasons: in the first place, the thermodynamic quantities require N to be large, lest they make no sense.[8] Secondly, the whole probabilistic approach loses its strict connection with reality, if it concerns a small number of objects. Although probabilities may remain in many cases the best theoretical tool to use, in general, they will describe non-thermodynamic phenomena.

Let us briefly consider these facts in terms of the microcanonical ensemble. Up to the analytic and combinatorial errors discussed above, let us approximate the microcanonical partition function as the following integral:

$$\Omega(U, V) = \int_{J_E} \mathrm{d}p\mathrm{d}q \frac{1}{N!h^{3N}} \tag{F.17}$$

J_E being the subset of $\mathcal{M}$ such that $E(\Gamma) \in [U - \delta E, U]$. The thermodynamic quantities are now defined as in the canonical case, replacing $\exp(-\beta H)$ by 1, Z by Ω, and the integral over the whole space with an integral over J_E. The temperature is still defined as $T = 2K(\mu)/3kN$ but this time one starts from the entropy rather than the free energy, introducing:

$$S = k \log \Omega(U, V) \tag{F.18}$$

which implies

$$\mathrm{d}S = \frac{k}{\Omega(U, V)}\left[\frac{\partial \Omega}{\partial U}(U, V)\mathrm{d}U + \frac{\partial \Omega}{\partial \mathrm{V}}(U, V)\mathrm{d}V\right] \tag{F.19}$$

---

[8] This does not forbid considering generalized quantities that go beyond the thermodynamic ones; on the contrary, it opens very interesting perspectives. The question is whether such quantities obey the thermodynamic relations, in a robust fashion with respect to parameters variations. That must be assessed on a case by case basis.

Now, denote the full average of K^{-1} as

$$\mathbb{E}\left[K^{-1}\right] = \frac{1}{\int_{J_E} \mathrm{d}p\mathrm{d}q} \int_{J_E} \frac{\mathrm{d}p\mathrm{d}q}{K(p)} \tag{F.20}$$

and the one conditioned to one particle being in a region dV near the boundary of V as

$$\mathbb{E}\left[K^{-1}\right]^* = \frac{1}{\int_{J'_E} \mathrm{d}p\mathrm{d}q} \int_{J'_E} \frac{\mathrm{d}p\mathrm{d}q}{K(p)} \tag{F.21}$$

where the integral in J'_E is the subset of J_E with $q_1 \in dV$. Then, assuming the following conditions (Gallavotti 1999):

- stability: for every configuration $q = (q_1, ..., q_N)$, there exists B such that

$$\phi(q) = \sum_{ij=1}^{N} 1/2\varphi(q_i - q_j) \geq -BN \tag{F.22}$$

- temperedness: there exist $c, x, R > 0$ such that

$$|\phi(q - q') \leq c|q - q'|^{-3-x} \quad \text{for } |q - q'| > R \tag{F.23}$$

one finds that

$$\mathbb{E}\left[K^{-1}\right] \simeq K(\mu)^{-1}, \quad \text{and} \quad \mathbb{E}\left[K^{-1}\right]^* = K(\mu)^{-1}(1 + \theta_N) \tag{F.24}$$

become the same, with $\theta_N \to 0$, in the $N \to \infty$ limit, see Gallavotti (1999) for the details.

The above discussion highlights ingredients that are eventually *sufficient* to justify the use of ensembles, but sufficient conditions are not *necessary* for a given result to hold. Therefore, the formalism may apply even if some condition is violated. For instance, large neutral systems made of hard core electrically charged particles can still formally obey a kind of standard thermodynamics. However, when conditions are violated, one must assess the applicability of the theory case by case. Gravitating systems fail that test.

F.2 Equivalence of Ensembles

Begin considering a system of non-interacting particles, $\varphi = 0$. Then, using polar coordinates and assuming h is negligible, the canonical and microcanonical partition functions write

$$\begin{aligned} Z(\beta, V) &= \frac{V^N}{N!h^{3N}}\left(\frac{2\pi m}{\beta}\right)^{3N/2} \\ \Omega(U, V) &= \frac{V^N}{N!h^{3N}}\left[(2mU)^{3N/2} - (2mU - 2m\delta E)^{3N/2}\right]\frac{\pi^{3N/2}}{\Gamma(3N/2)\,3N} \end{aligned} \tag{F.25}$$

where $\Gamma(x) = (x-1)!$ and $\pi^{3N/2}/[\Gamma(3N/2)\,3N]$ is the surface of the N-dimensional unit sphere. Taking the thermodynamic limit, which means

$$N \to \infty, \quad V \to \infty, \quad \frac{V}{N} = v, \quad \frac{U}{N} = u, \quad u, v = \text{fixed} \tag{F.26}$$

and using Stirling's formula to approximate the $N!$ yields

$$\begin{aligned} S = k \log \Omega(U, V) &= Nk \left[\log \frac{V}{N} + \frac{3}{2} \log \frac{U}{N} + \text{const} + O(\frac{\log N}{N}) \right] \\ F = -\frac{1}{\beta} \log Z(\beta, V) &= -\frac{N}{\beta} \left[\log \frac{V}{N} - \frac{3}{2} \log \beta + \text{const} + O(\frac{\log N}{N}) \right] \end{aligned} \tag{F.27}$$

where S is interpreted as the microcanonical entropy, and $F = U - TS$ as the canonical free energy. Therefore, considering both ensembles as orthodic, the pressure can be obtained as

$$\begin{aligned} \frac{p}{T} &= \left.\frac{\partial S}{\partial \mathrm{V}}\right|_U = Nk\frac{1}{V}\left(1 + O\left(\frac{1}{N}\right)\right) \\ p &= -\left.\frac{\partial F}{\partial \mathrm{V}}\right|_\beta = \frac{1}{\beta}\frac{V}{N} \end{aligned} \tag{F.28}$$

proving the equivalence of the ensembles, given the observables of interest, if k is the same in both expressions, and if it takes the well-known value of the Boltzmann constant k. More generally, let us introduce the following quantity:

$$\Omega_0(U, V) = \int_{E(p,q) \le U} \frac{\mathrm{d}p\mathrm{d}q}{N!h^{3N}}$$

so that $\Omega(U, V) = \Omega_0(U, V) - \Omega_0(U - \delta E, V)$ and

$$Z(\beta, V) = \beta \int_{U_0}^{\infty} \mathrm{d}E\ \Omega_0(E, V) e^{-\beta E}$$

Integrating by parts produces

$$Z(\beta, V) = \int_{U_0}^{\infty} \mathrm{d}E\ \frac{\partial \Omega_0(E, V)}{\partial E} e^{-\beta E} \tag{F.29}$$

as the boundary terms at U_0 and ∞ vanish, and the free energy takes the form:

$$F(\beta, V) = -\frac{1}{\beta} Z(\beta, V) = -\frac{1}{\beta} \left[\log \beta + \log \int_{U_0}^{\infty} \mathrm{d}E\ \Omega_0 e^{-\beta E} \right] \tag{F.30}$$

Now, the specific (per particle) quantities in the thermodynamic limit are expressed by

$$
\begin{aligned}
v &= \frac{V}{N} \\
f_c(\beta, v) &= \lim_{N\to\infty} \frac{F(\beta, v)}{N} \\
u_c(\beta, v) &= \lim_{N\to\infty} \frac{U(\mu)}{N} = \frac{\partial \beta f_c}{\partial ﬁ}(\beta, v) \\
T_c &= \frac{1}{k_B \beta} = \frac{2\,K(\mu)}{3k_B N} \\
p_c &= \lim_{N\to\infty} P(\mu) = -\frac{\partial f_c}{\partial V}(\beta, v) \\
s_c &= \frac{u_c - f_c}{T_c}
\end{aligned}
\tag{F.31}
$$

while in the microcanonical case one has

$$
\begin{aligned}
v_m &= \frac{V}{N} \\
f_m(u_m, v_m) &= -T_m s_m + u_m \\
u_m &= \frac{U(\mu)}{N} \\
{T_m}^{-1} &= \frac{\partial s_m}{\partial u_m} = \frac{2\,K(\mu)}{3k_B N} \\
p_m &= T_m \frac{\partial s_m}{\partial v_m}(u_m, v_m) \\
s_m &= \lim_{N\to\infty} \frac{k_B}{N} \log \Omega_0(U, V)
\end{aligned}
\tag{F.32}
$$

In particular, this means that we have two different expressions for the entropy. To the dominant order for large N, and ignoring that exchange of limits could be problematic, we can write

$$
\Omega_0(U, V) = \exp\left(\frac{N}{k_B} s_m(u, v)\right) \tag{F.33}
$$

Then, writing $E = Nu$ and using the steepest descent for large N, one obtains

$$
\begin{aligned}
Z(\beta, V) &= \beta \int_{U_0}^{\infty} \mathrm{d}E\, \Omega_0(E) e^{-\beta E} N\beta \int \mathrm{d}u\, e^{-\beta N u} e^{N s_m(u)/k_B} = \\
&\approx \text{const}\, \sqrt{N} \exp\left\{ N \max_u \left[-\beta u + \frac{1}{k_B} s_m(u) \right] \right\}
\end{aligned}
\tag{F.34}
$$

and

$$\beta = \frac{1}{k_B}\frac{\partial s_m}{\partial u}(u_0, v_m) \tag{F.35}$$

provided there is a single maximum, located at u_0. Then, the large N limit implies

$$\begin{aligned} u_c &= \lim_{N\to\infty} \frac{U(\mu)}{N} \\ &= \lim_{N\to\infty} \frac{1}{\int_{U_{\min}}^{\infty} e^{-\beta E}\Omega_0(E, V)\, dE} \int_{U_{\min}}^{\infty} \frac{E}{N} e^{-\beta E}\Omega_0(E, V)\, dE = u_0 \end{aligned} \tag{F.36}$$

because the leading contribution is given by the values next to u_0, and we can write

$$\frac{1}{\beta} = \left[\frac{1}{k_B}\frac{\partial s_m}{\partial u}(u_c, u_m)\right]^{-1} = k_B T_m(u_c, v_m) \tag{F.37}$$

Then, taking $v_c = v_m$ and u_c such that $T_c = T_m$, one obtains $u_c = u_m = u_0$ from the expression of T_c and T_m. Moreover, we have

$$\begin{aligned} f_c(\beta, v_m) &= -\frac{1}{\beta}\max_u \left[-\beta u + \frac{s_m(u, v_m)}{k_B}\right] = u_c - \frac{s_m(u_c, v_m)}{k_B \beta} \\ &= u_c - T_c\, s_m(u_c, v_m) = u_m - T_m s_m(u_m, v_m) = f_m(u_m, v_m) \end{aligned} \tag{F.38}$$

As free energy, internal energy and absolute temperatures are equal, and because canonical and microcanonical ensembles are orthodic, hence they lead to the standard thermodynamic relations, also the two entropies are equal. In other words, the microcanonical and the canonical ensembles are equivalent in the thermodynamic limit: $N, V \to \infty$ with fixed $v = V/N$.

The equivalence may be understood as the relation between local observables, i.e., observables concerning a macroscopic part of the system, much smaller than the total, and the total is isolated. Therefore, the total system is in a microcanonical state and plays the role of a heat bath for one of its parts, which is then canonical. Restricting to local observables, the two descriptions are the same if part and total are both sufficiently large. In the proper limit, different models yield the same thermodynamics.

The above derivation of the equivalence of ensembles relies on the uniqueness of the extremum, u_0, of $-\beta u + s_m(u, v_m)/k_B$. This is the case when dealing with strictly convex functions, but it also happens that there is a whole interval (u_-, u_+) in which the function is extreme. In that case, equivalence fails. However, in general, such an event concerns only special values of β, and the consequent lack of uniqueness, rather than being a hindrance, provides a way to treat also phase transitions within the same conceptual scheme of equilibrium statistical mechanics. If there is more than one equilibrium, phase transitions will appear as *instabilities* of the thermodynamic properties with respect to variations of the boundary conditions (Gallavotti 1999).

The question of the equivalence of ensembles is particularly interesting in the case of nonequilibrium systems, because the models that are appropriate are still an open question. At the moment, a general theory is not available; hence, different models are used with different degrees of success to describe different phenomena. Could one prove that under certain conditions the different descriptions lead to the same result, would be an important step toward a unified theory of nonequilibrium, analogously to the statistical mechanics of equilibrium phenomena. In fact, various attempts have been made in this direction, cf. Refs. Evans and Morriss (2021), Gallavotti (2014), Gallavotti et al. (2004), Cohen and Rondoni (1998).

Fluctuation Relation from Anosov Framework

G

tat tvam asi

Chandogya Upanishad

The derivation of the Fluctuation Relation of Ref. Evans et al. (1993) was framed in a rigorous mathematical setting by Gallavotti and Cohen (1995), introducing the following:

Chaotic Hypothesis: *A reversible N-particle system in a stationary state can be regarded as a transitive Anosov system, for the calculation of its macroscopic properties.*

In analogy to the ergodic hypothesis, which is too strong to strictly represent physically relevant systems, but nevertheless, it is successfully used if applied to a limited and special set of observables.

A flow $(\mathscr{M}, \Phi^t)$ is called Anosov if it is smooth on a smooth manifold $\mathscr{M}$, and it is strictly *hyperbolic*: i.e., $\forall x \in \mathscr{M}$ the following holds:

- **i.** the tangent space at x splits as $T_x\mathscr{M} = E_x^u \oplus E_x^s \oplus E_x^0$, the splitting is continuous with x, and E_x^0 is a one-dimensional subspace.
- **ii.** the subspaces E_x^s and E_x^u of the tangent space are invariant under the tangent map $D\Phi^t$: i.e., $\forall x \in \Omega$, $D\Phi^t(x)E_x^i \subset E_{\Phi^t(x)}^i$ with $i = s, u$.
- **iii.** $\exists\, C > 0$, $\theta \in (0, 1)$ such that $\forall t \geq 0$, $\| D\Phi^t \xi \| \leq C\,\theta^t \| \xi \|$, for all $\xi \in E_x^s$ and $\| D\Phi^{-t} \eta \| \leq C\,\theta^t \| \eta \|$, $\forall \eta \in E_x^u$.
- **iv.** The stable and unstable manifolds W_x^s, W_x^u are dense in $\mathscr{M}$ for every $x \in \Omega$ (transitivity), the periodic orbits are dense in $\mathscr{M}$, and the *physical* measure, i.e., the one corresponding to the frequency of visitations of the regions of $\mathscr{M}$, has density along the unstable directions (which is then called the SRB measure).

C. Mejía-Monasterio and L. Rondoni, *Nonequilibrium Statistical Mechanics I*, Theoretical and Mathematical Physics,
https://doi.org/10.1007/978-3-031-98249-1

For such systems, the following holds:

Theorem (Sinai, 1968): *Every transitive Anosov system admits Markov partitions.*

Which allows a construction of the SRB measure, attributing weights to the cells of the Markov partition of $\mathcal{M}$ and then taking the limit of finer and finer partitions. Equivalently, one may use the decomposition in unstable periodic orbits (Ruelle 1999). Given a transitive, reversible Anosov diffeomorphism and a Hölder continuous function $\phi : \mathcal{M} \to I\!R$; there is a unique ergodic measure μ maximizing $h(\mu) + \int \phi(x)\mu(dx)$, where h is the Kolmogorov-Sinai entropy and μ is called the Gibbs state for ϕ. Sinai proved that the invariant measure which gives the forward time statistics is the Gibbs state of $\phi = -\log J^u$. Denoting by $\text{Fix}\Phi^n = \{x \in \mathcal{M} : \Phi^n x = x\}$, the set of periodic points of period n and by $\mathcal{O}$ a continuous function, the probability measure μ_n defined by the averages

$$\mu_n(\mathcal{O}) = \frac{\sum_{\text{Fix}\Phi^n} \mathcal{O}(x)\Pi_{k=0}^{n-1} J^u(\Phi^k x)^{-1}}{\sum_{\text{Fix}\Phi^n} \Pi_{k=0}^{n-1} J^u(\Phi^k x)^{-1}} \tag{G.1}$$

is invariant for Φ and tends weakly to μ for $n \to \infty$: $\mu_n(\mathcal{O}) \to \mu(\mathcal{O})$. Moreover, by definition $\mu(e_\tau) = 1$ and, because of time reversibility, one has

$$J(i\Phi x) = J(x)^{-1}, \quad e_\tau \circ i \circ \Phi^\tau = -e_\tau, \quad \text{and} \quad J^s(i\Phi x) = J^u(x)^{-1} \tag{G.2}$$

since i exchanges the stable and unstable directions. To prove the steady-state Fluctuation Relation for the phase space contraction rate $(-\Lambda)$, observe first that, given $[p, q] \subset I\!R$, there are $a, b > 0$ for which

$$\frac{1}{\tau \mathbb{E}_{\text{SRB}}[-\Lambda]} \log \frac{\mu(x : e_\tau(x) \in [p, q])}{\mu(x : e_\tau(x) \in [-q - a/\tau, -p + a/\tau])} \leq q + \frac{b}{\tau} \tag{G.3}$$

where $\mathbb{E}_{\text{SRB}}[*]$ is the average with respect to the SRB measure. This result relies on a long list of properties, including Bowen's shadowing, topological mixing, a property of sums for Hölder continuous functions, the expansiveness of the dynamics, the continuity of the splitting of the tangent bundle, and the expression of μ in terms of periodic orbits. Furthermore, it is used the fact that the dynamics can be mapped into a one-dimensional systems with short range interactions, considering $\phi = -\log J^u$ and $\psi = -\log J/\mathbb{E}_{\text{SRB}}[\Lambda]$, so that $(1/\tau)\sum_{k=0}^{\tau-1} \psi(\Phi^k x) = e_\tau(x)$. Then, there are $p^* > 0$ and a real analytic and strictly concave function η in the interval $(-p^*, p^*)$ such that, for every other interval I which intersects $(-p^*, p^*)$, the following holds

$$\lim_{\tau \to \infty} \frac{1}{\tau \mathbb{E}_{\text{SRB}}[-\Lambda]} \log \mu(\{x : e_\tau(x) \in I\}) = \frac{1}{\mathbb{E}_{\text{SRB}}[-\Lambda]} \sup_{u \in I \cap (-p^*, p^*)} \eta(u) \tag{G.4}$$

where the Gibbs state of ϕ, μ, and time reversibility have been used. Combining this with (G.3), one obtains

$$\lim_{\tau \to \infty} \frac{1}{\tau \mathbb{E}_{\text{SRB}}[-\Lambda]} \log \frac{\mu(\{x : e_\tau(x) \in (p - \delta, p + \delta)\})}{\mu(\{x : e_\tau(x) \in (-p - \delta, -p + \delta)\})} \leq p + \delta \tag{G.5}$$

for $\delta > 0$ and $|p| < p^*$. Taking this result, and the one corresponding to $-p$, the fluctuation theorem for the phase space contraction rate is obtained.

Theorem (Ruelle 1999) *Let Φ be a $C^{1+\alpha}$, $\alpha > 0$, Anosov diffeomorphism of the compact connected manifold $\mathcal{M}$, and let μ be the corresponding SRB measure. Assume reversibility, with involution i, and consider the dimensionless phase space contraction rate e_τ with respect to an i-invariant Riemann metric on $\mathcal{M}$. Then, there exists $p^* > 0$ such that the Fluctuation Relation for the phase space contraction rate:*

$$p - \delta \leq \lim_{\tau\to\infty} \frac{1}{\tau \mathbb{E}_{\text{SRB}}[-\Lambda]} \log \frac{\mu(\{x : e_\tau(x) \in (p-\delta, p+\delta)\})}{\mu(\{x : e_\tau(x) \in (-p-\delta, -p+\delta)\})} \leq p + \delta \tag{G.6}$$

holds if $|p| < p^*$ and $\delta > 0$.

This means that the function $C(p; \tau, \varepsilon) = (1/\tau \mathbb{E}_{\text{SRB}}[-\Lambda]) \log \left[\pi_\tau(B_{p,\varepsilon})/\pi_\tau (B_{-p,\varepsilon})\right]$, tends to a straight line of slope 1 for growing τ, apart from errors that can be made as small as one likes, where $B_{q,\varepsilon}$ is the ball of radius ε centered in q.

Under the assumption that $-\Lambda$ coincides with the entropy production rate, the Fluctuation Relation can be used to obtain the Green-Kubo relations and the Onsager reciprocal relations, in the limit of small external drivings. This way, the Fluctuation Relation appears to be an extension of such relations to nonequilibrium systems. In order to obtain this result, Gallavotti assumes that the (Gaussian isokinetic, continuous time) system is driven by ℓ fields $F = (F_1, F_2, ..., F_\ell)$, that the phase space contraction rate vanishes when all drivings vanish, and that it can be expanded as $-\Lambda(X) = \sum_{i=1}^{\ell} F_i J_i^0(X) + O(F^2)$. In this way, the linear "currents" J_i^0, which are proportional to the forces F_i, are defined. If the decay of the Λ-autocorrelation function is sufficiently fast, one then has

$$\zeta(p) \equiv -\lim_{\tau\to\infty} \frac{\log \pi_\tau(p)}{\tau} = \frac{\mathbb{E}_{\text{SRB}}[-\Lambda]^2}{2C_2}(p-1)^2 + O((p-1)^3 F^3) \tag{G.7}$$

where

$$C_2 = \int_{-\infty}^{\infty} \mathbb{E}^c_{\text{SRB}}\left[\Lambda(\Phi^t X)\Lambda(X)\right] \mathrm{d}t \tag{G.8}$$

and $\mathbb{E}^c_{\text{SRB}}[*]$ denotes the cumulant. Thus, the validity of the Fluctuation Relation implies $\mathbb{E}_{\text{SRB}}[-\Lambda] = C_2/2 + O(F^3)$. Now, let the full (nonlinear) "currents" be defined by $J_i(X) = -\partial_{F_i}\Lambda(X)$, and the transport coefficients by $L_{ij} = \partial_{F_j}\mathbb{E}_{\text{SRB}} [J_i]\,|_{F=0}$. The derivatives with respect to the parameters F require a property of differentiability of SRB measures, proven for Anosov systems in Ruelle (1997).

Assuming this property, the validity of the Fluctuation Relation and time reversibility, one can write

$$\mathbb{E}_{\mathrm{SRB}}\left[-\Lambda\right] = \frac{1}{2}\sum_{i,j=1}^{\ell}\left(\partial_{F_j}\mathbb{E}_{\mathrm{SRB}}\left[J_i\right] + \partial_{F_i}\mathbb{E}_{\mathrm{SRB}}\left[J_j\right]\right)\Bigg|_{F=0} F_i F_j$$
$$= \frac{1}{2}\sum_{i,j=1}^{\ell}(L_{ij} + L_{ji})F_i F_j \tag{G.9}$$

to second order in the forces. Equating this with $C_2/2$ and considering $(L_{ij} + L_{ji})/2$ with $i = j$, one recovers the Green-Kubo relations. To obtain the symmetry $L_{ij} = L_{ji}$, Gallavotti extends the Fluctuation Relation to consider the joint distribution of $-\Lambda$ and its derivatives. He introduces the dimensionless current q, averaged over a long time τ, through the relation

$$\frac{1}{\tau}\int_{-\tau/2}^{\tau/2} F_j \partial_{F_j}\Lambda(\Phi^t X)dt = q(X)\; F_j \mathbb{E}_{\mathrm{SRB}}\left[\partial_{F_j}\Lambda\right] \tag{G.10}$$

and considers the joint distribution $\pi_\tau(p, q)$, with corresponding large deviation functional $\zeta(p, q) = -\lim_{\tau\to\infty}\frac{1}{\tau}\log\pi_\tau(p, q)$. The result is similar to the Fluctuation Relation:

$$\lim_{\tau\to\infty}\frac{1}{\tau\mathbb{E}_{\mathrm{SRB}}\left[-\Lambda\right]p}\log\frac{\pi_\tau(p, q)}{\pi_\tau(-p, -q)} = 1 \tag{G.11}$$

This makes the difference $(\zeta(p, q) - \zeta(-p, -q))$ independent of q, which leads to the desired result, $L_{ij} = L_{ji}$, in the limit of small F.

Assuming that the currents and transport coefficients, defined here in terms of the phase space contraction rate, do represent physical quantities, these results show that the FRs are consistent with Irreversible Thermodynamics, close to equilibrium. Hence they may be considered in the construction of a comprehensive nonequilibrium theory. However, some difficulties affect the present approach, the first being that Λ is proportional to the energy dissipation rate divided by the kinetic energy only in very special cases.

A local version of the Fluctuation Relation derived in this appendix has also been derived for an infinite chain of weakly interacting chaotic maps. Let V_0 be a finite region of the chain centered at the origin, $T_0 > 0$ be a time interval, and define

$$\mathbb{E}_{\mathrm{SRB}}\left[\Lambda\right] = \lim_{V_0,T_0\to\infty}\frac{1}{|V_0|T_0}\sum_{j=0}^{T_0-1}\Lambda_{V_0}(S^j X)\,, \quad p = \frac{1}{\mathbb{E}_{\mathrm{SRB}}\left[\Lambda\right]|V|}\sum_{j=-T_0/2}^{T_0/2}\Lambda_{V_0}(S^j X) \tag{G.12}$$

where $V = V_0 \times T_0$, $\Lambda_{V_0}(X)$ is the contribution to Λ given by V_0, and $|V|$ denotes the volume of the set V. Then, one obtains

$$\pi_V(p) = e^{\zeta(p)|V|+O(|\partial V|)}\,, \quad \text{with } \frac{\zeta(p) - \zeta(-p)}{p\mathbb{E}_{\mathrm{SRB}}\left[-\Lambda\right]} = 1 \;\text{ and } |p| < p^* \tag{G.13}$$

where $|\partial V|$ is the size of the boundary of V, $p^* \geq 1$ and ζ is analytic in p. The contribution of the boundary term $|\partial V|$ should decrease with growing V, leading to the Λ-FR in the limit of large (compared to microscopic scales) volume V_0 and long times T_0.

Some Properties of the Dissipation Function Ω

H

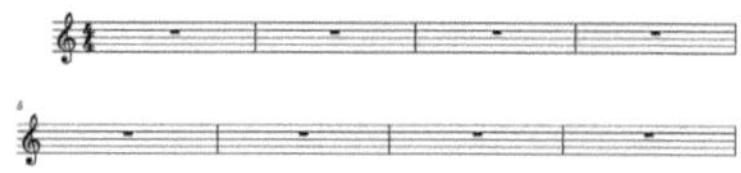

John Cage, 4'33"

Let us prove that the dissipation function satisfies $\Omega^{f_s}(\Gamma) = \Omega^{f_0}(S^{-s}\Gamma)$. First of all, Ω^{f_s} must exist, which requires f_s to be non-singular. This will be the case at all finite times for smooth dynamics, starting from a smooth initial f_0. If this is the case, we may write

$$\int_a^b \Omega^{f_s}(\Gamma)\, ds = \int_a^b \Omega^{f_0}(S^{-s}\Gamma)\, ds$$

on any interval $[a, b]$. In fact, we have

$$\begin{aligned}
\Omega^{f_s}_{a,b}(\Gamma) &= \ln \frac{f_s(S^a\Gamma)}{f_s(S^b\Gamma)} - \Lambda_{a,b}(\Gamma) \\
&= \ln \frac{f_0(S^{a-s}\Gamma)\exp\{-\Lambda_{0,s}(S^{a-s}\Gamma)\}}{f_0(S^{b-s}\Gamma)\exp\{-\Lambda_{0,s}(S^{b-s}\Gamma)\}} - \Lambda_{a,b}(\Gamma) \\
&= \ln \frac{f_0(S^{a-s}\Gamma)}{f_0(S^{b-s}\Gamma)} - \Lambda_{0,s}(S^{a-s}\Gamma) + \Lambda_{0,s}(S^{b-s}\Gamma) - \Lambda_{a,b}(\Gamma) \\
&= \ln \frac{f_0(S^{a-s}\Gamma)}{f_0(S^{b-s}\Gamma)} - \Lambda_{a,a+s}(S^{-s}\Gamma) + \Lambda_{b,b+s}(S^{-s}\Gamma) - \Lambda_{a+s,b+s}(S^{-s}\Gamma)
\end{aligned}$$

C. Mejía-Monasterio and L. Rondoni, *Nonequilibrium Statistical Mechanics I*, Theoretical and Mathematical Physics,
https://doi.org/10.1007/978-3-031-98249-1

$$= \ln \frac{f_0(S^{a-s}\Gamma)}{f_0(S^{b-s}\Gamma)} - \Lambda_{a,b+s}(S^{-s}\Gamma) + \Lambda_{b,b+s}(S^{-s}\Gamma)$$

$$= \ln \frac{f_0(S^{a-s}\Gamma)}{f_0(S^{b-s}\Gamma)} - \Lambda_{a,b}(S^{-s}\Gamma) = \Omega_{a,b}^{f_0}(S^{-s}\Gamma)$$

Since $a < b$ are arbitrary, it follows that $\Omega^{f_s}(\Gamma) = \Omega^{f_0}(S^{-s}\Gamma)$.

Another property of interest is the following. Given an initial probability density f_0 on the phase space $\mathcal{M}$ and its evolved density at time t, f_t say, we have

$$\mathbb{E}_t\left[\Omega^{f_t}\right] = \int \Omega^{f_t}(\Gamma) f_t(\Gamma) d\Gamma = \int \frac{\partial f_t}{\partial t}(\Gamma) d\Gamma = \frac{d}{dt}\int f_t(\Gamma) d\Gamma = \frac{d}{dt} 1 = 0$$

Also, because Ω^{f_t} is itself a phase function, an important result follows:

$$\frac{d}{ds}\mathbb{E}_s\left[\Omega^{f_t}\right]\Bigg|_{s=t} = \int \Omega^{f_t}(\Gamma) f_t(\Gamma) \Omega^{f_t}(\Gamma) d\Gamma = \mathbb{E}_t\left[\left(\Omega^{f_t}\right)^2\right] \geq 0$$

which also implies

$$\frac{\mathrm{d}}{\mathrm{d}t}\mathbb{E}_t\left[\Omega^{(0)}\right]\Bigg|_{t=0} = \mathbb{E}_0\left[\Omega^{(0)}\,\Omega^{(0)}\right] = \mathbb{E}_0\left[\left(\Omega^{(0)}\right)^2\right] > 0 \qquad \text{(H.1)}$$

Equation (H.1) requires f_t to be non-symmetric under time reversal, in the first instants of its evolution, even if there is no dissipation, and the state converges to a different equilibrium state. This is yet another way to see that evolving ensembles are not directly related to physical quantities, in general. A number of conditions, such as those allowing the derivation of the Boltzmann equation from Liouville's equation, are in general required.

Deterministic Thermostats

I

But to which side shall we incline? Reason can decide nothing here. There is an infinite chaos which separates us.

Thoughts, n. 233; Blaise Pascal

This appendix is devoted to a brief account of numerical simulation techniques known as deterministic thermostats, which have evolved in time to constitute theories of nonequilibrium systems. We follow in part Ref. Jepps and Rondoni (2010). Moreover, in the exercises Sect. 9.7.1 below, various examples are treated in detail.

Computer simulations of molecular systems have grown in time to become an indispensable research tool. Molecular simulation allows both qualitative and quantitative investigation of real-life systems, thanks to progress across many disciplines. For real-world systems in thermal equilibrium with their surroundings, Monte Carlo techniques provide a natural simulation setting that relies upon ergodicity: the measurement result $\overline{\mathscr{O}}$ is estimated as the weighted average $\mathbb{E}\left[\mathscr{O}\right]$, generated through an appropriate sampling of phase space. However, such ensemble averages cannot be calculated, in general, for nonequilibrium systems, since the relevant ensembles are usually not known. For these purposes, molecular dynamics is the prevailing technique, which also makes use of an ergodic hypothesis: for the assumed equivalence of time averages from any initial condition, and for the determination of thermodynamic properties such as entropy, temperature, or heat capacity.

Molecular dynamics takes the mechanical approach by solving the evolution dynamics of particles, but is naturally suited to simulations of systems *isolated* from their surroundings, rather than in thermal equilibrium with them. The dynamics must therefore be manipulated in order to generate the relevant equilibrium ensemble. This is achieved by introducing "thermostats," "barostats" and other numerical devices representing the coupling of the system with its environment. The extension of molecular dynamics to nonequilibrium systems has focused on the creation of

C. Mejía-Monasterio and L. Rondoni, *Nonequilibrium Statistical Mechanics I*, Theoretical and Mathematical Physics,
https://doi.org/10.1007/978-3-031-98249-1

behavior consistent with the target system, given the purpose of the simulation. The claim is not that the resulting equations of molecular motion are somehow a true representation of the underlying dynamics—only that the reported values are correct for properties we are interested in. The point is that any natural system can be studied from countless different perspectives, but it is neither possible nor interesting to consider more than a few at a time.

This is particularly true of nonequilibrium molecular dynamics (NEMD), whose original purpose was to compute transport coefficients. NEMD introduces artificial or "fictitious" forces that provide a convenient mechanical substitute for the thermodynamic forces (gradients in density, temperature, shear rate) driving real nonequilibrium behavior. The key advantage is that simulations driven by these fictitious forces are generally more efficient than those performed using thermodynamic boundary conditions. In NEMD, the thermostat dissipates precisely that energy introduced into the system by the fictitious forces. This procedure appears to generate ensembles that are in some sense the nonequilibrium counterparts of the equilibrium ensembles, as the simulated states appear to depend only on the macroscopic boundary conditions. Unlike the equilibrium case, however, our understanding of these ensembles, their distributions, and indeed their interpretation, is still limited. A glimpse of the physical appropriateness and relevance of thermostatted dynamics may, however, be obtained developing a theory of equivalence of nonequilibrium ensembles focusing, in particular, on equivalence of transport coefficients. In the following, we illustrate some of the available results in this research area.

I.0.1 Early Thermostats

The aim of the earliest thermostats was to provide a means of simulating thermal equilibrium, rather than providing a comprehensive dynamical theory on the subject. Thermostatting was performed with recourse to the equipartition theorem, as the disordered part of the kinetic energy is related to temperature, in equilibrium, and controlling it amounts to controlling the system temperature. The earliest thermostats (Woodcock 1971) involved the regular rescaling of each particle's momentum over the course of a simulation. But difficulties arose because of the effect of scaling on the time correlation functions, which depends on the specific scaling scheme, and because of the lack of time reversibility. The Berendsen thermostat constitutes a first answer to such problems, based on the following equations of motion:

$$\dot{\mathbf{q}} = \frac{\mathbf{p}}{m}, \quad \dot{\mathbf{p}} = \mathbf{F} - \frac{1}{2\tau}\left(1 - \frac{T_0}{T}\right)\mathbf{p} \tag{I.1}$$

While these equations remain time irreversible, the rescaling of the kinetic energy is made smooth, with a characteristic time scale τ for relaxation of the instantaneous kinetic temperature T to the target temperature T_0.

Time-reversibility appears not necessary for the purposes of calculating certain properties of molecular systems, as the widespread use of the Berendsen thermostat attests. However, our understanding of microscopic dynamics is that they are reversible; therefore, it is important to try to preserve it, when possible. Gaussian thermostats and the Nosé-Hoover thermostats are among those which keep the dynamics time reversal invariant.

I.0.2 Gaussian Thermostats

Gaussian thermostats are derived from Gauss' principle of least constraint (Gauß 1829):

Gauss Principle (1829): *Consider N point particles of mass m_i, subjected to frictionless bilateral constraints ϕ_i and to external forces $\mathbf{F}_i$. Among all motions allowed by the constraints, the natural one minimizes the curvature, defined by*

$$C = \sum_{i=1}^{N} m_i \left(\ddot{\mathbf{q}}_i - \frac{\mathbf{F}_i}{m_i} \right)^2 = \sum_{i=1}^{N} \frac{1}{m_i} \phi_i^2$$

Gauss' principle is equivalent to d'Alembert's principle, in terms of accelerations. It is not independent of coordinate systems, and therefore, it is not as generally applicable as the Lagrangian formulations of mechanics. However, it applies equally for non-holonomic and holonomic constraints, and in the case of holonomic constraints, it produces Hamiltonian equations of motion (Lanczos 1979).

On the other hand, nonequilibrium simulations require non-holonomic constraint, which leads to non-Hamiltonian equations of motion. For instance, fixing the kinetic energy of the system, $K = \sum_i \mathbf{p}_i^2/2m$, yields the isokinetic (IK) dynamics. Fixing the internal energy $H_0 = K + \Phi^{int}$, for particle interaction potential Φ^{int}, produces the isoenergetic (IE) dynamics. For an N-particle system subjected to an external field, these constraints can be implemented using Gauss' principle and yield the following equations of motion:

$$\begin{cases} \dot{\mathbf{q}}_i = \mathbf{p}_i/m \\ \dot{\mathbf{p}}_i = \mathbf{F}_i^{int}(\mathbf{q}) + \mathbf{F}_i^{ext}(\mathbf{q}) - \alpha(\Gamma)\mathbf{p}_i \end{cases} \qquad i = 1, ..., N \tag{I.2}$$

where $\mathbf{F}_i^{int}(\mathbf{q})$ and $\mathbf{F}_i^{ext}(\mathbf{q})$ denote inter-particle forces and external forces doing work on the system, respectively. The term $-\alpha(\Gamma)\mathbf{p}_i$, where $\Gamma = (\mathbf{q}, \mathbf{p})$, makes the dynamics dissipative for non-vanishing external forces, allowing the system to reach, in time, a steady state. If $\mathbf{F}_i^{ext}(\mathbf{q}) = C_i \mathbf{F}_e$, where C_i is a coupling constant for the ith particle, and $\mathbf{J} = \sum_{i=1}^{N} C_i \dot{\mathbf{q}}_i$, then the IK case corresponds to

$$\alpha(\Gamma) = \alpha_{IK}(\Gamma) = \frac{1}{2K} \left(\mathbf{J} \cdot \mathbf{F}_e + \sum_{i=1}^{N} \frac{\mathbf{p}_i}{m} \cdot \mathbf{F}_i^{int} \right) \tag{I.3}$$

while the IE case corresponds to

$$\alpha(\Gamma) = \alpha_{IE}(\Gamma) = \frac{1}{2K}\mathbf{J} \cdot \mathbf{F}_e \tag{I.4}$$

It is remarkable that these dynamics are time reversal invariant, and that α contains terms involving the product of thermodynamic force and conjugate flux, indicating that α could be associated with the energy dissipation of the system. It is through this relationship that the thermostat became an important ingredient of nonequilibrium theories. For equilibrium systems, the IE dynamics reduces to the usual Hamiltonian equations of motion, which has vanishing instantaneous dissipation, while the IK dynamics has zero time averaged dissipation.

IK and IE constraints are only two possible options in Gaussian thermostats, which have been extended to include a wide range of constraints including, for instance, isothermal-isobaric, isobaric-isoenthalpic, and constant stress ensembles. In the IE systems of many interacting particles, the link between the phase space variation rate, defined by

$$\Lambda \equiv \mathrm{div}_\Gamma \dot{\Gamma} = \sum_{i=1}^{N} \left[\frac{d}{d\mathbf{q}_i} \cdot \dot{\mathbf{q}}_i + \frac{d}{d\mathbf{p}_i} \cdot \dot{\mathbf{p}}_i \right] \tag{I.5}$$

and the instantaneous energy dissipation rate is obtained because Λ is proportional to α_{IE}. However, this is not a general property of thermostatted dynamics.

I.0.3 Nosé-Hoover Thermostat

S. Nosé and W.G. Hoover devised a popular thermostat that can be seen as a generalization of the Gaussian thermostats. Consider a system of particles with internal energy H_0 exchanging heat with a reservoir at temperature T. The interaction between particles and reservoir is represented by variables (s, p_S), and the whole system has a Hamiltonian expressed by

$$H = H_0 + 3NkT \ln s + \frac{p_S^2}{2Q} = \frac{\mathbf{p}^2}{2m} + \Psi(\mathbf{q}) + 3NkT \ln s + \frac{p_S^2}{2Q} \tag{I.6}$$

The ingenuity of this approach is that, under the transformed variables

$$\tilde{\mathbf{q}} = \mathbf{q}\,; \quad \tilde{\mathbf{p}} = \frac{\mathbf{p}}{s}\,; \quad \tilde{t} = \int_0^t \frac{d\tau}{s}\,; \quad \tilde{s} = s\,; \quad \xi = \frac{p_S}{Q} \tag{I.7}$$

the partition function takes the form

$$\begin{aligned} Z &= \int \frac{1}{s}\delta(H - E)\, \mathrm{d}\mathbf{q}\mathrm{d}\mathbf{p}\mathrm{d}s\mathrm{d}p_S \\ &= \left[\int \frac{Q}{3NkT} e^{-\beta E} e^{-\beta Q\xi^2/2} \mathrm{d}\xi\right] \int e^{-\beta H_0(\tilde{\mathbf{q}},\tilde{\mathbf{p}})}\, \mathrm{d}\tilde{\mathbf{q}}\mathrm{d}\tilde{\mathbf{p}} \end{aligned} \tag{I.8}$$

where $H(s_0) = E$ and $\beta = 1/kT$. The distribution is therefore *microcanonical* in the original frame $(\mathbf{q}, \mathbf{p}, s, p_S)$, but *canonical* with temperature T in the transformed frame $(\tilde{\mathbf{q}}, \tilde{\mathbf{p}})$. We can therefore simulate a canonical system $(\tilde{\mathbf{q}}, \tilde{\mathbf{p}})$ through the Hamiltonian dynamics concerning the variables $(\mathbf{q}, \mathbf{p}, s, p_S)$. The resulting equations of motion are quite straightforward:

$$\frac{\mathrm{d}\tilde{\mathbf{q}}}{\mathrm{d}\tilde{t}} = \frac{\tilde{\mathbf{p}}}{m} \, ; \quad \frac{\mathrm{d}\tilde{\mathbf{p}}}{\mathrm{d}\tilde{t}} = \mathbf{F} - \xi \tilde{\mathbf{p}} \, ; \quad \frac{\mathrm{d}\xi}{\mathrm{d}\tilde{t}} = \frac{1}{Q}\left(\frac{\tilde{\mathbf{p}}^2}{m} - gkT\right) \tag{I.9}$$

These equations differ from the Gaussian ones because the terms implementing the thermostat, α and ξ, play quite different roles: α depends on time only through the phase space point Γ, while ξ evolves according to its own equations of motion. On the other hand, the connection between phase space variation and energy dissipation associated with external fields case is given, in the Nose-Hoover case, by the thermostat term ξ, like it is given by α in the Gaussian case.

As with the Gaussian approach, Nosé-Hoover dynamics can be applied to produce different ensembles, including the canonical and isothermal-isobaric ensembles. Furthermore, Nosé's Hamiltonian is only one of a family of Hamiltonians which, coupled with an appropriate transformation to a new set of variables $(\mathbf{q}, \mathbf{p}) \mapsto (\tilde{\mathbf{q}}, \tilde{\mathbf{p}})$, produce canonical distributions. This is useful, because Nosé-Hoover dynamics are not ergodic, in general, especially when applied to low-dimensional systems.

I.0.4 Configurational Thermostats

The previous thermostats control the temperature by controlling the kinetic energy, which is the microscopic quantity most commonly associated with temperature. In fact, there is an infinite family of phase functions whose average at equilibrium is the system temperature, including functions of positions and not of momenta. For instance:

$$\frac{1}{kT} = \mathbb{E}\left[\nabla \cdot \frac{\nabla H}{\|\nabla H\|^2}\right] \tag{I.10}$$

expresses the temperature through a rather complicated, but legitimate formula. This allowed the development of the so-called configurational thermostats, which fix the configurational temperature implementing a *holonomic* constraint. This is achieved, for instance, by the following equations of motion:

$$\frac{\mathrm{d}\mathbf{q}}{\mathrm{d}t} = \frac{\mathbf{p}}{m} - \xi \nabla_{\mathbf{q}} T_C(\mathbf{q}) \, ; \quad \frac{\mathrm{d}\mathbf{p}}{\mathrm{d}t} = \mathbf{F} \tag{I.11}$$

Computationally, these equations that instantaneously fix the temperature are quite expensive. A better alternative is to consider a thermostatted equation of the form:

$$\frac{\mathrm{d}\mathbf{q}}{\mathrm{d}t} = \frac{\mathbf{p}}{m} - \xi \mathbf{F} \, ; \quad \frac{\mathrm{d}\mathbf{p}}{\mathrm{d}t} = \mathbf{F} \, ; \quad \frac{\mathrm{d}\xi}{\mathrm{d}t} = \frac{1}{Q}\left(\mathbf{F} \cdot \mathbf{F} - kT \nabla \cdot \mathbf{F}\right) \tag{I.12}$$

which are reminiscent of Nosé-Hoover equations.

I.1 Physical Suitability

As described in the previous section, thermostats are mathematical tools introduced to conveniently simulate thermodynamic phenomena from a microscopic, dynamical standpoint. For a particular system of interest, it is simply unfeasible to simulate the microscopic boundary interactions responsible for its thermodynamic state. This is as true of the exchange of energy through walls to maintain thermal equilibrium as it is of the transport processes that maintain a temperature gradient between two reservoirs. Experiments of driven systems show that different initial states may reach a final state that depends only on the driving mechanism, in particular on the boundary conditions. For equilibrium systems, the statistical mechanical theory of ensembles solves the problem, also thanks to the equivalence of ensembles in the thermodynamic limit.

For nonequilibrium transport processes, one needs methods to account for the coupling with the reservoirs, with computationally acceptable costs. Nonequilibrium molecular dynamics succeeded in this task, introducing artificial forces and an accompanying thermostat. Indeed, this has proven to be a highly successful technique for modeling nonequilibrium phenomena including rheological properties as well as fracturing, to mention quite diverse ones. Naturally, as with any model of a physical phenomenon, some aspects of reality are properly reproduced, while others are artifacts that do not reflect the behavior of real systems, as discussed below.

I.1.1 Non-Hamiltonian Equations

Given that the simplest and most common mechanical systems considered by physicists are naturally described using Hamiltonian equations of motion, one might object to the use of thermostatted equations of motion for the construction of a theory of nonequilibrium phenomena, on the basis that they contain artificial forces resulting in non-Hamiltonian dynamics.

There is strong justification, however, for the use of non-Hamiltonian dynamics to represent nonequilibrium dynamics. A system that exchanges energy with its environment should not be modeled by Hamiltonian equations of motion if that environment is not included in the description. The subsystem of interest is not bound to preserve its volume (so it cannot be Hamiltonian), even if the universe to which it belongs is Hamiltonian. These ideas are schematically represented in Fig. I.1 for a two-dimensional supposedly phase space volume preserving and reversible dynamics, where the subsystem phases correspond to the projection on the horizontal axis. While the subsystem dynamics retains the time reversal invariance of the whole system, it does not retain the preservation of volume. Moreover, physically dissipative systems, in which some current is established, cluster their trajectories inside the small (negligible indeed) nonequilibrium region of phase space corresponding to such currents.

Therefore, there is no reason in principle to avoid dynamical models, such as those of nonequilibrium molecular dynamics, that are time reversal invariant, and

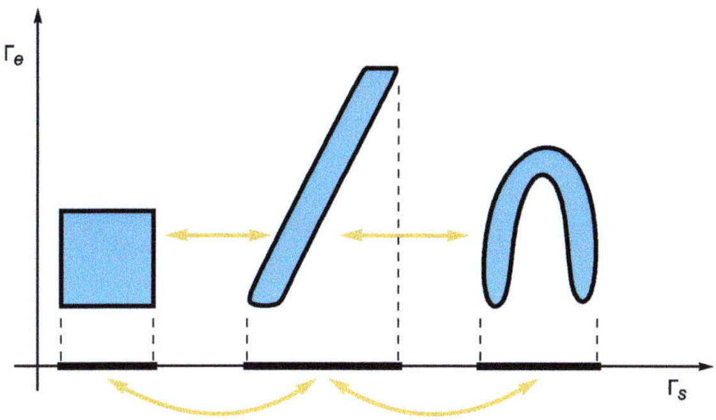

Fig. I.1 Three snapshots of the idealized evolution of phase space volume in time, made of system's coordinates Γ_s and environment coordinates Γ_e, and their projection on the system's space Γ_s. It is assumed that the dynamics are reversible and preserve volumes in phase space. Instead, in the system's space the volume fluctuates. However, as the backward evolution is allowed in phase space, it is also allowed in the system's space

that do not preserve volumes, allowing them to fluctuate in time, eventually shrinking them over sets of zero volume. The use of *artificial* forces in describing transport phenomena is not necessarily a problem for the NEMD techniques, also because the equivalence with thermodynamic driving forces has been demonstrated. The fact that they are not obtained through the ideal projection procedure described above, however, implies that they must be used *cum grano salis*. Indeed, problems may arise in describing aspects of the objects of interest directly related to the artificial forces of the models. For instance, the connections between temperature and kinetic energy, and between entropy and phase space volume, are well understood for equilibrium systems, but are much less clear for nonequilibrium steady states. However, the success of nonequilibrium molecular dynamics vindicates them in a wide context of research and applications.

I.1.2 Linear and Non-linear Response Theory

As argued earlier, one should not expect any mathematical model or molecular simulation to produce a perfect replication of microscopic dynamics, but should be content to realize specific objectives. In the case of nonequilibrium molecular dynamics, the original purpose was to obtain the linear transport coefficients, characterizing the response (induced flux) to an applied external field which is small or moderately large, on the thermodynamic scale. The linear response can be associated with equilibrium fluctuations, as shown in Chap. 7, implying that transport coefficients measured in the linear regime of a nonequilibrium system can be obtained from equilibrium calculations. The success of the nonequilibrium molecular dynamics in this endeavor is well established and often associated with greater computational efficiency, than afforded by other approaches.

Therefore, a theory has been developed to demonstrate that the response of a system to a real thermodynamic gradient is identical to the response to the artificial fields of nonequilibrium molecular dynamics, in the thermodynamic and small driving limit (Evans and Morriss 2021). As an example, let us mention the Evans-Gillan equations of motion:

$$\begin{aligned} \dot{\mathbf{q}}_i &= \mathbf{p}_i/m \\ \dot{\mathbf{p}}_i &= \mathbf{F}_i(\mathbf{q}) + (E_i - \bar{E})\mathbf{F}_e(t) - \tfrac{1}{2}\sum_{j=1}^{N} \mathbf{F}_{ij}(\mathbf{q}_{ij} \cdot \mathbf{F}_e(t)) + \tfrac{1}{2N}\sum_{j,k}^{N} \mathbf{F}_{jk}(\mathbf{q}_{jk} \cdot \mathbf{F}_e(t)) \end{aligned} \tag{I.13}$$

where E_i is the energy of particle i and $\bar{E} = \sum_i^N E_i/N$, which provide arguably the most efficient available model for heat flow in systems close to equilibrium. The external field $\mathbf{F}_e$ is artificial but induces a response that precisely mimics the effect of a real temperature gradient, transforming an inhomogeneous thermodynamic boundary condition (the different temperatures at the two ends of the system) that is hard to simulate, into a homogeneous mechanical force. The term $(E_i - \bar{E})\mathbf{F}_e(t)$ pushes particles with above-average energy in the direction of $\mathbf{F}_e(t)$, and pushes particles with below-average energy in the opposite direction. Consequently, this field induces an energy flow in the absence of temperature gradients and mass flows, efficiently simulating heat flow under a temperature gradient, provided the number of particles is sufficiently high, and artificial forces do not dominate the intermolecular interactions, so that their impact on the system behavior is by and large negligible. In fact, such conditions are those required by the Green-Kubo theory, cf. Chap. 7, which suggests that under appropriate limits the macroscopic response may be equivalently obtained from different microscopic dynamics, as analogously produced at equilibrium by the equivalence of ensembles. Of course, measurable differences may appear and eventually dominate as the conditions are more and more substantially violated, in particular as the number of particles gets small or the fields large. This will have to be assessed case by case.

I.1.3 Equivalence of Nonequilibrium Ensembles

Analogously to the pairs of conjugate thermodynamic variables generating conjugate equilibrium ensembles, related, e.g., to the entropy, there are nonequilibrium conjugate variables as well: thermodynamic forces X and corresponding fluxes J. Therefore, one may consider ensembles in which the quantities (N, V, T, X), for instance, are preserved, and the others, including J, fluctuate, or ensembles in which the quantities (N, V, T, J) are fixed and the others, including X fluctuate. In 1985, Evans and Morriss postulated a principle of equivalence of nonequilibrium ensembles, analogously to that concerning equilibrium ensembles (Evans and Morriss 2021). The principle states that linear responses of conjugate nonequilibrium ensembles are equivalent. This principle, later generalized also to nonlinear response, may provide a more formal basis for analyzing the suitability and range of applicability of the deterministic thermostatted dynamics, beyond the use of transport coefficients, notwithstanding the "artificial" nature of some of their ingredients.

The first result of this sort was related by Evans and Morriss to the equivalence of the Norton constant current (flux) electrical circuit and the corresponding Thevénin constant voltage (force) circuit. Voltage and current, being conjugate variables, suggest that the ensembles (N, V, T, J) and (N, V, T, E) may be regarded as equivalent, from the viewpoint of transport in the linear regime. To prove that this is, indeed, the case, two Gaussian multipliers are introduced, to hold the temperature and the current constant. The resultant equations of motion afford an expression for the required applied field, whose fluctuations are consistent with the flux fluctuations of the constant force ensemble. At zero frequency, the "Thevénin" fluctuation correlation function is proportional to the conductivity, while the "Norton" fluctuation correlation function is proportional to the resistivity. That this equivalence actually holds does not imply equivalence in every respect, because only one property has been directly involved: the proportionality between driving field and resulting current. However, this is the same for the equivalence of equilibrium ensembles. For instance, the energy does not fluctuate in the microcanonical case, while it does in the canonical case. Interestingly, the zero-frequency condition appears in some sense to be the nonequilibrium counterpart to the thermodynamic-limit condition: the zero-frequency contribution correctly represents the system behavior when the non-zero-frequency contributions are negligible, something which requires the number of particles, N, to be very large.

As an example, observe that the phase space variation rate of thermostatted systems cannot *in general* be thought of as proportional to the entropy production rate of the phenomenon under consideration, but taking an appropriate large N limit makes different models agree. A simple way of realizing that goes through an analysis of Gaussian IK and IE thermostatted particle systems, whose equations of motion are given by Eq. (I.2), with different Lagrange multipliers α given, respectively given by Eqs. (I.3) and (I.4), and correspondingly different phase space volumes variation rates Λ. For the time average of Λ in IK case, we have

$$\overline{\Lambda}_{IK} = \mathbb{E}\left[\Lambda_{IK}\right] = -(dN-1)\left(\mathbb{E}\left[\frac{\sum_{i=1}^{N}\frac{\mathbf{p}_i}{m}\cdot\mathbf{F}_i^{int}}{\sum_{i=1}^{N}\frac{\mathbf{p}_i^2}{m}}\right] + \mathbb{E}\left[\frac{\sum_{i=1}^{N}\frac{\mathbf{p}_i}{m}\cdot\mathbf{F}^{ext}}{\sum_{i=1}^{N}\frac{\mathbf{p}_i^2}{m}}\right]\right)$$

if the dynamics are ergodic, and $\mathbb{E}$ represents averaging with respect to the steady-state probability distribution. Moreover, assuming the system is macroscopic and near thermodynamic equilibrium, the kinetic energy expresses the fixed and uniform temperature, and the internal conservative forces give vanishing contribution to the phase space variation rate. Therefore, one can write

$$\overline{\Lambda}_{IK} = -\frac{1}{m}\frac{\mathbb{E}\left[\sum_{i=1}^{N}\mathbf{p}_i\cdot\mathbf{F}^{ext}\right]}{k_B T} \tag{I.14}$$

because a constant can be taken out of the averaging operator. Introducing the particle current density, $\mathbf{I} = \mathbb{E}\left[\sum_{i=1}^{N}\frac{\mathbf{p}_i}{m}\right]/V$, and dividing Λ_{IK} by the volume V of the system, to compare dynamical averages with macroscopic quantities, one obtains

$$-\hat{\Lambda}_{IK} = \frac{\mathbf{I}\cdot\mathbf{F}^{ext}}{k_B T} \tag{I.15}$$

for the phase space expansion rate, where the right-hand side of Eq. (I.15) is *formally* the expression for the entropy production rate in irreversible thermodynamics.

While Eq. (I.15) may not appear to depend on large N, in reality there is no concept of a thermodynamic quantity away from this limit. For small N we can only speak of a connection between Λ, the kinetic energy, and the dissipated energy in *statistical* rather than *thermodynamic* terms.

In the IE case, the temperature can again be (formally) defined by

$$\mathbb{E}\left[\sum_{i=1}^{N}\frac{\mathbf{p}_i^2}{m}\right] = dNk_BT \tag{I.16}$$

which turns concrete under the usual local thermodynamic equilibrium conditions. Now, Eq. (I.4) yields

$$\hat{\Lambda}_{IE} \equiv \frac{\mathbb{E}\left[\Lambda_{IE}\right]}{V} = -(dN-1)\mathbb{E}\left[\frac{\sum_{i=1}^{N}\frac{\mathbf{p}_i}{mV}\cdot\mathbf{F}^{ext}}{\sum_{i=1}^{N}\frac{\mathbf{p}_i^2}{m}}\right] \tag{I.17}$$

where the kinetic energy is not constant, hence it cannot be taken out of the averaging operator, unless some condition is verified. Assuming that fluctuations are negligible with respect to mean values, which requires very large N, one may argue that the average of the ratio on the right-hand side of (I.17) can be replaced by the ratio of the averages. On the other hand, speaking sensibly of temperatures and entropies requires the same conditions to be realized. We can thus neglect terms of order $O(1/N)$ to obtain

$$-\hat{\Lambda}_{IE} = \frac{\mathbf{I}\cdot\mathbf{F}^{ext}}{k_BT} \tag{I.18}$$

This shows that even in the IE case, entropy production and phase space expansion rates may be simply related only for large N. In general, one may only speak of energy dissipation.

Nevertheless, we have obtained $\hat{\Lambda}_{IE} = \hat{\Lambda}_{IK} + O(1/N)$, when density and energy density take a common value. This constitutes one example of dynamical (and possibly thermodynamic) equivalence of nonequilibrium models, which tells that different microscopic models lead to the same macroscopic results, under proper conditions and for specific observables. In turn, this explains why “artificial” nonequilibrium molecular dynamics models perform so well in computing properties of nonequilibrium states.

References

T.M. Apostol, *Mathematical Analysis* (Addison Wesley, Reading, 1981)

V.I. Arnold, A. Avez, *Ergodic Problems of Classical Mechanics* (Benjamin, 1968)

N.W. Ashcroft, N.D. Mermin, *Solid State Physics* (Harcourt College Publishers, Orlando, 1976)

E. Aurell, C. Mejía-Monasterio, P. Muratore-Ginanneschi, Optimal protocols and optimal transport in stochastic thermodynamics. Phys. Rev. Lett. **106**, 250601 (2011). https://doi.org/10.1103/PhysRevLett.106.250601

M. Baiesi, C. Burigana, L. Conti, G. Falasco, C. Maes, L. Rondoni, T. Trombetti, Possible nonequilibrium imprint in the cosmic background at low frequencies. Phys. Rev. Res. **2**(1), 013210 (2020)

M. Baldovin, A. Vulpiani, G. Gradenigo, Statistical mechanics of an integrable system. J. Stat. Phys. **183**, 41 (2021). https://doi.org/10.1007/s10955-021-02781-7

F. Balducci, M. Beau, J. Yang, A. Gambassi, A. Del Campo, Large deviations beyond the Kibble-Zurek mechanism. Phys. Rev. Lett. **131**(23), 230401 (2023)

J. Barbour, T. Koslowski, F. Mercati, Identification of a gravitational arrow of time. Phys. Rev. Lett. **113**, 181101 (2014)

R. Belousov, S. Savino, P. Moghe, T. Hiiragi, L. Rondoni, A. Erzberger, Poissonian cellular potts models reveal nonequilibrium kinetics of cell sorting. Phys. Rev. Lett. **132**(24), 248401 (2024)

L. Bertini, A. De Sole, D. Gabrielli, G. Jona-Lasinio, C. Landim, Macroscopic fluctuation theory for stationary non-equilibrium states. J. Stat. Phys. **107**, 635–675 (2002)

L. Bertini, A. De Sole, D. Gabrielli, G. Jona-Lasinio, C. Landim, Macroscopic fluctuation theory. Rev. Mod. Phys. **85**, 593–636 (2015)

T. Bodineau, B. Derrida, Current fluctuations in non-equilibrium diffusive systems: an additivity principle. Phys. Rev. Lett. **92**, 180601–1 – 6180601–4 (2004)

G. Boffetta, G. Lacorata, S. Musacchio, A. Vulpiani, Relaxation of finite perturbations: beyond the fluctuation dissipation relation. Chaos **13**, 3 (2003)

L. Boltzmann, *Theoretical Physics and Philosophical Problems: Selected Writings* (D. Reidel Publishing Company, 1974)

M. Bonaldi, L. Conti, P. De Gregorio, L. Rondoni, G. Vedovato, A. Vinante, M. Bignotto, M. Cerdonio, P. Falferi, N. Liguori, S. Longo, R. Mezzena, A. Ortolan, G.A. Prodi, F. Salemi, L. Taffarello, S. Vitale, J.-P. Zendri, Nonequilibrium steady-state fluctuations in actively cooled resonators. Phys. Rev. Lett. **103**, 010601 (2009). https://doi.org/10.1103/PhysRevLett.103.010601

S. Bonella, G. Ciccotti, L. Rondoni, Time reversal symmetry in time-dependent correlation functions for systems in a constant magnetic field. Eur. Lett. **108**(6), 60004 (2015)

C. Mejía-Monasterio and L. Rondoni, *Nonequilibrium Statistical Mechanics I*, Theoretical and Mathematical Physics,
https://doi.org/10.1007/978-3-031-98249-1

F. Bonetto, E.A. Carlen, R. Esposito, J.L. Lebowitz, R. Marra, Propagation of chaos for a thermostated kinetic model. J. Stat. Phys. **154**(1), 265–285 (2014)
L. Breiman, *Probability* (SIAM, 1968)
R. Brown, A brief account of microscopical observations made in the months of June, July, and August, 1827 on the particles contained in the pollen of plants; and on the general existence of active molecules in organic and inorganic bodies. Philos. Mag. **4**, 161 (1828)
P.R. Buenzli, R. Soto, Violation of the action-reaction principle and self-forces induced by nonequilibrium fluctuations. Phys. Rev. E **78**(2), 793–4 (2008)
H.B. Callen, *Thermodynamics and an Introduction to Thermostatistics* (Wiley, 1985)
C. Callender, Thermodynamic asymmetry in time. *The Stanford Encyclopedia of Philosophy* (2012). http://plato.stanford.edu/archives/fall2011/entries/time-thermo/
L. Campbell, W. Garnett, *The Life of James Clerk Maxwell* (MacMillan and Co., London, 1882)
G. Cantoni, Su alcune condizioni fisiche dell'affinità e sul moto browniano. Il Nuovo Cimento **27**, 156 (1867)
D. Carbone, P. De Gregorio, L. Rondoni, Time reversal symmetry for classical, non-relativistic quantum and spin systems in presence of magnetic fields. Ann. Phys. **441**, 168853 (2022)
D. Carbone, M. Hua, S. Coste, E. Vanden-Eijnden, Efficient training of energy-based models using Jarzynski equality. *Advances in Neural Information Processing Systems*, vol. 36 (2024)
H.B.G. Casimir, On Onsager's principle of microscopic reversibility. Rev. Mod. Phys. **17**, 343–350 (1945). https://doi.org/10.1103/RevModPhys.17.343
E. Cassirer, *An Essay on Man: An Introduction to a Philosophy of Human Culture* (Yale University Press, Veritas paperback edition, 2021)
M. Cencini, F. Cecconi, A. Vulpiani, *Chaos: From Simple Models to Complex Systems* (World Scientific, Singapore, 2009)
C. Cercignani, *Ludwig Boltzmann: The Man Who Trusted Atoms* (Oxford University Press, 2006)
C. Cercignani, R. Illner, M. Pulvirenti. *The Mathematical Theory of Dilute Gases* (Springer, 1994)
C. Cercignani, G.M. Kremer, *The Relativistic Boltzmann Equation: Theory and Applications* (Birkhäuser, Basel, 2002)
B. Cessac, J.-A. Sepulchre, Linear response, susceptibility and resonances in chaotic toy models. Physica D **225**, 13 (2007)
S. Chandrasekhar, Stochastic problems in physics and astronomy. Rev. Mod. Phys. **15**(1), 1 (1943)
J. Chen, T. Gao, Y. Li, J. Duan, Detecting the most probable transition pathway based on optimal control theory. Appl. Math. Model. **127**, 217–236 (2024)
S. Chibbaro, L. Rondoni, A. Vulpiani, *Reductionism, Emergence and Levels of Reality, The Importance of Being Borderline* (Springer, 2014)
S. Ciliberto, Experiments in stochastic thermodynamics: short history and perspectives. Phys. Rev. X **7**(2), 021051 (2017)
E.G.D. Cohen, L. Rondoni, Note on phase space contraction and entropy production in thermostatted Hamiltonian systems. CHAOS **8**, 357 (1998)
M. Colangeli, L. Rondoni, A. Vulpiani, Fluctuation-dissipation relation for chaotic non-Hamiltonian systems. J. Stat. Mech. **L04002** (2012)
P. Collet, J.-P. Eckmann, *Concepts and Results in Chaotic Dynamics: A Short Course* (Springer, Berlin, Heidelberg, 2006)
L. Conti, P. De Gregorio, G. Karapetyan, C. Lazzaro, M. Pegoraro, M. Bonaldi, L. Rondoni, Cef-fects of breaking vibrational energy equipartition on measurements of temperature in macroscopic oscillators subject to heat flux. J. Stat. Mech. (2013a)
L. Conti, P. De Gregorio, G. Karapetyan, C. Lazzaro, M. Pegoraro, M. Bonaldi, L. Rondoni, Effects of breaking vibrational energy equipartition on measurements of temperature in macroscopic oscillators subject to heat flux. J. Stat. Mech.: Theory Exp. **2013**(12), P12003 (2013b)
E. Cristiani, B. Piccoli, A. Tosin, *Multiscale Modeling of Pedestrian Dynamics*, vol. 12 (Springer, 2014)
G.E. Crooks, Path-ensemble averages in systems driven far from equilibrium. Phys. Rev. E **61**, 2361 (2000)

S.R. De Groot, P. Mazur, *Non-Equilibrium Thermodynamics* (Dover Publications, New York, 1984)
I. de Vega, D. Alonso, Dynamics of non-Markovian open quantum systems. Rev. Mod. Phys. **89**, 015001 (2017). https://doi.org/10.1103/RevModPhys.89.015001
G. Dematteis, L. Rondoni, D. Proment, F. De Vita, M. Onorato, Coexistence of ballistic and Fourier regimes in the β Fermi-Pasta-Ulam-Tsingou lattice. Phys. Rev. Lett. **125**(2), 024101 (2020)
C.-J. Ding, Y.-W. Lei, X.-Y. Wang, X.-L. Li, X.-Y. Li, Y.-G. Zhang, Y.-C. Xu, C.-S. Liu, X.-B. Wu, A deep learning interatomic potential suitable for simulating radiation damage in bulk tungsten. Tungsten 1–19 (2023)
J.L. Doob, *Stochastic Processes* (Wiley Classics, 1990)
J. Dunkel, P. Hänggi, Relativistic Brownian motion. Phys. Rep. **471**(1), 1–73 (2009)
A. Einstein, Über die von der molekularkinetischen theorie der wärme geforderte bewegung von in ruhenden flüssigkeiten suspendierten teilchen. Annalen der Physik **322**(8), 549–560 (1905). https://doi.org/10.1002/andp.19053220806
A. Einstein, *Investigations on the Theory of the Brownian Movement* (Dover Publications, 1956)
D.J. Evans, G. Morriss, *Statistical Mechanics of Nonequilibrium Liquids* (Cambridge University Press, 2021)
D.J. Evans, E.G.D. Cohen, G.P. Morriss, Probability of second law violations in shearing steady flows. Phys. Rev. Lett. **71**, 2401 (1993)
V. Eyert, J. Wormald, W.A. Curtin, E. Wimmer, Machine-learned interatomic potentials: recent developments and prospective applications. J. Mater. Res. **38**(24), 5079–5094 (2023)
M. Falcioni, L. Palatella, S. Pigolotti, L. Rondoni, A. Vulpiani, Initial growth of Boltzmann entropy and chaos in a large assembly of weakly interacting systems. Phys. A-Stat. Mech. Appl. **385**(1), 170–184 (2007)
E. Fermi, *Thermodynamics* (Dover Publications, 1956)
D.K. Ferry, J. Robert Barker, C. Jacobini, *Physics of Nonlinear Transport in Semiconductors*, vol. 52 (Springer, 2012)
R.P. Feynman, *Principles of Least Action in Quantum Mechanics* (Princeton University, 1942)
R.P. Feynman, R.B. Leighton, M. Sands, *The Feynman Lectures on Physics* (Addison-Wesley Reading, MA, 1964)
P. Fireman, The expansion of gas into a vacuum and the kinetic theory of gases. J. Phys. Chem. **6**, 463 (1902)
G.B. Folland, *Real Analysis* (Wiley, New York, 1999)
M.I. Freidlin, A.D. Wentzell, *Random Perturbations of Dynamical Systems* (Springer, 1998)
L.E. Fried, W.M. Howard, An accurate equation of state for the exponential-6 fluid applied to dense supercritical nitrogen. J. Chem. Phys. **109**(17), 7338–7348 (1998)
G. Gallavotti, *Statistical Mechanics: A Short Treatise* (Springer, Berlin, 1999)
G. Gallavotti, *Nonequilibrium and Irreversibility* (Springer, Heidelberg, 2014)
G. Gallavotti, E.G.D. Cohen, Dynamical ensembles in stationary states. J. Stat. Phys. **80**, 931 (1995)
G. Gallavotti, L. Rondoni, E. Segre, Lyapunov spectra and nonequilibrium ensembles equivalence in 2d fluid mechanics. Phys. D: Nonlinear Phenom. **187**, 338 (2004)
A.A. Gamba, L. Rondoni, Current fluctuations in the nonequilibrium Lorentz gas. PHYSICA. A **340**, 274–282 (2004)
Y. Gao, H. Fang, K. Ni, Y. Feng, Water clusters and density fluctuations in liquid water based on extended hierarchical clustering methods. Sci. Rep. **12**(8036), 1–11 (2022)
C. Gardiner, P. Zoller, *Quantum Noise: A Handbook of Markovian and Non-Markovian Quantum Stochastic Methods with Applications to Quantum Optics* (Springer, Berlin, 2004)
C.W. Gardiner, *Handbook of Stochastic Methods* (Springer, Berlin, 1985)
C.F. Gauß, über ein neues allgemeines grundgesetz der mechanik. Journal für die reine und angewandte Mathematik **1829**(4), 232–235 (1829). https://doi.org/10.1515/crll.1829.4.232
M. Ghil, V. Lucarini, The physics of climate variability and climate change. Rev. Mod. Phys. **92**(3), 035002 (2020)

G. Giacomin, S. Olla, E. Saada, H. Spohn, G. Stoltz, *Stochastic Dynamics Out of Equilibrium* (Springer, 2019)
C. Giardina, J. Kurchan, V. Lecomte, J. Tailleur, Simulating rare events in dynamical processes. J. Stat. Phys. **145**, 787–811 (2011)
C. Giberti, L. Rondoni, C. Vernia, Temporal asymmetry offluctuations in the nonequilibrium fpu model. Phys. D-Nonlinear Phenom. **228**, 64–76 (2007)
H. Grad, On the kinetic theory of rarefied gases. Commun. Pure Appl. Math. **2**, 331 (1949)
D. Grebenkov, R. Metzler, G. Oshanin, *Target Search Problems* (Springer, Cham, 2024). https://doi.org/10.1007/978-3-031-67802-8
M.S. Green, Markoff random processes and the statistical mechanics of time-dependent phenomena. J. Chem. Phys. **20**, 1281 (1952)
M.S. Green, Markoff random processes and the statistical mechanics of time-dependent phenomena. ii. irreversible processes in fluids. J. Chem. Phys. **22**, 398 (1954)
R.F. Greene, H.B. Callen, On the formalism of thermodynamic fluctuation theory. Phys. Rev. **83**(6), 1231 (1951)
F. Guerra, L.M. Morato, Quantization of dynamical systems and stochastic control theory. Phys. Rev. D **27**, 1774–1786 (1983). https://doi.org/10.1103/PhysRevD.27.1774
J.P. Hansen, I.R. McDonald, *Theory of Simple Liquids* (Academic Press Inc., London, 1986)
P.M. Harman (ed.), *The Scientific Letters and Papers of James Clerk Maxwell*, vol. I (Cambridge University Press, Cambridge, 1990)
R.J. Harris, H. Touchette, Phase transitions in large deviations of reset processes. J. Phys. A: Math. Theor. **50**(10), 10LT01 (2017)
T.L. Hill, *Statistical Mechanics* (Dover Publications, 1987)
T.L. Hill, A different approach to nanothermodynamics. Nano Lett. **1**(5), 273–275 (2001)
D.R. Hofstadter, *Gödel, Escher, Bach: An Eternal Golden Braid* (Basic Books, 1979)
K. Huang, *Statistical Mechanics* (Wiley, New York, 1987)
J.C. Hull, *Options, Futures, and Other Derivatives* (Pearson Global Editions, Harlow UK, 2021)
O. Jepps, L. Rondoni, Deterministic thermostats, theories of nonequilibrium systems and parallels with the ergodic condition. J. Phys. A **43** (2010)
D. Jou, G. Lebon, J. Casas-Vázquez, *Extended Irreversible Thermodynamics* (Springer, 2010)
M. Kac, *Probability and Related Topics in Physical Sciences* (American Mathematical Society, 1957)
J.B. Keller, Diffusion at finite speed and random walks. PNAS **101**, 1120–1122 (2004)
A.I. Khinchin, *Mathematical Foundations of Statistical Mechanics* (Dover Publications, 2013)
J.G. Kirkwood, Statistical mechanics of fluid mixtures. J. Phys. Chem. Phys. 300–313 (1935)
J. Klafter, I.M. Sokolov, Anomalous diffusion spreads its wings. Phys. World **18**(8), 29 (2005)
A.N. Kolmogorov, *Foundations of the Theory of Probability* (Chelsea Publishing Co., New York, 1956)
D. Kondepudi, I. Prigogine, *Modern Thermodynamics: From Heat Engines to Dissipative Structures*, 2nd edn. (Wiley, 1998)
A. Koyré, *Newtonian Studies* (The University of Chicago Press, 1968)
H.J. Kreuzer, *Nonequilibrium Thermodynamics and Its Statistical Foundations* (Claredon Press, 1981)
R. Kubo, Statistical-mechanical theory of irreversible processes. i. general theory and simple applications to magnetic and conduction problems. J. Phys. Soc. Jpn. **12**, 570 (1957)
R. Kubo, *Thermodynamics* (North-Holland Publishing, 1968)
R. Kubo, M. Toda, N. Hashitsume, *Statistical Physics II: Nonequilibrium Statistical Mechanics* (Springer, 1998)
R. Kuske, J.B. Keller, Large deviation theory for stochastic difference equations. Eur. J. Appl. Math. **8**, 567 (1997)
A. Lampo, M.A. García March, M. Lewenstein, *Quantum Brownian Motion Revisited: Extensions and Applications* (Springer, Cham, New York, 2019)
C. Lanczos, *The Variational Principles of Mechanics* (Dover Publications, New York, 1979)
L.D. Landau, E.M. Lifshitz, *Statistical Physics, Part 1* (Pergamon, New York, 1980)

O.E. Lanford, The hard sphere gas in the Boltzmann-Grad limit. Phys. A: Stat. Theor. Phys. **106**(1–2), 70 (1981)
P. Langevin, Sur la théorie du mouvement brownien. Compt. Rend. Acad. Sci. (Paris) **146**, 530–533 (1908)
P. Laplace, *Essai Philosophique sur les Probabilities, Courcier, Paris. The 5th edition (1825) was the last revised by Laplace. English translation in 1952 as Philosophical Essay on Probabilities* (Dover Publications, 1814)
H.E. Lieb, J. Yngvason, The physics and mathematics of the second law of thermodynamics. Phys. Rep. **310**, 1 (1999)
F. Lu, M. Maggioni, S. Tang, Learning interaction kernels in stochastic systems of interacting particles from multiple trajectories. arXiv preprint arXiv:2007.15174 (2020)
R. Luo, G. Benenti, G. Casati, J. Wang, Onsager reciprocal relations with broken time-reversal symmetry. Phys. Rev. Res. **2**, 022009 (2020). https://doi.org/10.1103/PhysRevResearch.2.022009
R. Mañe, *Ergodic Theory and Differentiable Dynamics* (Springer, 1987)
S. Machlup, L. Onsager, Fluctuations and irreversible processes. ii. systems with kinetic energy. Phys. Rev. **91**(6), 1512 (1953)
L. Maffioli, E.R Smith, J.P. Ewen, P.J. Daivis, D. Dini, B.D. Todd, Slip and stress from low shear rate nonequilibrium molecular dynamics: the transient-time correlation function technique. J. Chem. Phys. **156**(18) (2022)
U.M.B. Marconi, R. Puglisii, L. Rondoni, A. Vulpiani, Fluctuation-dissipation: response theory in statistical physics. Phys. Rep. **461**, 111–195 (2008)
D.A. McQuarrie, *Statistical Mechanics* (Harper & Row, 1975)
C. Mejía-Monasterio, G. Oshanin, G. Schehr, First passages for a search by a swarm of independent random searchers. J. Stat. Mech.: Theory Exp. **2011**(06), P06022 (2011)
R. Metzler, J. Klafter, The random walk's guide to anomalous diffusion: a fractional dynamics approach. Phys. Rep. **339**(1), 1–77 (2000). https://doi.org/10.1016/S0370-1573(00)00070-3
A.M. Monge, M. Manosas, F. Ritort, Experimental test of ensemble inequivalence and the fluctuation theorem in the force ensemble in dna pulling experiments. Phys. Rev. E **98**(3), 032146 (2018)
T. Morawietz, A. Singraber, C. Dellago, J. Behler, How van der Waals interactions determine the unique properties of water. PNAS **113**(30), 8368–8373 (2016)
S. Nazarenko, *Wave Turbulence*, vol. 825 (Springer, 2011)
I. Nordlund, Eine neue bestimmung der avogadroschen konstante aus der brownschen bewegung kleiner, in wasser suspendierten quecksilberkügelchen. Z. Phys. Chem. **87**, 40 (1914). https://doi.org/10.1515/zpch-1914-8703
I.V.G. Oliveira, C. Wang, G. Dong, R. Du, C.E. Fiore, A.L.M. Vilela, H.E. Stanley, Entropy production on cooperative opinion dynamics. Chaos Solitons Fractals **181**, 114694 (2024)
L. Onsager, S. Machlup, Fluctuations and irreversible processes. Phys. Rev. **91**(6), 1505 (1953)
L. Onsager, Reciprocal relations in irreversible processes. i. Phys. Rev. **37**, 405–426 (1931a)
L. Onsager, Reciprocal relations in irreversible processes. ii. Phys. Rev. **38**, 2265–2279 (1931b)
L.A. Openov, A.I. Podlivaev, Violation of the equipartition theorem for thermally insulated clusters of atoms with different masses. Phys. Solid State **50**, 1195–1200 (2008)
C. Paneni, D.J. Searles, L. Rondoni, Temporal asymmetry of fluctuations in nonequilibrium steady states: links with correlation functions and nonlinear response. J. Chem. Phys. **128**, 164515–1–164515–11 (2008)
R.F. Pawula, Approximation of the linear Boltzmann equation by the Fokker-Planck equation. Phys. Rev. **162**(1), 186 (1967)
L. Peliti, *Statistical Mechanics in a Nutshell* (Princeton University Press, 2011)
L. Peliti, S. Pigolotti, *Stochastic Thermodynamics: An Introduction* (Princeton University Press, 2021). ISBN 9780691215525
R. Penrose, *The Road to Reality: A Complete Guide to the Laws of the Universe* (Vintage, 2007)

R. Peredo-Ortiz, L.F. Elizondo-Aguilera, P. Ramírez-González, E. Lázaro-Lázaro, P. Mendoza-Méndez, M. Medina-Noyola, Non-equilibrium Onsager–Machlup theory. Mol. Phys. e2297991 (2023)

J. Perrin, L'agitation moléculaire et le mouvement brownien. Compt. Rend. (Paris) **146**, 967–970 (1908)

J. Perrin, Mouvement brownien et realite moleculaire. Ann. Chim. Phys. **18**, 1–114 (1909). https://doi.org/10.1051/jphystap:0191000900500

M.J. Pinheiro, On Newton's third law and its symmetry-breaking effects. Physica Scripta **84**(5), 055004–12 (2011)

H. Poincaré, *Science and Method* (Thomas Nelson and Sons, London, 1914)

A. Porporato, S. Calabrese, L. Rondoni, Thermodynamic Langevin equations. Phys. Rev. E **110**(5), 054136 (2024)

I. Prigogine, *From Being to Becoming: Time and Complexity in the Physical Sciences* (W. H. Freeman and Company, New York, 1980)

S. Redner, *A Guide to First-Passage Processes* (Cambridge University Press, Cambridge England, 2001)

F.H. Ree, Molecular interaction of dense water at high temperature. J. Chem. Phys. **76**(12), 6287–6302 (1982)

L.E. Reichl, *A Modern Course in Statistical Physics* (Wiley-VCH, 1998)

D.R.M. Renger, Anisothermal chemical reactions: Onsager-Machlup and macroscopic fluctuation theory. J. Phys. A: Math. Theor. **55**(31), 315001 (2022)

H. Risken, *The Fokker-Planck Equation* (Springer, Berlin, 1989)

L. Chris, G. Rogers, D. Williams, *Diffusions, Markov Processes and Martingales: Volume 2, Itô Calculus*, vol. 2 (Cambridge University Press, 2000)

L. Rondoni, C. Mejía-Monasterio, Fluctuations in nonequilibrium statistical mechanics: models, mathematical theory, physical mechanisms. Nonlinearity **20**(10), R1 (2007). https://doi.org/10.1088/0951-7715/20/10/R01

G.M. Rotskoff, E. Vanden-Eijnden, Neural networks as interacting particle systems: asymptotic convexity of the loss landscape and universal scaling of the approximation error. stat **1050**, 22 (2018)

G.M. Rotskoff, E. Vanden-Eijnden, Dynamical computation of the density of states and Bayes factors using nonequilibrium importance sampling. Phys. Rev. Lett. **122**(15), 150602 (2019)

D. Ruelle, General linear response formula in statistical mechanics, and the fluctuation-dissipation theorem far from equilibrium. Phys. Lett. A **245**, 220 (1998)

D. Ruelle, Smooth dynamics, new theoretical ideas in nonequilibrium statistical mechanics. J. Stat. Phys. **95**, 393 (1999)

D. Ruelle, Differentiation of SRB states. Commun. Math. Phys. **187**(1), 227–241 (1997)

D. Ruelle, *Elements of Differentiable Dynamics and Bifurcation Theory* (Elsevier, 2014)

L. Saint-Raymond, *Hydrodynamic Limits of the Boltzmann Equation* (Springer, 2009)

D.J. Searles, B. Johnston, D.J. Evans, L. Rondoni, Time reversibility, correlation decay and the steady state fluctuation relation for dissipation. Entropy **15**, 1503 (2013)

U. Seifert, Stochastic thermodynamics, fluctuation theorems and molecular machines. Rep. Prog. Phys. **75**(12), 126001 (2012)

K. Sekimoto, *Stochastic Energetics*. Lecture Notes in Physics (Springer, 2010)

S.I. Serdyukov, The Onsager–Machlup theory of fluctuations and time-dependent generalized normal distribution. J. Non-Equilib. Thermodyn. **48**(3), 243–254 (2023)

Y.G. Sinai, *Introduction to Ergodic Theory* (Princeton University Press, 1976)

G.E. Smith, R. Wolfe, Experimental verification of the Kelvin relation of thermoelectricity in a magnetic field. J. Phys. Soc. Japan **21**, 651 (1966)

H. Spohn, *Large Scale Dynamics of Interacting Particles* (Springer, 1991)

M. Suceska, M. Braithwaite, T.M. Klapötke, B. Stimac, Equation of state of detonation products based on exponential-6 potential model and analytical representation of the excess Helmholtz free energy. Propellants, Explos., Pyrotech. **44**(5), 564–571 (2019)

W. Sutherland, A dynamical theory of diffusion for non-electrolytes and the molecular mass of albumin. Phil. Mag. **9**(54), 781–785 (1905). https://doi.org/10.1080/14786440509463331

F.J. Tipler, *The Physics of Immortality: Modern Cosmology, God and the Resurrection of the Dead* (Doubleday, 1995)

M. Toda, R. Kubo, N. Saito, *Statistical Physics I: Equilibrium Statistical Mechanics* (Springer, 1998)

R.C. Tolman, *The Principles of Statistical Mechanics* (Oxford University Press, 1955)

M.E. Tuckerman, *Statistical Physics: Theory and Molecular Simulation* (Oxford Academic, 2023)

G.E. Uhlenbeck, L.S. Ornstein, On the theory of the Brownian motion. Phys. Rev. **36**, 823–841 (1930). https://doi.org/10.1103/PhysRev.36.823

N. van Kampen, The case against linear response theory. Physica Norvegica **5**, 279 (1971)

J.L. Vega, T. Uzer, J. Ford, Chaotic billiards with neutral boundaries. Phys. Rev. E **48**, 3414 (1993)

A. Vezzani, E. Barkai, R. Burioni, Single-big-jump principle in physical modeling. Phys. Rev. E **100**(1), 012108 (2019)

M. Viana, K. Oliveira, *Foundations of Ergodic Theory*. Cambridge Studies in Advanced Mathematics (Cambridge University Press, 2016)

C. Villani, A review of mathematical topics in collisional kinetic theory. Handbook of Mathematical Fluid Dynamics **1**, 71 (2002)

J. Vollmer, L. Rondoni, M. Tayyab, C. Giberti, C. Mejía-Monasterio, Displacement autocorrelation functions for strong anomalous diffusion: a scaling form, universal behavior, and corrections to scaling. Phys. Rev. Res. **3**, 013067 (2021a)

J. Vollmer, L. Rondoni, M. Tayyab, C. Giberti, C. Mejía-Monasterio, Displacement autocorrelation functions for strong anomalous diffusion: a scaling form, universal behavior, and corrections to scaling. Phys. Rev. Res. **3**(1), 013067 (2021b)

M. von Smoluchowski, Zur kinetischen theorie der brownschen molekularbewegung und der suspensionen. Annalen der Physik **326**(14), 756–780 (1906). https://doi.org/10.1002/andp.19063261405

A. Vulpiani, F. Cecconi, M. Cencini, A. Puglisi, D. Vergni, Large deviations in physics. *The Legacy of the Law of Large Numbers* (Springer, Berlin, 2014)

T. Wen, L. Zhang, H. Wang, E. Weinan, D.J. Srolovitz, Deep potentials for materials science. Mater. Futur. **1**(2), 022601 (2022)

N. Wiener, The average of an analytic functional and the Brownian movement. PNAS **7**, 294–298 (1921)

L.V. Woodcock, Isothermal molecular dynamics calculations for liquid salts. Chem. Phys. Lett. **10**, 257 (1971)

L. Yang, Z. Zhang, Y. Song, S. Hong, R. Xu, Y. Zhao, W. Zhang, B. Cui, M.-H. Yang, Diffusion models: a comprehensive survey of methods and applications. ACM Comput. Surv. **56**(4), 1–39 (2023)

J. Yin, A. Porporato, L. Rondoni, Nonequilibrium fluctuations of global warming. J. Clim. **37**(9), 2809–2819 (2024)

M.W. Zemansky, *Heat And Thermodynamics* (Tata McGraw Hill Education Private Limited, 2011)

Index

C. Mejía-Monasterio and L. Rondoni, *Nonequilibrium Statistical Mechanics I*, Theoretical and Mathematical Physics,
https://doi.org/10.1007/978-3-031-98249-1

MIX
Papier aus verantwortungsvollen Quellen
Paper from responsible sources
FSC® C105338

If you have any concerns about our products,
you can contact us on
ProductSafety@springernature.com

In case Publisher is established outside the EU,
the EU authorized representative is:
Springer Nature Customer Service Center GmbH
Europaplatz 3, 69115 Heidelberg, Germany

Printed by Libri Plureos GmbH
in Hamburg, Germany